Preservation of Surfactant Formulations

Preservation of Surfactant Formulations

Edited by

F.F. MORPETH
Zeneca Biocides
Wilmington
Delaware

SPRINGER-SCIENCE+BUSINESS MEDIA, B.V

First edition 1995

Originally published by Chapman & Hall in 1995
Softcover reprint of the hardcover 1st edition 1995

Typeset in 10/12pt Times by Cambrian Typesetters, Frimley, Surrey

ISBN 978-94-010-4272-7 ISBN 978-94-011-0621-4 (eBook)
DOI 10.1007/978-94-011-0621-4

A catalogue record for this book is available from the British Library
Library of Congress Catalog Card Number: 95–75580

Printed on permanent acid-free text paper, manufactured in accordance with ANSI/NISO Z39.48-1992 (Permanence of Paper).

Preface

Microbes are known to live in an enormous range of environments. Their ability to survive and proliferate in diverse industrial systems is often a surprise to those not exposed to these problems in their work. These systems contain a range of potential carbon sources, one common theme being surfactants. Surfactants are often not the components most prone to spoilage since some systems contain highly susceptible natural components, such as starch and xanthum gum, but the surfactant is a key part of the formulation, and its extensive breakdown usually means that the material is beyond recovery.

The aim of this book is to describe in detail all aspects of the preservation of surfactant containing materials.

The book should be viewed as being in three discrete sections.

- chapters 1–5 deal with and summarise essential background information
- chapters 6–11 discuss in detail various end use applications
- chapters 12–15 outline the regulatory and toxicology implication associated with the safe handling of preservatives

Given the format of the book there is inevitably some duplication of information in the middle section with different authors describing essentially the same phenomena but on different substrates. I hope the reader will find that although different chapters touch on the same topics the information around these areas is sufficiently different to justify their inclusion in this book and to be of interest. It should also demonstrate what can be the most useful source of information, the hard practical experience of the authors.

Much of the expertise in this area is based in industry. This is reflected in the spread of authors and indeed much of the information presented here has been gained from direct experience and is not found anywhere else.

As might be imagined from the comments above, the literature around the area covered in this book is sparse. Thus any comments on the content of this book or information about the subject will be gratefully received.

Finally I would like to thank Marla, Ross and Lawrence for their understanding during the long hours spent pulling this manuscript together.

F.F.M

Contributors

P.W. Austin	Zeneca Specialties Research Centre, Zeneca Specialties, Blackley, Manchester, M9 3DA, UK
D.K. Brannan	Abilene Christian University, ACU Box 7721, Abilene, Texas 37134, USA
M.E. Burt	Zeneca Biocides, 1800 Concord Pike, FOC 2W, Wilmington, Delaware 19897, USA
M.A. Cresswell	Vinamul Ltd, Mill Lane, Carshalton, Surrey SM2 2JU, UK
H. Gibson	Campden FDRA, Chipping Campden, Gloucestershire, GL55 6LD, UK
J.T. Holah	Campden FDRA, Chipping Campden, Gloucestershire, GL55 6LD, UK
K. Holland	National Starch & Chemical Ltd., Galvin Road, Slough, Berkshire, SL1 4DF, UK
D.A. Knowles	FORM-AK, 10 The Forstal, Hadlow, Tonbridge, Kent, TN11 0RT, UK
P.S.K. Lee	Zeneca Biocides, 1800 Concord Pike, CR & DL 208, Wilmington, Delaware 19897, USA
J.J. Lewis	ICI Surfactants, PO Box 90, Wilton Centre, Middlesborough, Cleveland, TS90 8JE, UK
G. Lloyd	Albright & Wilson Ltd., 210–222 Hagley Road West, Oldbury, West Midlands, B68 0NN, UK
J. Martin	Calgon Corp. PO Box 1346, Pittsburgh, Pennsylvania 15230–1346, USA
F.F. Morpeth	Zeneca Biocides, 1800 Concord Pike, Wilmington, Delaware 19897, USA
D.N. Munro	Zeneca FCMO, North of England Works, PO Box A38, Leeds Road, Huddersfield, Yorkshire, HD2 1FF, UK
J.A. Pierce	Zeneca Biocides, 1800 Concord Pike, CR & DL 248, Wilmington, Delaware 19897, USA

P.M. Prichard ECC International Technology Centre, PO Box 471, Kaolin Road, Sandersville, Georgia 31082, USA

R.A. Rodford Unilever Research, Environmental Safety Laboratory, Colworth House, Sharnbrook, Bedford, MK44 1LQ, UK

R.D. Vore Zeneca Biocides, 1800 Concord Pike, CR & DL 248, Wilmington, Delaware 19897, USA

G.F. White School of Molecular and Medical Biosciences, Biochemistry Unit, University of Wales, College of Cardiff, PO Box 911, Cardiff, CF1 3US, UK

Contents

1 An introduction to microbial spoilage

F.F. MORPETH

1.1 Growth of spoilage microorganisms

Microbes can survive and flourish in a range of aqueous systems. The dominant organisms vary with the nature of the components and parameters such as pH, temperature, oxidation–reduction potential and water activity. The influence of these parameters of microbial growth is considered below.

1.1.1 pH

Most microorganisms grow best at around neutral pH (6.5–7.5). Few thrive below pH 4 or above pH 9 although microbes can survive from less than pH 1 to greater than pH 11. In general bacteria tend to be more fastidious about pH than yeasts or fungi which can predominate in acidic systems less than pH 6, such as acidic lattices and some dyes. The pH of most of the systems described in this book is fixed to within about 0.2 to 0.3 of a pH unit. However, it can be possible to limit the susceptibility of a system by taking away from neutrality by 1 pH unit or more. This can be especially useful in those cases where a single organism has infected the manufacturing plant and is proving resistant to killing. Tipping the balance against the organism in this situation can sometimes have just enough of a detrimental effect on the microbe to enable the biocide to re-establish control.

1.1.2 Temperature

Specialized microorganisms have been reported to be able to grow from less than −35 to greater than 90°C. However, most spoilage microbes tend to grow best between 15 and 40°C, though significant growth can occur between 7 and 60°C. Thus, many manufacturing facilities suffer spoilage problems only during the summer months. Indeed some companies even change their product specification between May and September to increase the amount of biocide incorporated.

1.1.3 Oxidation–reduction potential

Microorganisms display a varying degree of sensitivity to the oxidation–reduction potential of their environment. The most significant example of this is how the organisms respond to the presence of oxygen. Bacteria may be classified as:

- Aerobes (able to grow only in the *presence* of oxygen).
- Facultative anaerobes (able to grow in the presence or absence of oxygen).
- Anaerobes (only able to grow in the *absence* of oxygen).
- In addition, low oxygen tension favors certain bacteria, known as microaerophiles.
- Fungi and yeast tend to be aerobes though there are examples of facultative anaerobic fungi and many facultative and anaerobic yeasts.

There is a natural order in which microorganisms infect and grow in a newly prepared bulk industrial material not protected by a biocide. In general the initial growth tends to be that of vigorous aerobes usually the pseudomonads. However if a bulk storage tank is not agitated then these organisms quickly deplete the oxygen from the system. This allows the development of an insidious threat, the sulphate-reducing bacteria. Once established these organisms can cause significant pitting corrosion to any system.[1] Thus when the recovery is attempted of a particularly badly contaminated system it is often sensible to vigorously aerate the system prior to biocide addition and continue this agitation throughout the recovery process. This has two effects. First, it vents off any toxic hydrogen sulphide accumulated via the action of the sulphate reducing bacteria (which can also inactivate many commonly-used preservatives). Second, it inhibits further growth of the sulphate-reducing bacteria. Thus the effect of oxygen can be seen easily in a single bulk storage tank with pseudomonads predominating when sampled from the top and sulphate reducers predominating when sampled at the bottom.

In some cases the actual redox potential of a system also can affect profoundly the ability of microbes to survive and grow. However, this is rare.

1.1.4 Water activity

Water activity A_w is a measure of the moisture content of a system. It may be defined as the ratio of the vapor pressure of a system to that of water at the same temperature. Most spoilage bacteria require an A_w of at least 0.91; yeast can usually grow down to 0.88 and fungi can grow at 0.8. However, there are examples of microbes able to grow in an A_w as low as 0.6. Water activity is generally not used as a means of controlling microbial growth in industrial systems. Indeed it is a parameter which is seldom measured or used. This could be a missed opportunity.

1.2 Presence of a suitable carbon source

Various microorganisms are able to utilize a wide range of carbon compounds to support their growth. Both the concentration and type of carbon source will, to a large degree, determine which organisms are able to survive in any given industrial system. Examples of carbon sources able to support vigorous microbial growth, making them particularly prone to spoilage are soluble cellulose thickeners; starch derivatives; casein and other protein based materials; and biodegradable sufactants.

1.3 Presence of antimicrobial components

Though many components of aqueous industrial systems can serve as carbon sources able to support microbial growth, some will actually have an antimicrobial activity. This can either lower the possibility that the system will become spoiled or it can select for spoilage by those organisms not susceptible to these components. Examples are methanol commonly present in polyacrylic acid and the unreacted monomers in a latex system. Thus, the system to be preserved will have a profound influence on the range of organisms able to grow in it. Those organisms able to establish themselves, in turn determine what spoilage problems are encountered.

Most wet-state spoilage microbes are bacteria. Among those identified in the technical laboratories of ZENECA Biocides are members of the genera *Aeromonas* sp., *Alcaligenes* sp., *Bacillus* sp., *Flavobacterium* sp., *Pseudomonas* sp. and *Serratia* sp. with *Pseudomonas* sp. accounting for at least half the cases of spoilage.

Fungi can also contaminate some systems. Most common are *Fusarium* sp., *Penicillium* sp. and *Geotrichum* sp.

1.4 Consequences of microbial spoilage

Examples of the types of problem which may be encountered by different industries are buildup of gas and foul odors, loss of performance of processing equipment, thinning and phase separation and discoloration

1.4.1 Buildup of gas and foul odors in paints, adhesives and related systems

Bacterial growth can, if uncontrolled especially in hot climates, generate sufficient gas to distend and/or blow the top off paint and latex containers. These problems are particularly insidious since they are usually the result of long term microbial action and are only apparent when the final user opens the product. Occurrences such as this lead to a serious loss of reputation for the producing company. They can also lead to damaging claims for compensation against any company who supplied contaminated materials to the end user.

1.4.2 Damage or loss of performance of processing equipment

Failure to protect aqueous systems can lead to long term damage of manufacturing equipment. For example, it has been estimated that something like 10% of all corrosion is of microbial origin. This is mainly caused by cathodic depolarization and the *in situ* production of hydrogen sulphide by sulphate-reducing bacteria.

Less extreme, but more widespread, is the buildup of microbial biofilms in processing lines and on heat exchangers. Biofilms should probably be regarded as the 'natural' state of bacteria outside the laboratory.[2] They consist of mixed populations of bacteria encased in a polysaccharide (usually alginate) gel which is immobilized on a surface. Biofilms can be anything from one layer of cells to inches thick. Indeed they can be thick enough to block process pipes and significantly affect the efficiency of heat exchangers. In addition to this, they represent a reservoir of microbial contamination. A system not protected by a biocide which suffers a biofilm buildup cannot usually be cleared by chemical means alone. It will normally require physical cleaning. This can be an expensive, time-consuming process, involving considerable downtime for a production facility.

1.4.3 Thinning and phase separation

These are particularly difficult problems since they are brought about by enzymes secreted by microbes which persist after the organism has been killed. Thus, if biocide is added too late in a process, it will stop microbial growth but will not necessarily prevent further spoilage. This is a particular problem in the paint and latex industries but can occur in any system where cellulose ethers or surfactants are major components.

1.4.4 Discoloration

Discoloration can occur either through growth of a colored microorganism or through the microbial modification of colored components in a dye or paint. Sulphide precipitates can also occur when sulphate-reducing bacteria are present in a system.

References

1. Postgate, J.R. (1979) *The Sulphate Reducing Bacteria*, Cambridge University Press, Cambridge–London–New York–Melbourne.
2. Costerton, J.W., Cheng, K.J., Geesey, G.G., Ladd, T.I., Nickel, J.C., Dasgupta, M. and Marrie, T.J. (1987) Bacterial Biofilms in Nature and Disease. *Ann. Rev. Microbiol.*, **41**, 435–464.

2 Chemical preservatives

F.F. MORPETH and P.W. AUSTIN

A wide range of chemical agents is available to preserve surfactant-containing formulations. The criteria for the selection of the preservative of choice in each area are dealt with in subsequent chapters. The purpose of this chapter is to document the available preservatives and summarize their chemical, physical and antimicrobial properties. There have been some excellent books on preservatives which have covered in detail many of the more mature products.[1,2] In general many of these agents are in decline for safety and environmental reasons. Thus, this chapter will concentrate on the more 'modern' preservatives. Other agents will be referred to in passing and sufficient references given to enable the interested reader to pursue the subject. The reader is also referred to a recent book by Paulus which offers additional information to that given in this chapter.[3]

2.1 Isothiazolinones

The isothiazolinones as a class are probably the most commonly used industrial, agrochemical and cosmetic preservatives. The structures of those isothiazolinones in common use as preservatives are shown in Figure 2.1.

Benzisothiazolin-3-one. Benzisothiazolin-3-one (BIT) is an excellent broad spectrum antimicrobial. It is most commonly encountered in the Proxel range of preservatives by ZENECA Biocides. As such it is supplied either as a solution in caustic and glycol or as a neutral dispersion. The recommended use levels for BIT are in the range of 100 to 500 ppm. The physical-chemical properties of BIT are summarized in Table 2.1.

Some of the advantages of BIT are:

- good broadspectrum antimicrobial activity.
- excellent temperature stability.
- compatible with a wide range of industrial systems.
- excellent stability to both acid and alkaline pH.
- fully compatible with amines.

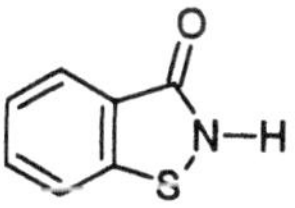

Benzisothiazolin-3-one

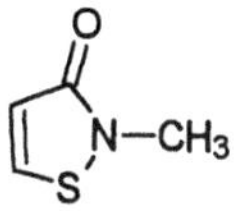

2-Methyl-isothiazolin-3-one

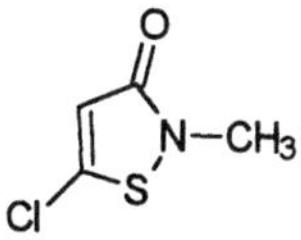

5-Chloro-2-methyl-isothiazolin-3-one

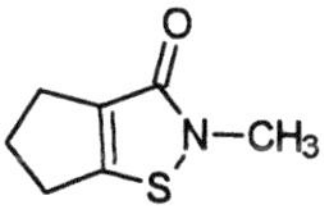

2-Methyl-4,5-trimethyleneisothiazolin-3-one

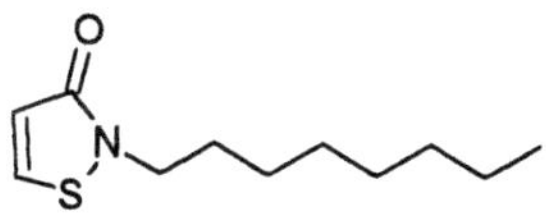

2-Octylisothiazolin-3-one

Figure 2.1 Structures of isothiazolinones in common use as preservatives.

Table 2.1 Physical-chemical properties of BIT

Property	Result
Colour	Colourless, although formulations may be straw coloured to brown
Melting point	157–158°C
Dissociation constant	$pK_a = 7.5$ at 25°C
Vapour pressure	5.8×10^{-5} Pa at 20°C
Solubility	Soluble in hot water. Forms water-soluble salts with alkalis and amines. Soluble in most organic solvents except petroleum ether.
Partition coefficient	Log $P = 1.3$ at 25°C
Density	1.45 g ml^{-1} at 20°C
Stability	Good heat resistance. Stable over a wide pH range

- chlorine/halogen free.
- comparatively low toxicity to higher forms of life.
- biodegradable by soil organisms, resulting in lack of accumulation in the environment.

The weaknesses of BIT tend to be common to all isothiazolinones. They are:

- inactivation by simple thiols. This is an inevitable consequence of its mode of action as a broad spectrum thiol reagent.[4–7]
- inactivation by oxidation/reducing agents.

BIT may be readily analysed by HPLC. Alternately in those systems where there are no other competing chromophores BIT's near UV spectrum (E_{max} = 318 nm) may be used to estimate its concentration. A rapid semi-quantitative method which is often used in production environments to confirm that BIT has been added to a material is to look for BIT fluorescence on irradiation with a fluorescence lamp. A key factor in the analysis of BIT in complex systems such as a polymer emulsion or a paint is the initial separation of the BIT. This may be done either by separating the system by centrifugation or by removing the BIT by dialysis. In both procedures the extraction is optimized by converting BIT into its highly soluble ammonium salt via the addition of 3M ammonia prior to centrifugation or dialysing against ammonia.

Mixture of 5-chloro-2-methyl-isothiazolin-3-one (CMIT) and 2-methyl-3-isothiazolin-3-one (MIT). Kathon is the tradename for a range of preservatives based on these actives and sold by Rohm and Haas Ltd. It is supplied in many formulations but mainly as either a 15% or a 1.5% solution. Rohm and Haas also supply the material in several grades of varying purity and containing various additives. Generic copies of this product are available in most countries (but not the USA) via a number of other manufacturers and formulators. The ratio of the two active components in Kathon is typically 3:1 in favour of the chlorinated component. Indeed most of the antimicrobial activity of Kathon can be ascribed to the chlorinated component which is about four to five hundred times more active than its unchlorinated analogue.[6] Physical-chemical properties of CMIT/MIT are summarized in Table 2.2

The main strengths of the CMIT/MIT mixture of biocides are:

- its extremely high activity
- excellent compatibility with many classes of commonly used industrial chemicals
- low VOC
- water-based formulation

Table 2.2 Physical-chemical properties of CMIT/BIT

Property	MIT	CMIT
Appearance	Colourless, extremely hygroscopic crystals	Slightly yellow crystals, darkening on storage
Melting point	50–51°C	54–55°C
Boiling point	93°C at 0.03 mm	
Solubility	30 g l^{-1} in H_2O. Soluble in a variety of organic solvents	Approximately 5 g l^{-1} in H_2O. Soluble in a wide variety of organic solvents
Stability	The free base is converted almost instantaneously into an oil on exposure to moist air. The hydrochloride is much less deliquescent.	The free base is not very stable on storage, tending to go dark and gummy. The hydrochloride is more stable, but not completely so.
	Decomposition on heating starts at 55°C.	Decomposition on heating starts at 55°C.
	Reacts rapidly with sodium bisulphite, and readily with oxidizing agents, thiols and (more slowly) amines.	Reacts rapidly with sodium bisulphite and readily with thiols, amines and alkali. Sensitive to oxidizing agents, but less so than most other isothiazolinones.

- increased stability to most oxidizing agents compared with other currently available isothiazolinones

Drawbacks are:

- high inherent toxicity, especially the 15% material
- instability at alkali pH. 5-Chloro-2-methyl-isothiazolinone tends to degrade rapidly above pH 9 though problems can occur if long term stability is required at a pH of 8.5 or even less.[8]
- lack of temperature stability. At temperatures above 45°C 5-chloro-2-methyl-isothiazolinone will slowly degrade over the course of several days.[9] Above 60°C this inactivation becomes quite rapid.
- inactivation by some nitrogen-containing compounds. Amine bases such as ammonia, triethanolamine, diethanolamine, monoethanolamine and some amine dispersing agents are able to react, in some cases rapidly, and inactivate 5-chloro-2-methyl isothiazolinone.[7,10]
- inactivation by some thiols. Simple thiols such as 2-mercaptoethanol and dodecyl mercaptan[7] are able to react with and inactivate CMIT/MIT (and indeed all isothiazolinones).
- inactivation by reducing agents. CMIT/MIT are rapidly inactivated by reducing agents such as sodium metabisulphite.

- a requirement for a stabilizer. This is required for many outlets for CMIT/MIT and is usually the nitrate of a divalent metal ion such as magnesium, copper or calcium though in the past formaldehyde has been used. There is an extensive patent literature on various stabilizers for CMIT/MIT.

2-Methyl-4,5-trimethyleneisothiazolin-3-one. 2-Methyl-4,5-trimethylene-isothiazolin-3-one (MTI) is a new antimicrobial agent recently introduced by ZENECA Biocides under the tradename of Promexal X50. A summary of the physical-chemical properties of MTI is given in Table 2.3.

MTI's main strengths are:

- high activity
- excellent compatibility with many classes of commonly-used industrial chemicals
- stable at alkaline pH
- good temperature stability
- water-based product
- no stabilizer required
- readily lost from the environment by a combination of biochemical and photochemical degradation
- chlorine free.

The drawbacks of MTI are essentially the same as those of benzisothiazolin-3-one.

2-Octyl-isothazolin-3-one. 2-Octyl-isothiazolin-3-one (OIT) differs from the other isothiazolinones in common use in that it is predominantly a fungicide. Its physical-chemical properties are outlined in Table 2.4.

Its main use is as a dry film fungicide for paint and plastic dry films. However, it is occasionally used either by itself or in a mixture with a suitable bactericide in those systems (usually low pH and high solids) which are particularly susceptible to fungal spoilage.

Table 2.3 Physical-chemical properties of MTI

Property	Result
Appearance	Colourless, crystalline solid
Melting point	122–123°C
Solubility	Very soluble in water. Soluble in the majority of organic solvents.
Thermal stability	Decomposition starts at about 216°C
Chemical stability	Reacts with reducing agents such a bisulphite and oxidizing agents. Also reacts with thiols and amines

Table 2.4 Physical-chemical properties of OIT

Property	Result
Appearance	Colourless to pale yellow oil
Boiling point	~125°C at 0.2 mm
Solubility	480 ppm in water. High solubility in the majority of organic solvents.
Partition coefficient log P	3.35
Vapour pressure	2.7×10^{-7} Pa at 20°C
Stability	Inactivation by ammonia, primary and secondary amines increasing with increasing pH. Reacts with reducing agents such as bisulphite and thiols and oxidizing agents.

2.1.1 Toxicology

All isothiazolinones are irritants and sensitizers.[11] The potential to cause allergic dermatitis is probably the biggest drawback of this class of molecule. The occurrence of sensitization due to CMIT is particularly well documented due to its widespread use in personal care products.[12]

2.1.2 Mode of action

The mechanism of action of BIT against one bacterium, *Staphylococcus aureus*, was investigated by Fuller and coworkers.[4] Briefly they found that BIT was a non-specific thiol reagent which reacted with a range of thiol-dependent processes associated with the cytoplasmic membrane. The general reaction of BIT with thiols is illustrated in Scheme 2.1.

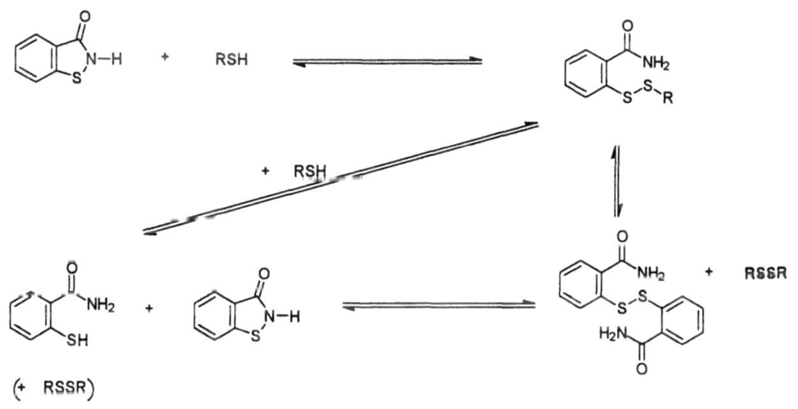

Scheme 2.1

BIT was shown to react rapidly with thiols, setting up a complex series of equilibria, depending on the concentrations involved. The initial reaction is to form an unsymmetrical disulphide, which slowly rearranges to the two symmetrical disulphides. In the presence of excess thiol, the unsymmetrical disulphide reacts further to give 2-mercaptobenzamide and the disulphide. The symmetrical disulphide from BIT is in equilibrium with BIT and 2-mercaptobenzamide, the reaction being particularly favoured at alkaline pH. When *Staphylococcus aureus* was challenged with sub-MIC levels of BIT, almost all of the BIT was detected as 2-mercaptobenzamide in the supernatant liquors, thus indicating that this was the preferred route for detoxification by the bacteria under these conditions.

This work was taken up and developed with other isothiazolinones, namely 2-methyl-isothiazolin-3-one and 5-chloro-2-methyl-isothiazolin-3-one.[5–7] These studies suggested a fundamental difference in the mode of action of 5-chloro-2-methyl-isothiazolin-3-one able to explain its much greater efficacy. The basis of this difference is shown in Scheme 2.2.

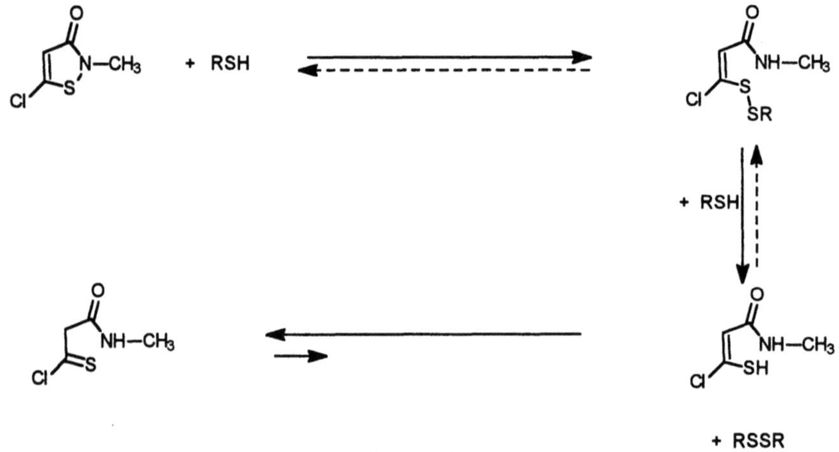

Scheme 2.2

Reaction of the isothiazolinone with thiols is thought to proceed initially as for BIT with the formation of the mixed disulphide, although these compounds are poorly characterized (except in the case of the hindered *t*-butylmercaptan) and the reversibility of the reaction is less well known. Further reaction with thiols initially produces the mercaptoacrylamide which rapidly tautomerizes to the thermodynamically more stable yet highly reactive thioacyl chloride. This highly reactive species is thought to be able to react rapidly and irreversibly with nucleophilic species such as amines, water, alcohols and thiols.[5–7] This suggests that the greater antimicrobial activity of CMIT depends not so much on the greater activity with thiols but rather on the extreme activity of one of its breakdown products. Thus, while the mercaptoamide structure in Scheme 2.2 is an innocuous breakdown product with BIT or 2-methyl-isothiazolin-3-one, with 5-chloro-*N*-methyl-isothiazolin-3-one it is a precursor of a highly active intermediate, explaining both the increased antimicrobial activity and its increased toxicity.

2.2 Other nitrogen–sulphur compounds

Dithio-2,2′-bis(benzmethylamide). This agent is structurally related to the isothiazolinones being the two-electron reduction product of 2-methyl BIT. Dithio-2,2′-bis(benzmethylamide) is sold by ZENECA Biocides as a neutral 25% dispersion under the trade name of Densil P. Densil P is an unusual preservative in that it is effective against both bacteria and fungi; its physical-chemical properties are summarized in Table 2.5.

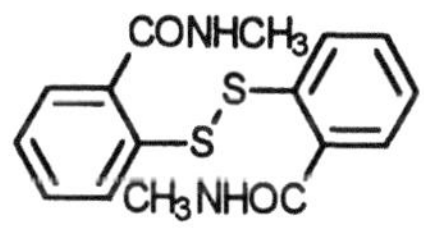

Typical use levels are 100–500 ppm active ingredient (AI). Its main drawbacks are its lack of solubility which is more pronounced at neutral to alkali pH and its lack of key regulatory clearances.

2-Mercaptopyridine N*-oxide.* The structures of the two commercially available forms of 2-mercaptopyridine *N*-oxide, (pyrithione) the sodium salt and the zinc complex, of this agent are shown at the top of page 14.

Table 2.5 Physical-chemical properties of Dervil P

Property	Result
Appearance	Colourless, crystalline solid
Melting point	220–221°C
Solubility	200 ppm in water. 1000 ppm in methyl ethyl ketone (MEK). 10 ppm in white spirit.
Heat stability	Very good.
Chemical stability	Hydrolyses in alkaline medium to *N*-methylbenzisothiazolinone and *N*-methylthiosalicylamide. Readily oxidized by peracids at alkaline pH, more slowly at pH 6. Reacts with bisulphite.

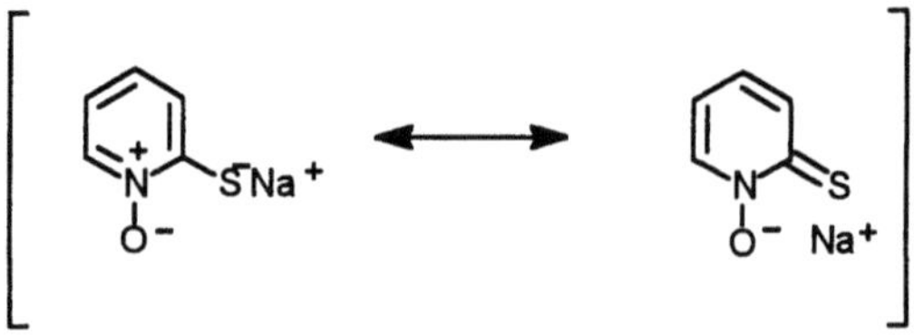

Sodium Pyrithione

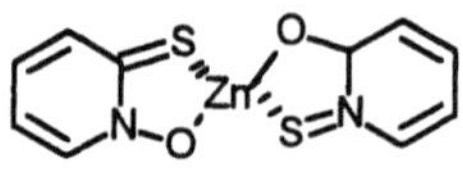

Zinc Pyrithione

Both forms of this agent are broad spectrum antimicrobials displaying activity against bacteria and fungi. However, they are usually considered to be fungicides with a significant antibacterial activity. Their physical-chemical properties are outlined in Table 2.6.

Table 2.6 Physical-chemical properties of sodium pyrithione and zinc pyrithione

Property	Sodium pyrithione	Zinc pyrithione
Appearance	White to pale yellow crystalline solid with a slight odour	White to pale yellow crystalline solid with a slight characteristic odour
Melting point	250°C (decomposition)	240°C (decomposition)
Solubility	>500 g l^{-1} in water. Soluble in ethanol, ethylene, glycol, dimethyl sulphoxide.	Very low solubility in water (~20 ppm), but solubility increased by the addition of amines such as diethylene triamine. Moderately soluble in dimethyl formamide and dimethyl sulphoxide.
pH of 2% aqueous solution	8	Insoluble
Stability	Solid is stable although can go yellowish on prolonged exposure to light. Aqueous solutions are stable between pH 4.5 to 9.5. At lower pHs converted to free pyrithione which is unstable to air (oxidation to the disulphide) and light. At more alkaline pHs slowly converts to the sodium salt of pyrithione sulphonic acid. Oxidizing agents convert it via the disulphide to the sulphinylsulphide, sulphonyl sulphide and eventually the sulphonic acid. Strong reducing agents reduce the *N*-oxide to give 2-mercaptopyridine. Forms highly coloured complexes with transition metal ions.	At acid pH the zinc is lost to give pyrithione which is unstable. Under alkaline conditions it reacts similarly to sodium pyrithione, although the very low solubility undoubtedly helps to protect it. Oxidizing and reducing agents affect it as for sodium pyrithione. The zinc atom is readily replaced by other transition metal ions that form stronger complexes, and these are usually strongly coloured.

The main difference between the two forms of pyrithione is their aqueous solubility and stability. That is, zinc pyrithione is stable and insoluble while sodium pyrithione is soluble but unstable particularly in the presence of light or oxygen. Sodium and zinc pyrithione are sold under the trade name, Omadine, by Olin Corporation.

The major use of sodium pyrithione is in the preservation of metal working fluids. Zinc pyrithione is predominantly sold as a personal care preservative though it is also used as a tank-side preservative for metal working fluids. Sodium pyrithione has only limited use in cosmetics due to concerns over irreversible effects on eyes.

Apart from the concerns over eye damage due to pyrithione, sodium and zinc pyrithione have a reasonably good toxicology profile. Both are skin irritants but are not sensitizers. Zinc pyrithione may be used over the range pH 4.5 to 9.5. Outside this range it breaks down to yield free pyrithione. However, at pH 9.5 pyrithione is fully ionized and is rapidly converted to the corresponding sulphinic acid which has no antimicrobial activity. At neutral pH zinc pyrithione is an extremely heat stable molecule tolerating temperatures of 100°C for 120 h and only decomposing at 240°C.

2.3 Organobromine compounds

2-Bromo-2-nitropropane-1,3-diol. 2-Bromo-2-nitropropane-1,3-diol commonly known as Bronopol is a formaldehyde condensate but it is considered separately here since it is known to kill microbes by a mechanism not related to its formaldehyde-release properties. The main suppliers of Bronopol are Boots and Angus.

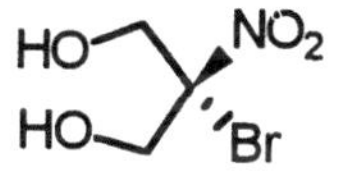

The physical-chemical properties of Bronopol are given in Table 2.7.

Traditionally Bronopol's main use has been as a preservative for personal care products. However, in recent years it has found an increasing application as a preservative in acidic polymer emulsions and adhesives. Bronopol's use as a preservative in more alkali systems is limited since, as discussed below, above pH 8 it can break down to give formaldehyde losing activity in the process. The decomposition of Bronopol in solution at elevated pH has been studied by various authors. In a recent study[13] a reaction scheme (Scheme 2.3) was proposed to account for the many observed decomposition products.

Table 2.7 Physical-chemical properties of Bronopol

Property	Result
Appearance	Colourless, odourless crystalline solid
Melting point	130°C
Vapour pressure at 20°C	1.68×10^{-5} hPa
Solubility	250 g l^{-1} in water. Soluble in alcohols. Very low solubility in petroleum ether
Stability	Stable as the solid. Aqueous solutions are relatively stable on the acid side but increasing tendency to decompose (releasing formaldehyde and many other products) as the pH is increased. Discoloration and corrosion can occur when aqueous solutions of Bronopol are in contact with iron and aluminium. Alcoholic solutions of bronopol are photochemically very unstable, readily going yellow to brown when exposed to light.

It can be seen that the decomposition of Bronopol is quite complicated, but the significant points are the release of formaldehyde and nitrite, the latter giving rise to the possibility of nitrosamine formation in the presence of amines.[14,15]

2.3.1 Mode of action

Various studies[16,17] have indicated that the antimicrobial activity of Bronopol is primarily due to its interaction with essential thiols within the

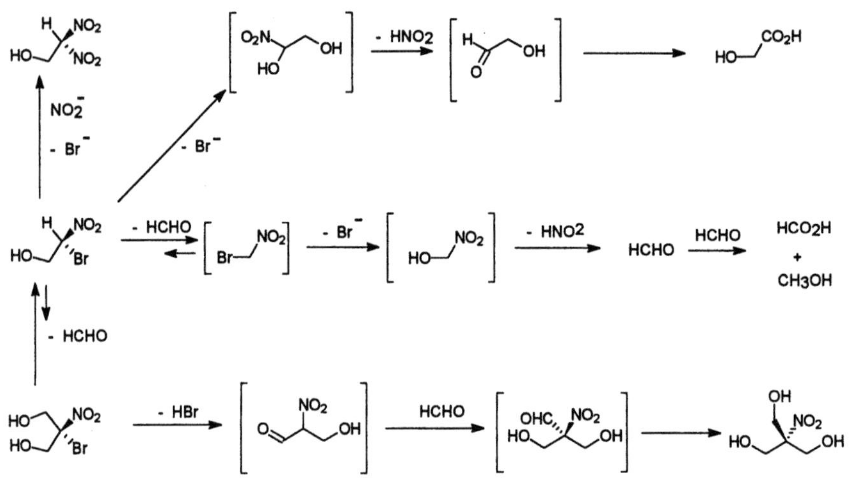

Scheme 2.3

cell, although there is a secondary action on the cell membrane which is not thought to be very significant. Although Bronopol can release up to two moles of formaldehyde, this plays at best only a minor role in its antimicrobial activity, formaldehyde-resistant and formaldehyde-sensitive strains of *Pseudomonas aeruginosa* being equally sensitive to the action of Bronopol.[18] Although Bronopol is thus essentially a thiol reagent, it differs in some respects from the isothiazolinones considered earlier, and there have been several studies on the interaction of *gem*-bromonitro compounds with thiols or thiolate anions.[17,19,20] Several different reaction types can take place, as can be seen from the accompanying schemes (Reaction paths 1–4) in which $R_2C(Br)NO_2$ is a generalized *gem*-bromonitro compound.

$$\left.\begin{array}{lcl} R_2C(Br)NO_2 + R^1S^- & \longrightarrow & R_2C(Br)NO_2^{\dot{-}} + R^1S^\bullet \\ R_2C(Br)NO_2^{\dot{-}} & \longrightarrow & R_2CNO_2^{\bullet} + Br^- \\ 2\,R^1S^\bullet & \longrightarrow & R^1SSR^1 \\ 2\,R_2CNO_2^{\bullet} & \longrightarrow & R_2C(NO_2){-}C(NO_2)R_2 \end{array}\right\}\ \text{Reaction path 1}$$

Scheme 2.4

Reaction path 1 (Scheme 2.4) involves the initial attack of the thiolate anion to form a radical and a radical anion and is reported to give almost quantitative yields of the dimeric products and bromide ion when performed anaerobically under argon.

$$\left.\begin{array}{lcl} R_2C(Br)NO_2 + R^1S^- & \longrightarrow & R_2C(Br)NO_2^{\dot{-}} + R^1S^\bullet \\ 2\,R^1S^\bullet & \longrightarrow & R^1SSR^1 \\ R_2C(Br)NO_2^{\dot{-}} + O_2 & \rightleftharpoons & R_2C(Br)NO_2 + O_2^{\dot{-}} \end{array}\right\}\ \text{Reaction path 2}$$

Scheme 2.5

Reaction path 2 (Scheme 2.5) is thought to be important under oxic conditions where the initially formed radical anion of Bronopol reacts with oxygen to generate superoxide and regenerate Bronopol.

$$\left.\begin{array}{lcll} R_2C(Br)NO_2 + R^1S^- & \longrightarrow & R_2C(Br)NO_2^{\dot{-}} + R^1S^\bullet & \text{(initiation)} \\ R_2C(Br)NO_2^{\dot{-}} & \longrightarrow & R_2CNO_2^{\bullet} + Br^- & \\ R_2CNO_2^{\bullet} + R^1S^- & \longrightarrow & R_2C(SR^1)NO_2^{\dot{-}} & \text{(propagation)} \\ R_2C(SR^1)NO_2^{\dot{-}} + R_2C(Br)NO_2 & \longrightarrow & R_2C(SR^1)NO_2 + R_2C(Br)NO_2^{\dot{-}} & \end{array}\right\}\ \text{Reaction path 3}$$

Scheme 2.6

Reaction path 3 (Scheme 2.6) is the $S_{RN}1$ reaction (radical chain nucleophilic substitution) where the initially formed radical anion takes part in chain propagation reactions. Termination is via radical–radical dimerizations.

$$R_2C(Br)NO_2 + R^1S^- \longrightarrow R_2C\bar{N}O_2 + R^1SBr$$
$$R^1SBr + R^1S^- \longrightarrow R^1SSR^1 + Br^-$$
$$R_2C\bar{N}O_2 + R_2C(Br)NO_2 \longrightarrow R_2C(NO_2){-}C(NO_2)R_2 + Br^-$$

Reaction path 4

Scheme 2.7

Reaction path 4 (Scheme 2.7) is the so-called 'halogenophile' reaction and does not involve a radical anion.

1,2-Dibromo-2,4-dicyanobutane. This agent is predominantly sold as the 'Tektamer 38' range of preservatives by Calgon Corporation and is available only as a racemic mixture of the two enantiomers (it is not known if there is any difference in activity between the two enantiomers). The main forms are Tektamer 38 Powder, a 98% active agent powder, which is recommended for addition to systems which require vigorous agitation and Tektamer 38 A.D a 25% dispersion. Probably the major use of 1,2-dibromo-2,4-dicyanobutane is in Europe as a personal care preservative. The major suppliers in this market are Schuelke and Mayer who sell 1,2-dibromo-2,4-dicyanobutane as a solution in 2-phenoxyethanol under the tradename of Euxkyl K400. 1,2-Dibromodicyanobutane is also widely used as a preservative for adhesives, slurries, polymer emulsions and some metal working fluids. Physical-chemical properties of this molecule are given in Table 2.8.

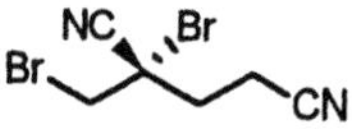

The key advantage of 1,2-dibromo-2,4-dicyanobutane is its good toxicology profile especially with regard to its low skin-sensitizing potential. The widespread use of this agent in cosmetics has resulted in some reports of sensitization but much fewer than competing materials. It also shows good compatibility with many industrial components particularly oxidation and reduction chemicals.

The weaknesses of 1,2-dibromo-2,4-dicyanobutane are its instability in the presence of amines and high pH. It is recommended that above pH 8 its stability in various systems is always checked. The breakdown products are not well characterized, but are presumed to be due to attack of nucleophiles present in the solution at elevated pH. Elimination of the

Table 2.8 Physical-chemical properties of 1,2-dibromodicyanobutane

Property	Result
Appearance	White to yellow crystalline powder
Melting point	51–53°C
Solubility	3.8 g l^{-1} in water. Soluble in most organic solvents
Stability	Reacts readily with a variety of nucleophiles. However, aqueous solutions are sufficiently stable up to pH 8. Relatively stable to oxidizing and reducing agents

tertiary bromine substituent is also expected to be a favoured reaction and the resulting substituted acrylonitrile should be susceptible to Michael addition reactions and would also activate the primary bromine substituent. 'Nu' represents a nucleophile (Scheme 2.8).

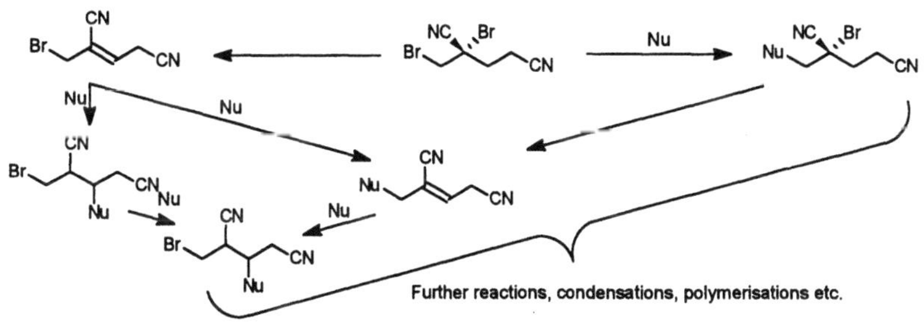

Scheme 2.8

Of course, where water is the nucleophile, the initially-formed alcohols could react further with unreacted bromo-compounds to give ethers etc. 1,2-Dibromo-2,4-dicyanobutane rapidly breaks down above 50°C. In solution, the breakdown pathway would presumably be similar to the scheme above. Under anhydrous conditions the thermal degradation pathway probably starts with the loss of HBr and goes on to give further degradation products (see Ref. 21) which comments on the thermal instability of the corresponding dichloro compound on attempted distillation at 12 Torr. At elevated levels these breakdown products can cause significant discoloration of many industrial products.

Little has been published in the open literature on the mode of antimicrobial action of 1,2-dibromodicyanobutane. However, based on the known chemistry of this class of molecules it most probably reacts as an alkylating agent. Thus it would be expected to react similarly to other alkylating agents such as chloroacetamide etc. with available cellular nucleophiles (see Scheme 2.8).

2.4 Thiocyanates

While there are many examples of antimicrobial thiocyanates, two compounds containing this active group are used as industrial antimicrobials. They are methylene bisthiocyanate (MBT) and 2-(thiocyanomethylthio)-benzothiazole (TCMTB). The structures of these agents are shown below.

MBT

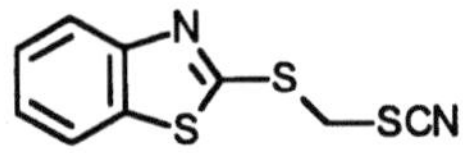

TCMTB

MBT and TCMTB are potent broad spectrum antimicrobials.[1] A summary of their physical properties is given in Table 2.9.

The thiocyanates find only limited use as wet-state preservatives the main use of MBT being as a water treatment chemical and TCMTB as a fungicide for timber and leather.

Organic thiocyanates are potent sulphydryl reagents and two reaction pathways (Scheme 2.9) have been characterized, depending on whether cyanide or thiocyanate is the leaving group.

Scheme 2.9

Table 2.9 Physical-chemical properties of MBT and TCMTB

Property	MBT	TCMTB
Appearance	Yellow crystalline solid with a characteristic odour	Yellow oil (or low melting solid) with a characteristic smell
Density	2.0 g ml^{-1}	1.38 g ml^{-1}
Melting point	105°C	
Vapour pressure at 25°C		4.75×10^{-6} hPa
Solubility	5 g l^{-1} in water. Soluble in most organic solvents	0.033 g l^{-1} in water. Soluble in acetone, DMF etc.
Stability	Stable in acidic systems at ambient temperatures. Decomposes in alkaline solution above pH 7.5. Unstable above 100°C	Decomposition in alkaline media (>pH 8) and on heating (>60°C)

It is considered that the first pathway is more important for aliphatic thiocyanates while the second becomes significant for aromatic thiocanates which are activated by electron-withdrawing groups.

2.5 Dithiocarbamates

Dithiocarbamates have the general structures shown below.

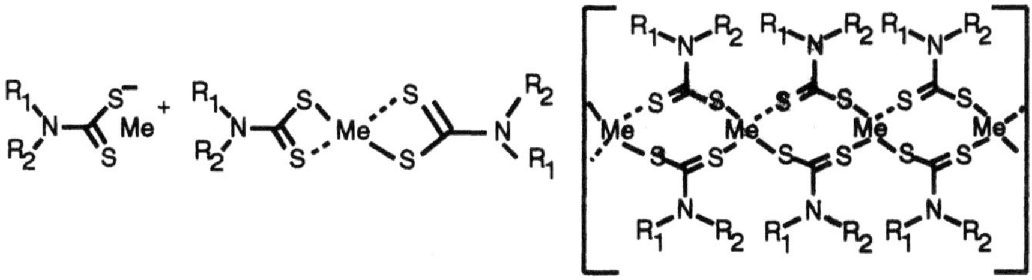

Where Me' is a cation, if the cation is polyvalent, then there will, of course, be more than one dithiocarbamate anion present to satisfy overall neutrality. If the metal is a transition metal, then neutral complexes may be formed, although these may be polymeric.

As a class the dithiocarbamates have been largely superseded by more modern preservatives such as the isothiazolinones. They are predominantly fungicides and their main non-agrochemical use has been in water treatment. However, they are occasionally used as in can preservatives, in systems prone to fungal spoilage, where their main disadvantage is a tendency to discolour some systems. Dithiocarbamates are believed to kill microorganisms via their thiol-modifying activity (in the case of those where one of R_1,R_2 is hydrogen, where they decompose to generate isothiocyanates) or via their metal-complexing abilities.

2.6 Organoiodine compounds

*3-Iodopropargyl-*N*-butylcarbamate.* 3-Iodopropargyl-*N*-butylcarbamate is sold almost exclusively as Troysan Polyphase by the Troy Corporation.

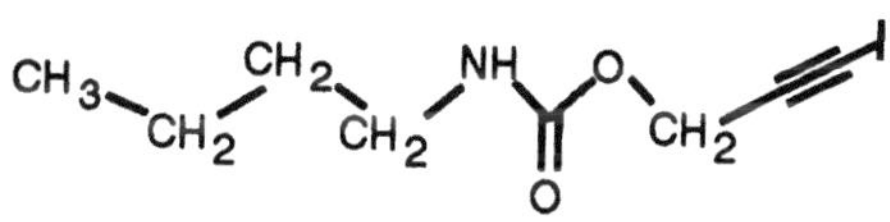

Its main use has been as a fungicide for paint films and timber. However it is also used, usually at the same time as an antibacterial, to provide protection to those systems which are particularly prone to fungal as well as bacterial spoilage. It has the chemical-physical properties shown in Table 2.10.

Very little information has been published about the mode of action of 3-iodopropargyl-*N*-butylcarbamate. The carbon–iodine bond is weak and readily broken (this accounts for the characteristic yellowing that plagues this class of compound). However, the antimicrobial activity (MIC) is unaltered in the presence of radical scavengers suggesting that the release of iodine radicals is not of major importance in its antimicrobial activity. The iodopropargyl group will react with a number of thiols, oxidizing them to disulphides and being reduced itself to the uniodinated propargyl derivative, thus acting as a source of 'positive iodine'. It must be emphasized that the iodine substituent in iodopropargyl compounds is very reluctant to act as a leaving group in nucleophilic substitution reactions.

$$2\ R{-}{\equiv}{-}I + 2\ R^1SH \rightarrow 2\ R{-}{\equiv}{-}H + R^1SSR^1 + 2I^-$$

Diiodomethyl-p-*tolylsulphone*. Diiodomethyl-*p*-tolylsulphone is also predominantly a fungicide which like 3-iodopropargyl-*N*-butylcarbamate has found application in boosting the antifungal activity of antibacterial preservatives. The properties of diiodomethyl-*p*-tolylsulphone are summarized in Table 2.11.

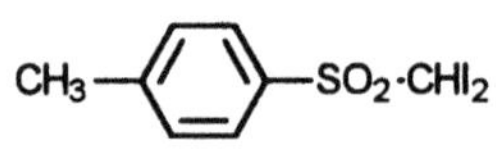

Diiodomethyl-*p*-tolylsulphone displays activity predominantly against fungi and yeast. It does have some activity versus Gram positive bacteria but is almost inactive against Gram negative microorganisms.

Table 2.10 Physical-chemical properties of Troysan Polyphase

Property	Result
Melting point	65–66°C
Solubility	156 ppm in water. Very soluble in most organic solvents.
Thermal stability	Decomposes above 192°C
Chemical stability	Formulations of Troysan need a stabilizer to prevent disproportionation (giving mixtures of mono-, di- and triiodoallylbutyl carbamate together with uniodinated species). Photochemical loss of iodine.

Table 2.11 Physical-chemical properties of diiodomethyl-*p*-tolylsulphone

Property	
Appearance	Fine, tan-coloured powder
Melting point	157°C
Solubility	~1 ppm in water. Soluble in many organic solvents such as acetone
Density	2.2 g ml^{-1}
Stability	Stable between pH 4–10. Photochemical loss of iodine

There has been very little published information on the mode of action of diiodomethylsulphones. In contrast to other aliphatic iodine compounds such as iodoacetamide (where the iodine substituent acts as a leaving group in nucleophilic substitution reactions) the iodine substituents in diiodomethylsulphones are virtually inert to normal substitution reactions and can probably best be considered as a source of 'positive iodine' which can act as an oxidizing agent, particularly for thiols (see section on the iodopropargyl compounds).

$$2\ ArSO_2CHI_2 + 2RSH \longrightarrow 2\ ArSO_2CH_2I + RSSR + 2I^-$$

2.7 Aldehydes and aldehyde-release agents

Formaldehyde and formaldehyde-release agents. Formaldehyde in aqueous/methanol solution (formalin) is one of the oldest most potent antimicrobials available. It displays many desirable features, being biodegradable and displaying a potent sporicidal activity. However, formaldehyde is not suitable in most cases for use as a preservative since it is rapidly lost from systems due to its volatility and unwanted side reactions. Indeed since formaldehyde is a well known cross-linking reagent its presence can cause gelling and severe compatibility problems in many systems. In addition there are serious toxicology concerns around formaldehyde and indeed free formaldehyde gas is significantly more toxic than cyanide!

To overcome this, use has been made of the ability of formaldehyde reversibly to form stable and in general safe adducts with itself and a very wide variety of compounds possessing active hydrogen atoms (alcohols, phenols, ketones, ammonia, amines, hydroxylamines, amides, hydantoins, ureas, thiols, thioureas, amino acids, amino alcohols, aliphatic nitro compounds etc.). The initially formed adducts can further react to give more condensed molecules so that there is an almost limitless number of formaldehyde addition and condensate molecules potentially available. These reaction types are shown in Scheme 2.10, but it must be emphasized that these are only a small fraction of the potential starting materials and reactions possible, and the reader is recommended to consult some of the many excellent reviews available.[22]

All these many addition and condensation reactions are potentially reversible but the position of equilibrium and the speed of the reaction to re-establish equilibrium once the equilibrium is disturbed (e.g. by removing formaldehyde from the system by reacting with some biological target) varies with the structure in question. Where the compound is very stable, such as phenol–formaldehyde condensates, there is usually insufficient formaldehyde ever available to provide an antimicrobial effect. As a general rule, unless the formaldehyde bound up within the molecule is sufficiently available to be detected by the Tannenbaum reagent[23] or the Nash reagent[24] then the molecule is very unlikely to have significant activity as a formaldehyde-release biocide, although of course there may be another mode of action. A further characteristic of formaldehyde-release biocides is their differing activities on formaldehyde-sensitive and -resistant strains of the same bacteria, which should parallel that of formaldehyde.[18] In view of the very large number of formaldehyde-release biocides that have been developed, it is impossible to include them all. Table 2.12 gives the structures of some of the more common ones, together with the number of moles of formaldehyde/mole of biocide potentially available (the potential formaldehyde being highlighted), and the release rate (although this is described as ‘fast’ or ‘slow’, there is, of course, a gradation between them all, and some compounds have both a fast and a slow component of the formaldehyde release).

As shown above formaldehyde is capable of reacting rapidly with a range of nucleophiles and this is undoubtably its mechanism of action. A full consideration of the toxicology of formaldehyde and its condensates may be found in chapter 13.

Glutaraldehyde. Glutaraldehyde is a potent antimicrobial whose use is increasing as formaldehyde and formaldehyde-release agents decline due to regulatory and toxicology pressures. A summary of its physical-chemical properties is given in Table 2.13.

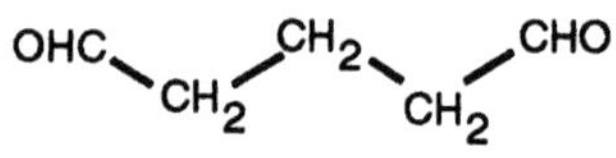

The antimicrobial action of glutaraldehyde has been the subject of several excellent reviews.[1,25,26] The main advantages of glutaraldehyde are

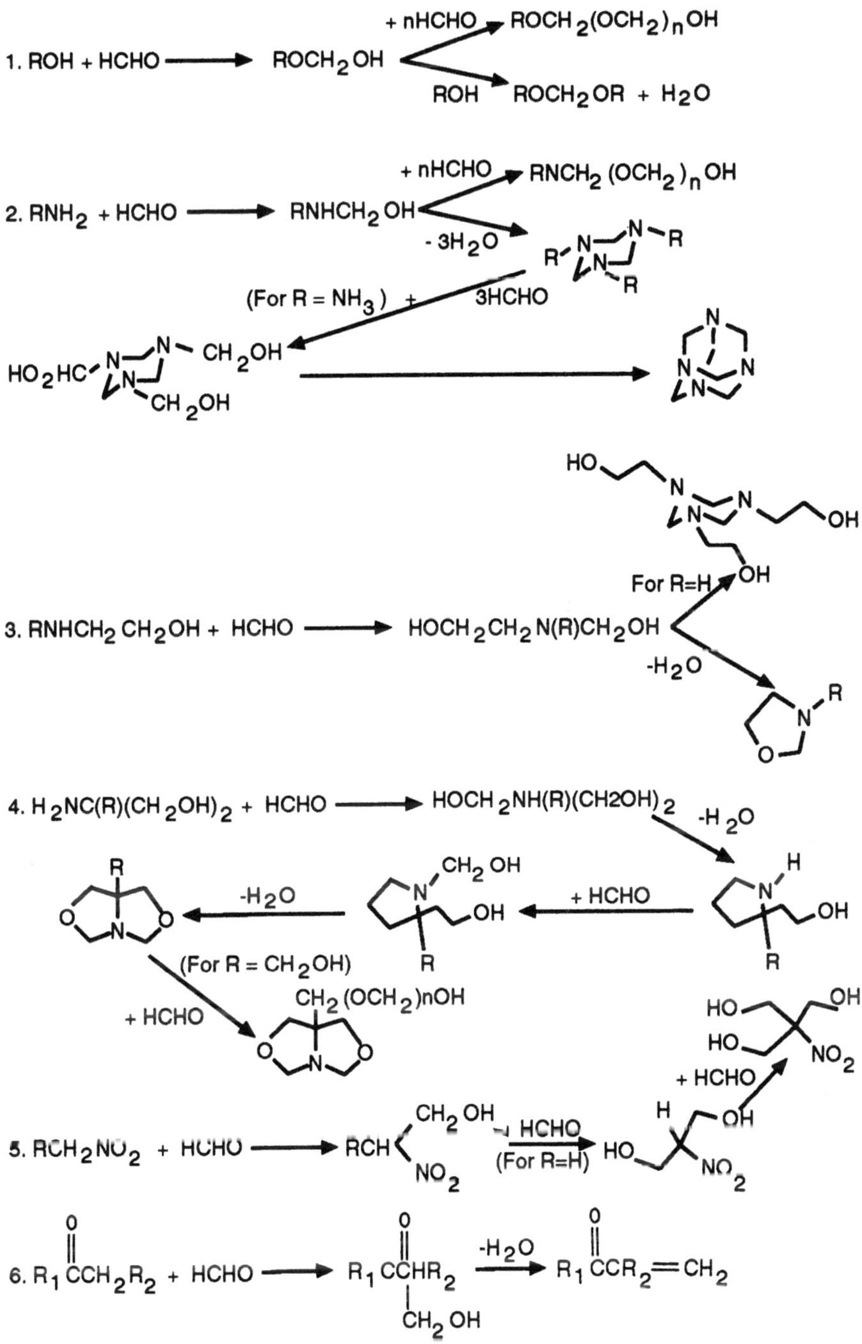

Scheme 2.10

Table 2.12 Structures of more common formaldehyde-release biocides

Structure	Moles HCHO/mole biocide	Ease of HCHO release
$PhCH_2O\mathbf{CH_2OH}$	1	Fast
$PhOCH_2CH_2O\mathbf{CH_2OH}$	1	Fast
$O_2NC(\mathbf{CH_2OH})_3$	3	Slow
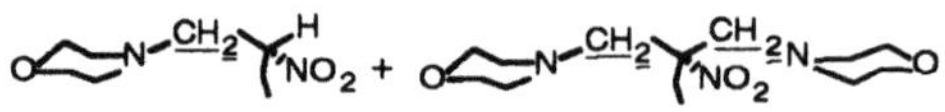	1 and 2	Slow
	6	Negligible at pH above 6, appreciable in acid solution
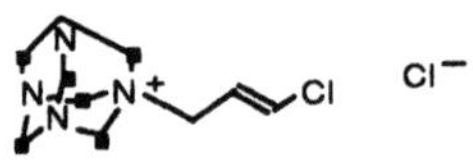	6	Slow
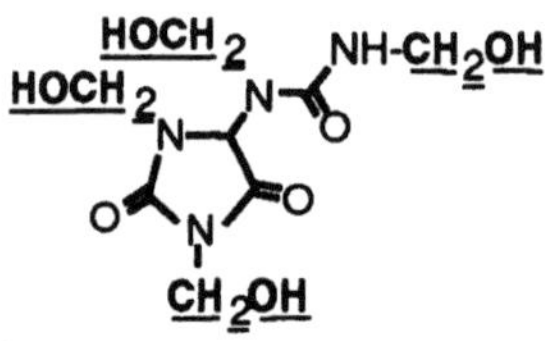	4	Slow
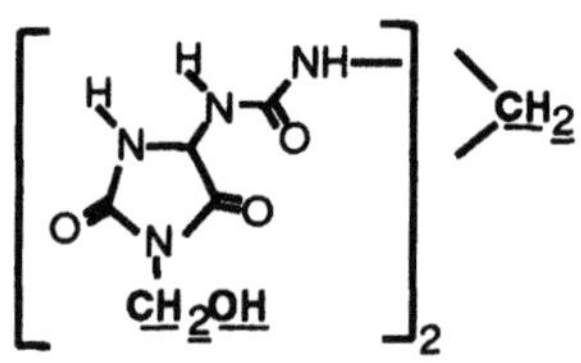	3	Slow
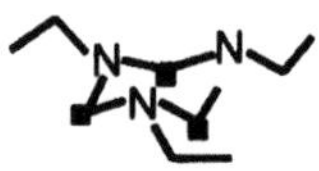	3	Fast
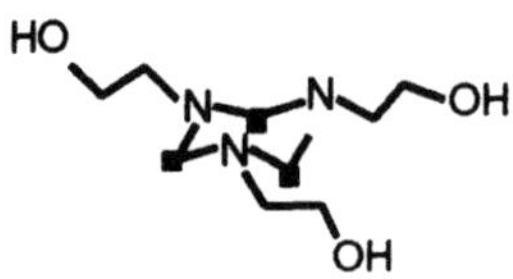	3	Fast
$CH_2(OCH_2)nOH$ 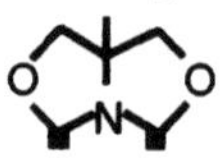	2 + *n*	Fast and slow

Table 2.12 *continued*

Structure	Moles HCHO/mole biocide	Ease of HCHO release
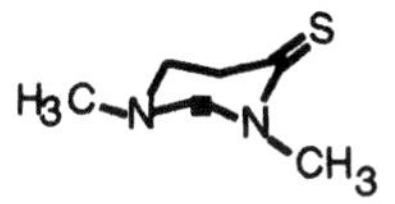	1	Fast. Mode of action not solely dependent on HCHO release
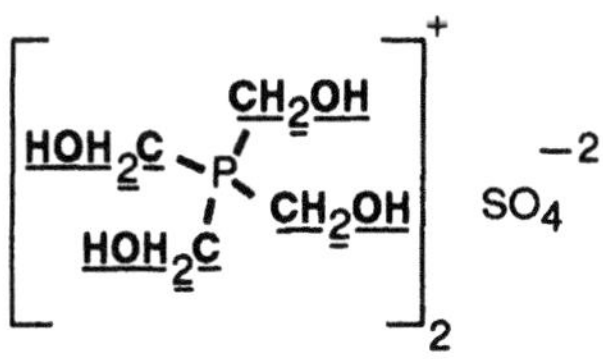	4	Slow

Table 2.13 Physical-chemical properties of glutaraldehyde

Property	Result
Appearance	Colourless oil, smell reminiscent of succindialdehyde
Boiling point	187–189°C (with decomposition) at atmospheric pressure, 106–108°C at 50 hPa, 83–85°C at 15 hPa.
n_D^{25}	1.43
Solubility	Soluble in water with a tendency to polymerize, depending on the pH. Soluble in most organic solvents
Stability	Characteristic reactions of an aldehyde (reduction, oxidation, condensation etc.). Solutions in water are stable at acid pH but rapidly decompose as the pH is increased, particularly in the presence of amines

its rapid rate of kill, broad spectrum of activity and sporicidal activity. Disadvantages are its odour, unacceptable in many applications, and instability at neutral to alkaline pH.

2.8 Phenols

The phenolics are an established class of antimicrobials with a long record of use as disinfectants. They are also used, albeit to a limited degree, as preservatives in some systems. Chlorinated phenols such as pentachlorophenol are no longer in common usage due to environmental and toxicology concerns.

2.9 Organic acids and salts

Used extensively in the food industry these agents are making a comeback in some of the areas covered in this book. This is despite their much lower efficacy than many of the other agents referred to in this chapter and reflects their excellent toxicology profile. Their use overall and in cosmetics is covered in chapter 7 of this book.

2.10 Mercury and other inorganic elements

For many years mercurials were among the most widely used industrial preservatives due to their high activity, low cost, and lack of colour and odour. While their use is declining steadily they are still popular in those markets where they are allowed. A review of the history and use of mercurials antimicrobial agents may be found in Ref 1. The two most commonly used mercurials are phenylmercuric acetate in its various forms and di(phenylmercury)dodecyl succinate. The main difference between the two agents is that phenylmercuric acetate is soluble in organic solvents but essentially insoluble in water while the converse is true of di(phenylmercury)dodecyl succinate.

Many other inorganic agents have been used historically as preservatives, for example tributyltinoxide, 10,10′-oxybisphenoxy arsine and barium metaborate, and indeed they still continue to be used in other industrial antimicrobial outlets. However, they do not come within the remit of this volume, though interested readers may refer to several chapters in Block.[1]

References

1. Block, S. (ed) (1991) *Disinfection, Sterilization and Preservation*, Lea and Febinger, Philadelphia.
2. Albert, A. (1985) *Selective Toxicity: Physical Chemical Basis of Therapy*, Chapman & Hall, London.
3. Paulus, W. (1993) *Microbicides for the Protection of Materials*, Chapman & Hall, London.
4. Fuller, S.J., Denyer, S.P., Hugo, W.B., Pemberton, D., Woodcock, P.M. and Buckley, A.J. (1985) The mode of action of 1,2-benzisothiazolin-3-one on *Staphylococcus aureus*. *Lett. Appl. Microbiol.*, **1**, 13–15.

5. Collier, P.J., Austin, P.W. and Gilbert, P. (1990) Uptake and distribution of some isothiazolone biocides into *Escherichia coli* ATCC 8739 and *Schizosaccharomyces pombe* NCYC 1354. *Internat. J. Pharmaceut.*, **66**, 201–206.
6. Collier, P.J., Ramsey, A.J., Austin, P.W. and Gilbert, P. (1990) Growth inhibitory and biocidal activity of some isothiazolone biocides. *J. Appl. Bacteriol.*, **69**, 569–577.
7. Collier, P.J., Ramsey, A.J., Waigh, R.D., Douglas, K.T., Austin, P.W. and Gilbert, P. (1990) Chemical reactivity of some isothiazolone biocides. *J. Appl. Bacteriol.*, **69**, 578–584.
8. Barman, B.N., and Preston, H.G. (1992) The effects of pH on the degradation of isothiazolone biocides. *Tribology Internat.*, **25**, 281–287.
9. Barman, B.N. (1994) Influence of temperature on the degradation of isothiazolinone biocides in aqueous media and in a metal working fluid concentrate. *Tribology Internat.*, **50**, 351–355.
10. Brown, S.A. personal communication.
11. Botham, P.A., Hilton, J., Evans, C.D., Lees, D. and Hall, T.J. (1991) Assessment of the relative skin sensitising potency of three biocides using the murine local lymph node assay. *Contact Dermatitis*, **25**, 172–177.
12. DeGroot, A.C. and Herxheimer, A. (1989) Isothiazolinone preservative: cause of a continuing epidemic of cosmetic dermatitis. *The Lancet*, 314–316.
13. Challis, B.C. and Yousaf, T.I. (1991) The reaction of geminal bromonitroalkanes with nucleophiles. Part 1. The decomposition of 2-bromo-2-nitropropane-1,3-diol ('Bronopol') in aqueous base. *J. Chem. Soc. Perkin Trans.*, **2**, 283–286.
14. Schmeltz, I. and Wenger, A. (1979) 2-Bromo-2-nitropropane-1,3-diol as a nitrosating agent for diethanolamine: a model study. *Food Cosmetic Toxicol.*, **17**, 105–109.
15. Holland, V.R. (1981) BNPD and nitrosamine formation. *Cosmetic Technol.*, May Issue, 31–36.
16. Stretton, R.J. and Manson, T.W. (1973) Some aspects of the mode of action of the antibacterial compound Bronopol (2-bromo-2-nitropropan-1,3-diol). *J. Appl. Bacteriol.*, **36**, 61–76.
17. Shepherd, J.A., Waigh, R.D. and Gilbert, P. (1988) Antibacterial action of 2-bromo-2-nitropropane-1,3-diol. *Antimicrob. Agents Chemother*, **32**, 1693–1698.
18. Sondossi, M., Rossmoore, H.W. and Wireman, J.W. (1986) The effect of fifteen biocides on formaldehyde-resistant strains of *Pseudomonas aeruginosa*. *J. Ind. Microbiol.*, **1**, 87–96.
19. Bowman, W.R. and Richardson, G.D. (1981) Reactions of thiolate anions with 2-substituted-2-nitropropanes. *Tet. Lett.*, **22**, 1551–1564.
20. Amrollah-Madjdabi, A., Begelmans, R. and Lechevalier, A. (1987) Sur la réactivité de *gem*-bromonitroalcanes en présence de thiolates. *Tet. Lett.*, **28**, 4525–4528.
21. Fritz, H., Weis, C.D., Winkler, T. (1976) 20.Halogenierte pyridine die herstellung von 3-halogenmethylpyridinen aus dimerem acrylnitril. *Helv. Chim. Acta*, **59**, 179–190.
22. Rossmore, H.W. and Sordossi, M. (1988) Applications and mode of action of formaldehyde condensate biocides. *Adv. Appl. Microbiol.*, **33**, 223–277.
23. Tannenbaum, M. and Bricker, C.E. (1951) Microdetermination of free formaldehyde. *Anal. Chem.*, **23**, 354–357.
24. Nash, T. (1953) The colorimetric estimation of formaldehyde by means of the Hantzsch reaction. *Biochem. J.*, **55**, 416–421.
25. Snyder, R.W. and Cheatle, E.L. (1965) Alkaline glutaraldehyde; an effective disinfectant. *Am. J. Hosp. Disin.*, **22**, 321–327.
26. Gorman, S.P. and Scott, E.M. (1980) The state of the glutaraldehyde molecule in relation to its biocidal activity. *J. Pharm. Pharmacol.*, **32**, 131–132.

3 Control of microbes through plant hygiene

H. GIBSON and J.T. HOLAH

3.1 Introduction

There is awareness in the food industry that control of factory hygiene is essential in the prevention of microbial contamination. This chapter covers the important considerations in controlling microbial contamination from environmental sources in the processing environment, in an industry sector in which hygiene is critical. The research and experience gained from the food industry and presented in these guidelines are of relevance to the manufacture of surfactant and surfactant based materials, particularly for surfactants used in the food industry or liable to spoilage, and should therefore be used in conjunction with risk assessments.

Contamination of the product may arise from four sources: the constituent raw materals, surfaces, people (and other animals) and the air. Control of the raw materials is addressed, for example, by specification, positive release of materials after sampling, quality assessment/quality control (QA/QC) procedures and supplier audits. This route of contamination is the only non-environmental route of contamination.

Water is often a raw material in many types of production, or may be used for conveying and cooling. In these cases treatment is required to produce water of an acceptable microbiological quality, for example by chlorination, filtration or ozonation.[1] This article describes control of the environmental routes of product contamination; discussion of water as a source of contamination is beyond the scope of this chapter.

Microbes on product contact surfaces can contaminate product directly whilst microorganisms on environmental surfaces can contaminate product indirectly through transfer by vectors such as personnel, utensils, pests, cleaning systems and air movement. The air of the processing environment acts as both a source of contamination and a transport medium, moving contamination from non-product contact surfaces to product contact surfaces.

Knowledge of the sources, vectors and routes of contamination is essential for the prevention of contamination. The environmental surfaces are controlled through the design, construction and operation of the plant. Control of contamination from product contact surfaces and equipment is dependent upon the hygienic design of the equipment and the use of effective cleaning and disinfection regimes. The personnel route of contamination is controlled through training and specific entry procedures for the production area. The production environment air is managed by air control systems is conjunction with control of practices that generate air-borne contamination. Finally management commitment is essential.

3.2 Design and construction of the premises

3.2.1 Site selection

Selection of a suitable site for new premises must involve consideration of the availability of services, i.e. water supply and effluent disposal. The premises should be located in areas which are not liable to flooding, smoke, dust or odours. The immediate surroundings of the premises should be well kept and free from refuse, rubbish, waste materials and overgrown vegetation. Ideally the site should be surrounded with a 1 m wide hard surface. The yard surfaces and roads of the site must have a suitable impervious surface with adequate drainage.

Activities such as agriculture, sewage treatment and rubbish tipping may present potential sources of contamination. Consideration of the control of birds, rodents and insects is important. In addition to preventing the access of pests to the process area, it is important that the areas around the site are kept free of rubbish and spilled product.

3.2.2 Factory structure

The types of external walls used include concrete, profiled metal sheeting, brick and block cavity walls, used in conjunction with profiled metal cladding and composite panels.

Pests may eat or contaminate ingredients, packaging materials and finished product. In addition to the associated health hazards, rodents can cause damage to buildings and electrical cabling, whilst birds can damage mortar. Pest control is therefore an inseparable part of profitable and hygienic production.

The major controls of pests are their exclusion from the premises and the prevention of spread within the building. The building should be constructed to prevent the entrance and harbouring of vermin, pests and birds. This is achieved by using insect screens, insectocutors, screens on air intakes and exits, air curtains or plastic strip barriers on doorways, and traps on drains. In addition foundations should be constructed to prevent rodents from burrowing underneath. Rodents are able to squeeze through very small (6–10 mm) holes to gain access to buildings. Pest proofing of doors is difficult because of the requirement for movement of staff and forklift trucks etc. However, all external doors should be self closing and fit closely so that gaps should be less than 3 mm.

3.2.3 Production area

3.2.3.1 Walls. The type of finish to be applied will depend on the particular area. In wet processing areas, the finish must be impervious to moisture and withstand the water pressure and chemicals used in the cleaning regime. In contrast, in dry and possibly dusty environments the walls must have a hard, smooth antistatic surface. The European Council Directive on the hygiene of foodstuffs gives general guidance for food premises.[2] In addition relevant Directives or Regulations for specific processing environments should be consulted for particular requirements. The Meat Products Directive,[3] for example, requires that walls must be covered with a light coloured coating. A hygienic, durable and easily cleanable coating is required for all processing areas.

The types of finishes available include liquid coatings, reinforced liquid coatings, cladding sheet, composite panels and tiles.

There are a range of liquid coatings including emulsion paints, oil-based one-pack epoxy and polyurethane paints, two-pack epoxy and polyurethane paints, chlorinated rubber paints and fungicidal and mould resistant paints. Emulsion paints are suitable for walls and ceilings in 'dry' areas and are cleanable rather than washable. Oil-based (one-pack) epoxy and polyurethane paints can present taint and toxicity hazards. Chlorinated rubber paints are particularly suitable for use in wet areas or the non-product contact surfaces of equipment involved in wet processing. Fungicidal and mould resistant paints are used in areas with high levels of humidity or condensation. The non-leaching fungicidal paints are appropriate as the leaching varieties may pose a potential taint hazard.

Reinforced liquid coatings which are based on glass fibres mixed with an epoxy resin result in an easily cleaned smooth surface which is resistant to impact damage, abrasion, a range of chemicals and heat.

Ceramic wall tiles are particularly suitable for areas where frequent washdowns occur and abrasion and heat resistance are required. Tiles should be fully vitrified glazed, light in colour and possess Group 1 classification for water absorption (for example in the UK in accordance with BS6431 (1983)[4] and any fixing or grouting material should be mould resistant (for example meet the requirements of BS5980 (1980)).[5]

Wall-to-floor stops are required to prevent doors damaging wall surfaces, and wall corners should be protected by non-corrosive metal or PVC angles. Galvanised steel or stainless steel rails should be used if trolleys are likely to damage wall surfaces.

3.2.3.2 Floors. Floors should be durable, non-absorbent, antislip, without crevices, capable of being effectively cleaned and should slope sufficiently for liquids to drain. In certain cases, floors must also be resistant to acids, grease and salts. Satisfactory drainage is obtained by ensuring sufficient fall to drainage points. Normally a fall of 1 in 60 is adequate, but this may vary depending on the surface texture of the flooring and type of operation.

There are several types of flooring materials, but all are dependent on the structural slab, the membrane, the screed and movement joints. The types of flooring include concrete, ceramic tiles and resins. Concrete is not generally suitable as although it is resistant to chemicals such as alkalis, mineral oil and many salts, it is attacked by acids, vegetable and animal oils, sugar solutions and some salts. In addition it is porous. The properties of concrete can be improved by the addition of a polymer which improves the resistance to liquid penetration.

Ceramic tiles are suitable as a durable and hygienic flooring material with a range of antislip surfaces; however, it is important that they are properly laid and grouted. Hygiene requirements dictate that only fully vitrified tiles are used because of their very low water absorption.

Various resin-based systems are available for application to the concrete substrate such as epoxy, polyurethane, polyester and methacrylate. These systems provide hygienic seamless floors.

3.2.3.3 Ceilings. Ceilings may be either solid or suspended. Suspended ceilings are preferable as all the supporting members and services are physically separated from the processing area and services taken down directly to the equipment. The ceiling should be smooth and easy to clean. Special attention must be paid to the ceiling finish above equipment generating steam so that mould growth is minimised.

3.2.3.4 Lighting. Suitable and sufficient lighting must be provided throughout the premises including store rooms and passageways. A lighting level of 500 lx is suitable for work areas and 700 lx in inspection and quality control procedures. All glass tubes in the processing area must be protected by covers to protect the glass and contain it in the event of breakage.

3.2.3.5 Ventilation. In areas where large quantities of steam and moist air are involved, adequate ventilation is required for the comfort of the personnel and to reduce microbial growth and corrosion of surfaces. Section 3.4 describes air control.

3.2.3.6 Factory layout. The factory layout should be such that processing operations are as direct as possible because the straight line flow minimises the possibilities of contamination of processed or semi-processed product by unprocessed or raw materials and is more efficient in terms of handling. This also allows easy separation of different processing operations. Premises should be designed so that there is physical separation of storage and processing operations whilst maintaining efficient product flow. Separation of processes reduces the risk of contamination. Separation of personnel is an essential part of the separation of processes with respect to the avoidance of cross contamination. Processing areas in which operations of high microbiological risk are undertaken should be limited to essential personnel.

Maintenance, hygiene and technical support, quality assurance and management personnel, who are required to move between high and low risk areas, must change their clothing and wash their hands.

3.2.3.7 Services. Services such as pipework should be installed in such a way that they do not constitute a hygiene hazard. Pipes should not pass over product lines, in order to prevent possible contamination from condensation, leakage, lagging or dust. Pipework must be arranged so that it can be cleaned by having adequate clearance from walls or ceilings. Pipe insulation should be crevice free, with sealed joints, and have a durable cleanable outer surface. Water supply pipelines should be free from 'dead legs' which could allow stagnation and an increase in the number of microorganisms.

Any drainage system must be of suitable construction for the particular effluent, adequate in size, cleaned regularly and capable of operating efficiently under the maximum load imposed. Effluent flow should always be directed from high risk areas.

3.3 Control of product contact surfaces

3.3.1 Hygienic design

Processing equipment of poor hygienic design can result in contamination incidents that can be both costly and damaging to the processor, but also increase cleaning costs and downtime. Hygienic processing equipment should be easy to maintain to ensure that it will perform as expected to prevent microbiological problems. Equipment must be cleanable and protect the product from contamination.

Hygienic design should be based on an appropriate combination of mechanical, process and microbiological requirements. Hygienic requirements should be taken into account at the initial design stage as upgrading existing designs to meet hygienic requirements may be prohibitively expensive and possibly unsuccessful. In addition to maintaining food quality and safety, compliance with hygienic design requirements may increase the life expectancy of the equipment and reduce maintenance needs and therefore reduce manufacturing costs.

In June 1989, a European Communities Directive relating to machinery was adopted.[6] This is chiefly concerned with safety, but there is a section concerning the essential health and safety requirements of agri-foodstuffs machinery. This states that equipment intended to prepare and process foodstuffs must be so designed and constructed as to avoid any risk of infection, sickness or contagion. Several hygienic rules are listed that refer to the materials of construction, surface finish, joints, fasteners, internal angles, drainage, deadspaces, prevention of contamination by lubricants etc. and installation. A harmonised European standard is being produced as a consequence.

Some of the major food processors have recognised the importance of hygienic design and produced their own standards and guidelines; however, these documents have not been generally available. Various organisations have, however, been producing guideline documents. The UK Food Manufacturers Federation in conjunction with the Food Machinery Association produced a booklet on the 'Hygienic Design of Food Plant' with particular reference to tanks, pumps and pipework. In addition, since 1982 the Campden Food and Drink Research Association has been working with an industrial working party of manufacturers and users of equipment to produce guidelines on hygienic design. Since 1992 the European Hygienic Equipment Design Group (EHEDG) has been publishing guideline documents on the hygienic design and operation of food plants and in particular a summary of the principle designations.[7]

3.3.1.1 Surfaces. Surfaces must be cleanable and must not present a toxicological hazard by leaching of components into the product. Product contact surfaces must be inert, be made of non-absorbent materials and must satisfy particular roughness requirements. All product contact surfaces must be easily accessible for visual inspection and manual cleaning, or it must be demonstrated that routine cleaning eliminates the possibility of contamination from bacteria. If cleaning-in-place (CIP) techniques are used it must be demonstrated that the results achieved without dismantling are satisfactory.

3.3.1.2 Draining and layout. The exterior and interior of all equipment and pipework should be self draining and easily cleanable. Horizontal surfaces must be avoided to prevent harbouring of product soils and cleaning liquids. Surfaces should slope so that any liquid would flow away from the main product area.

3.3.1.3 Installation. In design, construction, installation and maintenance it is important to avoid dead spaces which trap product, prevent effective cleaning and possibly allow microbial growth. Whenever possible, the risk of condensation on equipment, pipework and the fabric of the building should be avoided as the presence of condensation is particularly associated with microbial growth on surfaces. Any clearance between equipment and walls, floors and ceilings should be adequate for cleaning and inspection (a minimum of at least 1 m).

Noise suppression may be required to provide acceptable working conditions. It is important to select and install materials that are not prone to microbiological or infestation problems.

The equipment should be designed, installed and maintained so that foreign body contamination, e.g. by lagging material, loose nuts and bolts, does not occur. The connection of different items of equipment to form a line can cause major problems such as hold up of the product due to unsatisfactory transfer which may subsequently re-enter the main product flow. This can result in organoleptic or microbiological problems.

3.3.1.4 Construction materials. Materials used in the construction of product contact surfaces must be inert to the product under operating conditions, as well as to the detergents and disinfectants used in cleaning. The materials must also be corrosion resistant, non-toxic, mechanically stable and have a robust surface finish.

Austentic stainless steels are the most commonly used construction material for food processing equipment. Type AISI 304 is most often used in environments where fluids do not contain any chlorides. If chlorides are present, the corrosion resistant-type AISI 316 steel which contains molybdenum (or possibly titanium) is more suitable. Selection of the type of stainless steel mainly depends on the corrosive properties of the process and cleaning fluids as well as on welding requirements.

Hard plastics are also used as materials of construction but only certain plastics are approved for food contact and are inert to cleaning chemicals. Polypropylene is susceptible to attack by relatively low levels of sodium hypochlorite. Plastics are not as abrasion resistant as stainless steel and their cleanability is detrimentally affected. Polytetrafluoroethylene can be used for gaskets and seals but has the disadvangage that it suffers from cold flow under compression and is difficult to clean as it may not be permanently impermeable to microorganisms. Conveyor belts are particularly difficult to clean and can be a common source of contamination, especially if the belting has a carcass of an absorbent material such as canvas or cotton. As a consequence polyester is now widely used for the carcass of belting because it is stable and non-absorbent.

3.3.1.5 Internal design. Of major importance in the hygienic design of equipment is the internal design. The surface finish can affect the cleanability as rougher surfaces can be more difficult to clean. A maximum

Table 3.1 Examples of surface treatments and roughness (*R*a) of stainless steel

Treatment	*R*a (μm)
Cold rolling	0.2–0.5
Hot rolling	>4
Glass bead blasting	1.0–1.2
Bright-annealing	0.4–1.2
Electropolishing	Dependent on the original finish

surface roughness *R*a value of 0.8 μm is recommended for surface finish unless there is evidence that a rougher surface is acceptable. Stainless steel sheet, as supplied directly by the steel mill, typically has a roughness *R*a value of less than 0.5 μm. Electropolishing has relatively little effect on the surface finish, although eletropolishing can improve cleanability. Table 3.1 lists several surface treatments and the associated surface roughness.

Metal-to-metal joints should not be used in product contact areas as they are uncleanable and may retain high numbers of microorganisms which may subsequently contaminate product. Permanent joints should instead be continously welded and the weld ground and polished to a surface roughness value similar to that of the surrounding material to avoid cleaning problems. This may not always be possible, and in these cases a gasket of a suitable material should be used; however, it is important that a flush, continuous, crevice free surface is obtained. In certain cases non-toxic solders may also be used but it is important to ensure their integrity with respect to cleaning chemical action.

There should be no internal crevices or 'dead spaces' in which product can be retained or from which product residue cannot be easily removed by cleaning. To facilitate cleaning, the internal radii of tanks, hoppers etc. should be as generous as technically possible.

Exposed screw threads, nuts, bolts, and rivets must be avoided where possible as they are difficult to clean. In addition, any fastener used in or above product areas must be secure so as to avoid the possibility of product contamination.

3.3.2 Cleaning and disinfection

Provided that the process environment and production equipment have been hygienically designed, cleaning and disinfection are the major controls of the environmental routes to product contamination. Cleaning and disinfection, when performed correctly, are cost effective, easy to manage and can reduce the risk of microbial and foreign body contamination.

3.3.2.1 Principles of cleaning and disinfection. The cleaning and disinfection procedure removes undesirable material (or soil) from the surfaces, including microorganisms, product residues, foreign bodies and cleaning chemicals. This soil may be derived from normal production, spillages, line jams, maintenance of equipment, packaging or general contamination from the environment (dust and dirt).

Product residues are normally easy to visualise and are characterised by their chemical composition, e.g. fat, protein or carbohydrate. In addition, the process or environmental factors may influence the ease of soil removal. The moisture content of the soil, the soil temperature and the time period before cleaning takes place affect the ease of removal.

Microorganisms may be present in the soil, or can attach to surfaces and form biofilms. Holah and coworkers[8,9] described the attachment of stainless steel coupons to production lines to assess the buildup of microorganisms in a variety of processing environments and found levels of 10^3–10^7 cells cm^{-2}. Microorganisms are able to attach and grow on the materials used in the factory environment; these include stainless steel, aluminium, nylon, polypropylene, polycarbonate, PVC and Teflon.

Understanding the characteristics of the soil in terms of chemical and microbial components is essential for a successful and economic cleaning and disinfection programme. The cleaning and disinfection programme can be divided into a number of phases:[10–12]

1. Wetting and penetration of the soil and equipment surface by the cleaning solution,
2. Reaction of the cleaning solution with the surface and soil. This facilitates dissolution of soluble organics and minerals, emulsification of fats, peptisation of organic materials and removal of solid soil particles from the surface,
3. Prevention of redeposition of the soil removed from the surface,
4. Reaction of the disinfectant solution with the residual microbes, and possibly removal of the cells by rinsing.

There are four factors involved in a cleaning and disinfection programme which are used in combination to achieve the four phases described above. These factors are:

- Chemical energy
- Mechanical or kinetic energy
- Temperature or thermal energy
- Time

Chemical energy is important for the cleaning and disinfection phases. In the cleaning phase the chemicals break down the soils to facilitate removal from the surface. In the disinfection phase, the chemicals reduce the viability of the microbes remaining after cleaning.

Mechanical or kinetic energy is employed physically to remove soils from the surface and may include manual brushing, scraping, automated scrubbing, pressure jet washing or the circulation of fluid in clean-in-place systems.

Temperature affects cleaning and disinfection in several ways. First, the chemical effects increase linearly with temperature. Second, temperatures above the melting points of fats and oils facilitate their removal.

Increasing the time component, for example by the use of soak tanks, or foams and gels to extend the contact time, can increase the efficiency of cleaning and disinfection.

3.3.2.2 Cleaning chemicals. The determination of the correct detergent for any cleaning process is subject to a number of selection criteria. These criteria include plant design, cleaning techniques available, the type of soil present, the way the soil is formed, the nature of the production process and the water chemical composition. There is no single cleaning agent that fulfils all the requirements of an efficient cleaning programme, and for this reason cleaning solutions or detergents are formulated to contain a number of components with specific abilities. Typically a detergent may contain some of the following components: water, surfactants, inorganic alkalis, inorganic and organic acids and sequestering agents.

Water is the basic ingredient of most cleaning systems and provides a cheap readily available transport medium for rinsing and dispersing soils. Salts, sugars and other water-soluble components will be dissolved by water, and at temperatures above the melting point of fats, water will aid emulsification. Water is, however, a poor wetting agent and cannot dissolve nonionic compounds.

Surfactants are widely used in detergent formulations for wetting soils, soil penetration, soil suspension and to aid rinsing by reduction of surface tension. Surfactants are composed of a long non-polar (hydrophobic) tail and a polar (hydrophilic) head. They may be anonic, cationic or nonionic depending on their ionic charge, but anonics and nonionics are most common. If a surfactant is added to a drop of water the polar end of the molecule interacts with the water droplet in such a way as to reduce the surface tension of the droplet and thereby increase wettability and the spread of the droplet. In the case of fats and oils, and non-polar tail interacts with the oil and the molecule orientates with the hydrophilic part around the fat micelle, consequently aiding the dispersion of fats from the surface.

Alkalis (and particularly sodium hydroxide) are useful cleaning agents as they are relatively cheap, saponify fats, break down proteins and may also be biocidal at higher concentrations. Sodium hydroxide, a strong alkali, exhibits high levels of saponification and protein disruption. This alkali is, however, corrosive and hazardous to operatives. Weak alkalis may be used

Table 3.2 Cleaning procedures and solubility characteristics recommended for a range of soil types (modified from Elliot)[20]

Soil type	Solubility characteristics	Cleaning procedure recommended
Sugars, organic acids, salt	Water soluble	Mildly alkaline detergent
High protein	Water and alkali soluble Slightly acid soluble	Chlorinated alkaline detergent
Starchy foods	Alkali soluble, slightly water soluble	Mildly alkaline detergent
Fatty acids	Alkali soluble Water insoluble	Mildly alkaline detergent or strong alkali
Heat precipitated water hardness, milk stone, protein scale	Acid soluble Alkali and water insoluble	Acid cleaner used on a periodic basis

as they are less hazardous but also less effective. The major disadvantage of alkalis is precipitation of hard water ions (calcium and magnesium).

Inorganic and organic acids are not used as frequently as alkalis; however, they are effective at solubilizing mineral scales and are used in CIP applications.

Sequestering or chelating agents are used to prevent the precipitation of water hardness salts by affecting the physical structure of the precipitate or complexing with the salts to form water-soluble complexes. Sequestering agents may be organic or inorganic. Organic sequesterants are usually based on polyphosphates, whilst inorganic chelating agents include the sodium salts of ethylenediaminetetraacetic acid (EDTA) and nitrilotriacetic acid (NTA).

The detergent chosen for a particular application will depend on the soil to be removed and its solubility characteristics. Table 3.2 summarises the recommended detergents for a range of soil types.

3.3.2.3 Disinfectants. The cleaning phase has been shown to result in the most significant reduction in the numbers of bacteria present on the surfaces;[12] however, significant levels of viable cells are likely to remain even after cleaning. The aim of the disinfection stage is to remove or reduce the viability of these remaining microbes. Temperature is the ideal disinfectant as it penetrates the surface, is non-corrosive, is non-selective in terms of vegetative microorganism types and leaves no residue. High temperatures are often successfully used as disinfectants in CIP systems; the use of hot water on open surfaces is uneconomic, hazardous and impractical. In these cases chemical biocides are used.

A number of factors must be considered when choosing a disinfectant product:

1. Compatibility with other chemicals, such as any residual detergent that may be left on the surface,

2. Compatibility with the construction materials,
3. Hard water and soil tolerance,
4. Non-tainting – disinfectants should not affect the product by causing taints and therefore phenolics cannot be used in food applications,
5. Residual activity,
6. Rinsing ability,
7. Antimicrobial range,
8. Cost,
9. Toxicity, irritancy and environment impact.

There are two main types of disinfectants: oxidising and non-oxidising biocides. The oxidising type of disinfectants include halogen-releasing components (commonly chlorine), peroxy acids and hydrogen peroxide. These disinfectants are generally fast acting, bactericidal products with a wide spectrum of activity. The non-oxidising biocides include quaternary ammonium compounds (Quats or QACs), amphoterics and biguanides, and they are slower acting with a narrower spectrum of activity and may be bacteriostatic at certain concentrations.

The most commonly used disinfectants are chlorine-releasing components, QACs, iodine compounds, amphoterics and peracetic acid. Table 3.3 lists some of the characteristics of these products.

Table 3.3 Characteristics of the common disinfectants

Characteristic	Chlorine	QACs	Iodophor	Amphoteric	Peracetic acid
Microorganism control					
Gram positive	++	++	++	++	++
Gram negative	++	+	++	++	++
Spores	+	–	+/–	–	++
Yeast	++	++	++	++	++
Resistance acquisition	–	+	–	+	–
Inactivation by					
Organic matter	++	+	+	+	+
Water hardness	–	+	–	–	–
Detergency properties	–	++	+	+	–
Surface activity	–	++	+/–	++	–
Foaming potential	–	++	+	++	–
Problems with					
Taints	+/–	–	+/–	–	+/–
Stability	+/–	–	–	–	+/–
Corrosion	+	–	+	–	–
Safety	+	–	+	–	++
Other chemicals	–	+	+	–	–
Potential environmental impact	++	–/+	–/+	–/+	–
Cost	–	++	+	++	+

Key: –, No effect/no problem; +, effect; ++, large effect.

Chlorine is the cheapest disinfectant and is available as hypochlorite (fast acting) or slow releasing forms such as chloramines and dichlorodimethylhydantoin. Quaternary ammonium compounds are amphipolar cationic detergents derived from ammonium salts substituted with bromine or chlorine. Iodophores are soluble complexes between elemental iodine and nonionic surfactants. Amphoterics are based on the amino acid glycine, frequently with an imidazole group.

Disinfection efficiency is controlled primarily by four factors: interfering substances (such as organic matter), pH, concentration and contact time.

Any organic matter or cleaning chemical residues may protect the microbes from disinfectant penetration or may react with the disinfectant resulting in inactivation. For these reasons it is important to remove all cleaning residues and soil before disinfection.

Biocides have optimum activity within certain pH ranges, depending upon the disinfectant type.

Disinfectant concentration is important as the relationship between microbial death and disinfectant concentration is sigmoidal. Consequently at low concentrations the microbial population is difficult to kill but as the disinfectant concentration is increased a point is reached where the majority of the population is reduced. Above this concentration, a proportion of the cells may survive regardless of increasing biocide concentration. It is, therefore, essential to use the disinfectant at the correct concentration as changes in concentration may significantly reduce the activity, whilst increasing concentration may not produce any enhanced effect.

To be effective, biocides must reach the bacterial cell, and bind to and traverse the cell envelope to reach the target site.[13] For this reason sufficient contact time between the disinfectant and microbes is essential for biocide efficiency.

3.3.2.4 Cleaning techniques. Cleaning of open surfaces can be undertaken by hand using simple tools such as brushes, but specialist equipment is usually required for large areas to dispense chemical and/or provide mechanical energy.

Manual cleaning is limited due to the fact that only certain chemicals and low temperatures can be used for operator safety. Small items of equipment can be subjected to higher temperatures and chemical concentrations by soaking in tanks for extended periods.

To clean large areas of open surface, equipment is used to disperse chemicals and/or provide mechanical energy. Chemicals may be applied as mists, foams or gels and the mechanical energy supplied by water jets or scrubbing actions. The use of foams and gels prolongs the contact time between the chemical and the soil on the surface.

Closed systems such as liquid processing equipment can be cleaned in place (CIP). This involves circulation of water, detergents and disinfectants through the processing equipment whilst assembled and the mechanical energy is provided by the turbulent flow of the liquid through the pipework. CIP systems are ideally used to clean process equipment that has been specifically designed for CIP cleaning.

3.3.2.5 Cleaning and disinfection procedures. Cleaning and disinfection regimes are designed to be efficient with water and chemicals, to allow use of selected chemicals under optimum conditions, to operate safely, to be easily managed and to reduce manual labour. The cleaning procedure covers the stage at which cleaning is implemented and the sequence of cleaning and disinfection.

The main phases in the cleaning and disinfection programme are:

1. Production period practices. Good housekeeping practices should operate so that product is removed from production lines during breaks, possibly followed by controlled manual cleaning. In addition sound sanitation practices should be used for major product spillages during production.
2. Preparation of equipment. Equipment should be dismantled as far as is practical or necessary for cleaning, and dismantled equipment stored on racks etc. Machinery should be isolated electrically and electrics etc. covered for protection.
3. Removal of gross soil. Loosely adhered or gross product debris should be removed by brushing, vacuum, scraping, etc. Soil on floors should be picked up rather than rinsed to drain by hoses.
4. Pre-rinse. Loosely adhered small debris is removed with a low pressure cold water rinse. The use of hot water aids the removal of fats, but too high a temperature may coagulate proteins.
5. Cleaning. Cleaning chemicals, temperature and mechanical energy are applied to the surface to remove the adhered soil.
6. Inter-rinse. Detached soil and cleaning chemicals should be rinsed from the surfaces using low pressure cold water.
7. Disinfection. Chemical disinfectants are applied to reduce the viability of the remaining microbes to a level deemed to pose no significant risk.
8. Post-rinse. Disinfectants should be rinsed from the surface using low pressure cold water, although some disinfectants are surface active and are intended to be left on the surfaces.
9. Interproduction cycles. Certain procedures may be undertaken to prevent the growth of microorganisms on production contact surfaces before the next production period; these include removal of excess water and equipment drying. The process area may alternatively be fogged.

The sequence of cleaning and disinfection determines the order in which product contact surfaces (equipment) and environmental surfaces (walls, floors, drains etc.) are cleaned and disinfected to ensure that once product contact surfaces are disinfected they are not recontaminated. A typical sequence would be as follows:

i. remove gross soil from production equipment
ii. remove gross soil from environmental surfaces
iii. rinse down environmental surfaces (from top to bottom)
iv. rinse down equipment (from top to bottom) and flush to drain
v. clean environmental surfaces
vi. rinse environmental surfaces
vii. clean equipment
viii. rinse equipment
ix. disinfect equipment
x. fog

3.3.2.6 Monitoring the effectiveness of the cleaning and disinfection system. Due to their role in the control of the environmental routes of product contamination, cleaning and disinfection programmes require regular monitoring as part of a structured quality assurance system. This evaluation normally involves an immediate sensory evaluation and an historical assessment of the programme's effectiveness by detecting remaining microorganisms or chemicals.

Sensory evalution involves a visual inspection of the surfaces, smelling for products or offensive odours and feeling for greasy or encrusted surfaces.

Microbiological (or chemical) techniques may be applied if product residues are not detected. Traditional microbiological techniques involve sampling the microorganisms present on the surface using, for example, sterile cotton or alginate swabs and sponges. The microorganisms are then suspended in a suitable recovery or transport medium for dilution and plating. Other sampling techniques include self prepared or commercial agar contact plates that are pressed onto the surface. These methods give an historical evaluation of the hygiene of the surfaces. Relatively recently, rapid methods have been developed to assess surface hygiene, such as epifluorescent microscopy and adenosine triphosphate (ATP) analysis. These techniques give a result in a timescale relevant to process control such that the cleaning and disinfection regime can be repeated if necessary.

For epifluorescence microscopy, microorganisms are sampled by swabbing or rinsing and then collected onto filters; the microorganisms present on the filter are enumerated using the direct epifluorescent filter technique (DEFT).[14]

ATP bioluminescence is based on detecting the levels of adenosine triphosphate (ATP) present in plant, animal and microbial cells, using an enzyme system that produces light in proportion to the amount of ATP present. The light output can be measured in a luminometer in approximately 5 min. In the assessment of surface hygiene, analysis of total ATP is most suitable as after cleaning and disinfection any residues, whether microbial or product in origin, should have been removed.

Surface assessment methods only sample a very small proportion of the production plant surface and are therefore analogous to 'end product analysis' of a production batch. These assessment techniques should, therefore, monitor the effectiveness of a structured hazard analysis critical control point (HACCP) approach.[12] In a similar manner to production processes, critical control points (CCPs) require identification and control and vary between processing areas and types, but typical cleaning CCPs may include: detergent and disinfectant concentrations, chemical solution temperature, chemical contact times, degree and time of application procedures, equipment settings (i.e. pressure), cleaning equipment maintenance and chemical stock rotation. The use of HACCP in relation to cleaning and disinfection would put emphasis on control of the programme as it was happening and should ensure that an acceptable level of surface hygiene was consistently achieved.

3.4 Control of production environment air

3.4.1 Aim

Control of the production environment air may be for several purposes.

- prevention of contamination or deterioration of the product
- temperature control
- control of relative humidity
- control of dust
- control of vapours.

Air control systems do not guarantee freedom from air-borne contamination as operations within the production area such as movement of personnel, machinery and packaging, and cleaning will influence the level of air-borne contamination.

Air-borne particles cover the range of sizes from sub micron to about 100 microns (Figure 3.1), i.e. from viruses to large spores. The size of the air-borne particle has a direct effect upon its stability in aerosol form and particles of 2–3 μm or less remain suspended in air almost indefinitely. Larger particles sediment at rates dependent on their size and density.

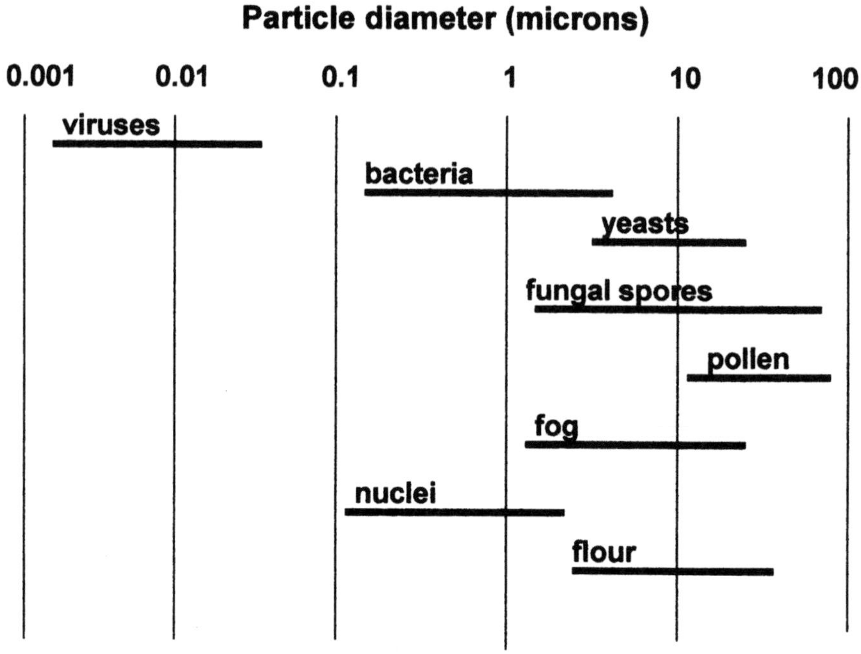

Figure 3.1 Comparison of the sizes of air-borne particles.

3.4.2 Sources of air-borne contamination

Microorganisms may be introduced into the processing area air from a number of sources and by several mechanisms. Contamination outside the processing area may enter via:

- doors and windows
- ventilation systems (therefore high efficiency particulate air (HEPA) filters should be fitted)
- wheels of trolleys and fork lift trucks, and shoes and clothing of personnel
- surface of outer packaging material or on the raw materials.

Air-borne contamination may be created within the processing area by the production practices.

- Movement of dry powders such as flour, spices, dried milk etc., generates significant levels of dust particles, some of which may have microorganisms attached to them.
- Processes involving liquid handling such as vegetable washing are likely to create aerosols.
- Cleaning systems and techniques such as high pressure/low volume

spraying, low pressure/high volume spraying, and floor scrubbers create aerosols containing viable microorganisms over significant distances from the site of use.[15]

Table 3.4 shows the levels of organisms present in the air related to the process taking place. Floors and drains have been shown to be a source of high levels of air-borne microorganisms, particularly when flooded.[16] Figure 3.2 shows the effect of flooding on the levels of air-borne bacteria.

- Dusty surfaces also contribute to the amount of air-borne contamination. Re-entrainment has been shown to occur at air velocities normally existing in the processing environment, and is dependent on surface roughness and air turbulence.
- Humans contribute to the presence of air-borne contamination. Skin particles are constantly shed from the body and may carry viable bacteria.[17] In addition hairs are lost from the head at a rate of 100 per day[18] and may also carry viable microorganisms. Speaking, coughing and sneezing generate aerosols containing microorganisms.
- The use of compressed air and vacuum systems can redistribute dust and microorganisms.

3.4.3 Control of air-borne contamination

Air control systems are mainly effective in controlling air-borne contamination entering the production area, but have little effect on contamination arising from other sources. Air control systems are very effective at controlling the air entering the area and can effectively combat condensation. Air filtration systems are also effective in controlling the contamination from compressed lines and exhausts from vacuum systems.

3.4.3.1 Air filtration. Filtration is the most commonly used technique for controlling air-borne contamination. Frequently two or more filters will be incorporated in the filter unit (e.g. primary, secondary and tertiary) to filter progressively decreasing particle sizes.

Air filters are categorised on the basis of arrestance and efficiency on a scale from 1 to 14 on the Eurovent 4/5 grading system. Arrestance is a measure of the filter's ability to capture and retain a known weight of synthetic dust fed into the filter at a specific rate to a final static pressure drop. Efficiency is a measure of the discoloration to filter test discs up and downstream of the filter. Table 3.5 summarises the grades of filters available.

The air quality required is dependent on the type of production area. For areas requiring good manufacturing practice (GMP) quality air, final filtration to EU4–5 is sufficient. For medium care areas, final filtration to EU5–7 would suffice, whereas high care areas require filtration to EU7–10

Table 3.4 Mean levels of microorganisms per 60 l of air for different production processes

Process		Mean cfu/60 l air	
		Total viable count	Yeasts/Mould count
Vegetable freezing		552.5 (n = 103)	316.3 (n = 104)
Cleaning		366.1 (n = 226)	225.0 (n = 210)
Meat filling		295.3 (n = 59)	34.6 (n = 42)
Waste disposal		280.0 (n = 6)	—
Engineering areas		278.0 (n = 6)	1.4 (n = 5)
Conveyors	total	207.8 (n = 26)	45.0 (n = 26)
	meat	282.0 (n = 26)	68.8 (n = 16)
	vegetable	146.7 (n = 6)	15.2 (n = 6)
	pizzas	3.3 (n = 4)	0.5 (n = 4)
Slicing		187.5 (n = 112)	83.5 (n = 112)
Raw materials receipt		183.6 (n = 16)	128.0 (n = 11)
Tray washing		145.5 (n = 94)	85.5 (n = 86)
Aseptic filling		89.5 (n = 54)	20.1 (n = 48)
Primary packaging		47.2 (n = 239)	13.6 (n = 213)
Stores		35.9 (n = 18)	19.9 (n = 18)

cfu = colony forming units.

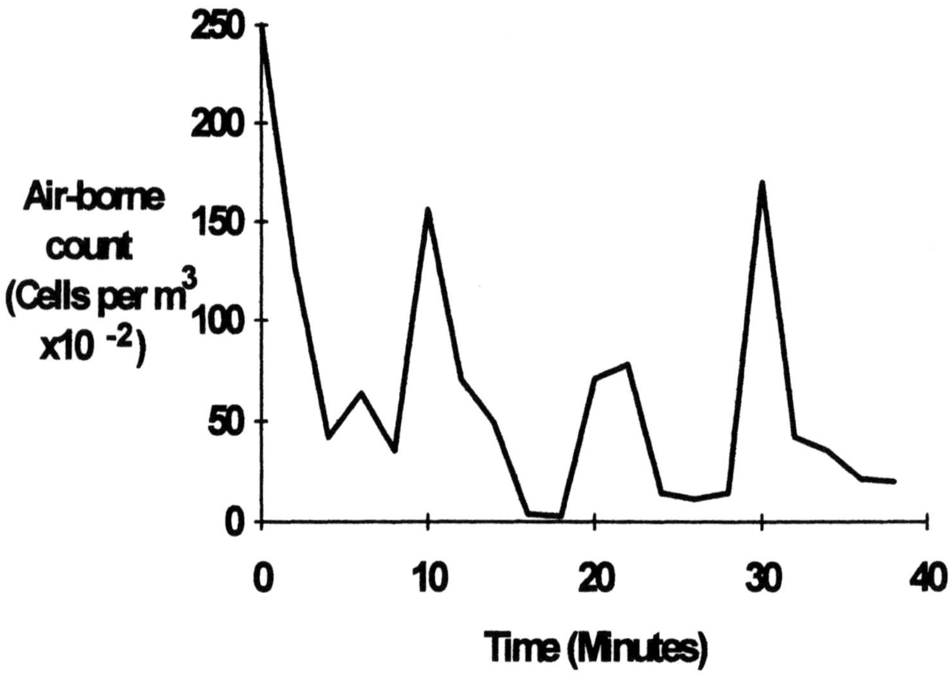

Figure 3.2 Effect of flooding at 10 min intervals on the air-borne bacteria counts above drains in four food production areas (adapted from Heldman).[16]

Table 3.5 Air filter types and their particle retention capabilities

EU grading Eurovent 4/5	General air filtration description	Approaching 100% retention of particles
EU1–EU4 (5)	Primary filters	>5 µm
EU6–EU7	Secondary filters	>2 µm
EU8–EU9	Secondary filters	>1 µm
EU10–EU14	Semi HEPA and HEPA filters	>0.5 µm

3.4.4 Methods of reduction of air-borne contamination

As aerosols that are produced in a food processing environment can potentially remain air-borne indefinitely, methods are required to reduce levels of air-borne microorganisms. The most widely used method to date is disinfectant fogging. The spraying of a disinfectant mist into the air can only be performed outside production hours due to the health risk to workers and the possibility of contamination of product. Alternative methods for the disinfection of air-borne particles include the use of negative ions, ozone, ultraviolet light and electric fields.

3.4.5 Monitoring of air quality

Air velocity is measured using vane anemometers which require calibration. Air pressure is measured by electronic differential pressure gauges and again calibration is required. Electronic particle counters are available but these instruments are designed for 'clean room'-type facilities and the sensors may become overwhelmed and fouled by processing environments.

There are a number of microbiological methods available for the estimation of air-borne contamination, although different methods of sampling will provide different estimates of air-borne contamination. The microbiological techniques include sedimentation, impingement techniques and filtration methods.

The sedimentation method involves exposing a standard Petri dish containing non-selective agar for a defined exposure time. This method is simple and easy, but does not measure the number of viable organisms per unit volume of air. It relies on the deposition of particles over a period of time and this can be greatly influenced by air currents.

Impingement methods involve purpose built samplers that draw a set volume of air over an agar plate or strip. These samples are portable and allow the sample volume to be adjusted.

Filtration methods involve placing a cellulose, glass fibre, or membrane filter in a suitable holder and drawing a known volume of air through the membrane using a vacuum pump. The membrane is then either placed directly on an agar plate, or the organisms are recovered from the filter and plated out. The filtration method is most suitable for desiccation-resistant mould and bacterial spores.

3.5 Control of the personnel route of contamination

People can be a source of contamination and/or can transfer contamination from other sources to the product.

Bacteria found on the skin fall into two categories: transients, which are organisms deposited onto the skin and do not multiply, and residents, which readily multiply on the skin (e.g. *Staphylococcus aureus*). The concentration of bacteria on the skin is normally 10^5–10^6 bacteria per cm^2 depending on the area of the body. Skin particles are constantly shed from the body due to friction with clothing and drying of skin particles. During undressing, for example, it is estimated that 5×10^5 scales become airborne, 5–10% of which may carry viable microbes.[17]

Hands are the principal agents in transferring pathogens to product and handling should be kept to a minimum. Microbiological examination of food handlers often demonstrates the presence of a large number of potentially pathogenic organisms.[19] Consequently regular and thorough hand washing is essential.

Before any staff are allowed to commence work they should be advised of their personal obligations with respect to hygiene and should complete a detailed medical questionnaire. An experienced medical officer should then decide whether a medical examination is required prior to employment. A high standard of personal hygiene is therefore required to minimise the risk of contamination to product. A list of the basic personal hygiene requirements is given below (for guidance only).

1. Protective clothing, footwear and headgear issued by the company must be worn.

   Protective clothing must be worn by all personnel in processing areas, and is designed to protect the product from contamination and the wearer's own clothing. The protective clothing should be in good repair, laundered regularly and changed when soiled. The clothing should be light coloured and may be colour coded if segregation of personnel is required, for example in high and low risk operations. The clothing must cover the operative from neck to knee, have tight cuffs and no outside pockets.

   A fine hairnet may be appropriate in addition to the protective headgear, depending on the processing environment. Hairclips and grips should not be worn.
2. Beards and moustaches must be kept short and trimmed and a protective cover worn when appropriate.

3. Personnel must wash their hands thoroughly before commencing work, after using toilets, after handling waste food, on returning to the production area after leaving for any reason and as frequently as necessary during the day. Hands should be washed in hot water (45°C) with a suitable non-perfumed bactericidal soup using non-hand operable sinks. The hands should then be dried effectively and in certain circumstances an alcohol-based hand sanitiser should be used.
4. Fingernails should be kept short and clean. The wearing of make up, false eyelashes, false nails and nail varnish should be prohibited.
5. Watches and jewellery must not be worn (except plain wedding bands and possibly 'sleeper' earrings).
6. Personal items must be left in lockers outside the production area.
7. Food and drink must not be taken into or consumed in areas other than the tea bars, restaurants etc. Sweets and chewing gum must not be consumed in production areas.
8. Smoking and the taking of snuff must be prohibited. If smoking is permitted at all, designated areas should be provided that are separated physically from production areas.
9. Spitting is forbidden.
10. Superficial injuries such as cuts, grazes etc. must be covered by a waterproof dressing that is distinctly coloured and metal detectable.
11. Personnel suffering from heavy colds, stomach disorders, diarrhoea, skin conditions, etc. must report the illness to the medical officer or supervisor. The person should be excluded from work involving food handling until pronounced fit. Staff returning from foreign travel or who have been in contact with infected persons should also inform the medical officer.

Management is responsible for ensuring that all personnel are aware of and understand the relevant factory regulations. Regular training programmes should be conducted to stress the importance of personal hygiene and behaviour and to issue up-dates on legal obligations.

Further reading

Imholte, T.J. (1984) *Engineering for Food Safety and Sanitation*, Technical Institute Foods Engineering, Crystal, Minnesota.

Sprenger, R.A. (1993) *Hygiene for Management*, Highfield Publications, Doncaster.

References

1. Troller, J.A. (1993) *Sanitation in Food Processing*, Academic Press Inc., San Diego, California.
2. Council Directive, 93/43/EEC (1993) on the hygiene of foodstuffs. *Official J. European Communities*, L175, 1–11.

3. Council Directive, 92/4/EEC (1992) for the intra-community trade in meat products. *Official J. European Communities*, L57, 1–14.
4. BS6431: Part 1 (1983) (EN87) *Ceramic Floor and Wall Tiles*, British Standards Institution, London.
5. BS5980 (1980) *Specifications for Adhesives for Use with Ceramic Tiles and Mosaics*, British Standards Institute, London.
6. Council Directive, 89/392/EEC (1989). Relating to machinery. *Official J. European Communities*, L183, 9–26.
7. European Hygienic Equipment Design Group (1993). Hygiene equipment design criteria. *Trends Food Sci. Technol.*, **4**, 225–229.
8. Holah, J.T., Betts, R.P. and Thorpe, R.H. (1989) The use of epifluorescent microscopy to determine surface hygiene. *Internat. Biodeterioration*, **25** 147–153.
9. Holah, J.T. and Kearney, L.R. (1992) Introduction to biofilms in the food industry. In *Biofilms: Science and Technology*, Melo, L.F., Bott, T.R., Fletcher, M. and Capdeville, B. (eds), Kluwer Academic Publishers, Dordrecht, The Netherlands, pp. 395–402.
10. Jennings, W.G. (1965) Theory and practice of hard surface cleaning. *Adv. Food Res.*, **14**, 325–459.
11. Koopal, L.K. (1985) Physico-chemical aspects of hard surface cleaning. *Netherlands Milk Dairy J.*, **39**, 127–154.
12. Holah, J.T. (1992) Cleaning and disinfection. In *Chilled Foods: A Comprehensive Guide*, Dennis, C. and Stringer, M. (eds), Ellis Horwood, London, pp. 319–341.
13. Klemperer, R. (1982) *Tests for Disinfectants: Principles and Problems in Disinfectants: Their Assessment and Industrial Use*, Scientific Symposia Ltd, London.
14. Pettipher, G.L., Mansell, R. and McKinnon, C.M. (1980) Rapid membrane filtration epifluorescent microscopy technique for direct enumeration of bacteria in raw milk. *Appl. Environ. Microbiol.*, **39**, 423–429.
15. Holah, J.T., Taylor, J.H. and Holder, J.S. (1993) *The Spread of* Listeria *by Cleaning Systems, Part II.* Technical Memorandum No. 673. Campden Food and Drink Research Association.
16. Heldman, D.R. (1974) Factors influencing air borne contamination of foods: A review. *J. Food Sci.*, **39**, 962–969.
17. Noble, W.C. (1981) Microbiology of human skin. *Major Problems in Dermatology*, **2**, 5353–655.
18. Haynes, P.R. (1985) *Food Microbiology and Hygiene*, Elsevier Applied Science Publishers, Barking, UK.
19. Kerr, K.G., Birkenhead, D., Seale, K., Major, J. and Hawkey, P.M. (1993) Prevalence of *Listeria* spp on the hands of food workers. *J. Food Protection*, **56**, 525–527.
20. Elliot, R.D. (1980) Cleaning and sanitation. In *Principles of Food Processing Sanitation*, Katsuyama, A.M. (ed.) The Food Processors Institute, USA, pp. 91–129.

4 An introduction to surfactants

J.J. LEWIS

4.1 Why are surfactants of interest?

Use of surfactant chemicals to modify the surface or interface between two phases leads to a number of very interesting effects. These include:

- The ability to form a stable emulsion of oil and water, which allows the production of mayonnaise for salad dressing, hand creams and lotions for personal grooming, or to emulsify bitumen for road laying.
- The ability to generate foam for use in bubble baths, toothpaste, or polyurethane foam for refrigerator insulation.
- The ability to disperse solid particles in a liquid matrix, and so give rise to cream cleansers for bathrooms, or emulsion paints.
- The ability to wet a surface not normally easy to wet, and so to provide leaf penetration for insecticides, or in combination with the other properties listed above, to provide detergency for the domestic washing of clothes.

The above list of effects is by no means exhaustive, but is included to show that there is a very wide variety of domestic and industrial uses for these products. The major uses, however, are for systems which deal with the interface between solid and liquid, liquid and liquid, and liquid and gas phases. In general, water is one of the liquid phases for these pairs of interactions.

This short introduction to surfactants seeks to illustrate the elegant nature of these molecules in their applications. This will be done by describing the reasons for their modes of action, the uses these actions can be employed for, some of the chemical syntheses and composition of commercially available materials and an approach to formulating with these materials.

The intention of this chapter is to provide a first impression of the main principles of surfactancy, and hopefully to provoke an interest in knowing more about what is an exceedingly diverse field of chemistry and physical phenomena. One of these physical phenomena is that of emulsification, where surfactants can be used to combine materials such as oil and water into a relatively stable, inhomogeneous mixture which may be difficult to achieve otherwise.

The question may be asked: What properties of surfactant molecules allow this emulsification to take place readily? In order to answer this question, it is worth asking another question first: Why are oil and water not miscible?

4.2 How surfactants work

4.2.1 Molecular interactions and interfacial tension

Whether a substance is a solid, liquid or gas at a particular pressure and temperature depends upon the strength of the intermolecular interactions between molecules. If we consider the types of cohesive energies which are operating in formulated products, the most significant of these are:

- Van der Waals' forces, either dispersion (arising from the relative dissymmetry of electron clouds of two non-polar molecules in close proximity) or permanent (arising from permanent dipoles in polar molecules).
- Hydrogen bonding, which arises by the formation of a 'hydrogen bridge' between two electronegative atoms or molecules.
- Ionic forces, which arise due to permanent ionic charge interactions between atoms or molecules.

The respective strengths of these three forces are of the order of 1:5:150 respectively, although this is system dependent.

For materials which only rely on weak molecular interactive forces, such as propane or butane, the result is that at room temperature and pressure these materials are gases. If the molecule is larger, such as pentane or hexane, the sum of these small cohesive forces over a bigger compound is enough to make these materials liquids. However, water is a much smaller molecule, but due to its polar nature and shape, and consequent hydrogen bonding, it is a liquid under the same conditions. Common salt, sodium chloride, is an ionic material, solid at room temperature, and only melting at 800°C. However, when salt is dissolved in water, the strong ionic bonding is replaced by hydrogen bonding of the anion and the cation with water. The combination of hydrogen bonding and electroneutrality in solution allows the ionic lattice of the salt to be replaced by a multitude of weaker forces.

If a paraffinic oil is mixed with water, a number of cohesive forces are in operation. For oil–oil, or oil–water, only dispersion forces are present. For water–water, the far stronger hydrogen bonding forces are in play. In consequence, water preferentially bonds to water, and the oil collects as a separate layer due to a difference in density. This immiscibility of oil and water is therefore the most thermodynamically stable state.

A further phenomenon now arises. For molecules in the bulk of the water, hydrogen bonding occurs in all the usual directions, and the net forces on these molecules is zero. However, at the interface between oil and water, water molecules are pulled towards the bulk by the hydrogen bonding, but only pulled away by the weak dispersion forces. Consequently, the entire interfacial area is subject to a force trying to contract it to the smallest possible area. If a single drop of water were present in the oil, it would be spherical. In the case of a water–air interface, the interfacial tension is much greater, as the dispersion forces of air are very much weaker than oil, due to the much lower density. The special case of interfacial tension against air is known as surface tension.

Most commonly encountered room temperature liquids have surface tensions against air or their vapours that lie in the range 10–80 dyn cm^{-1} (Table 4.1). Water has a value at the high end of this range (72–73 dyn cm^{-1}), while hydrocarbons are toward the lower end (20–30 dyn cm^{-1}).

Of course, the above is a simple picture of what is really going on. The interface is treated as a static entity, whereas there is actually a rapid interchange of molecules. A water molecule may typically have an average residence time at the interface of 3 μs at room temperature! Not surprisingly, temperature has a large effect on interfacial tension due to the mobility of surface molecules.

4.2.2 What does a surfactant do?

A surfactant is a chemical species that is active, and has a preference for, a surface or interface, which is the boundary between two phases. When the surfactant is used as an emulsifier, the surfactant will lower the energy differences between the two phases to allow a relatively stable mixture of oil and water to form. It should be recognised at this stage that the most thermodynamically stable state remains the separation of the emulsion or foam, but that surfactants can maintain the metastable state required for sufficient time in order to allow advantageous exploitation.

There are many compounds which, when dissolved in water, can

Table 4.1 Interfacial tension for different liquids

Liquid	Interfacial tensions (dyn cm^{-1})	
	vs Air	vs Water
Water	72.8	–
Octane	21.8	50.8
Benzene	28.9	35.0
n-Octanol	27.5	8.5

Table 4.2 Effect of organic additions on surface tension

System	Surface tension (dyn cm^{-1})
Water	73
Acetone in water (20%)	42
Acetone in water (50%)	30
EAS in water (0.0004%)	48
EAS in water (0.004%)	31

decrease the interfacial tension, but surfactants are generally the most cost effective and efficient way of achieving these effects. This is due to their ability to concentrate at the interface, rather than to remain in the bulk of the solution. Table 4.2 illustrates this point, comparing the reduction of interfacial tension of an air/water system by acetone, and an ethoxylated alcohol surfactant (EAS, lauryl alcohol plus 5 ethylene oxide).

Table 4.2 shows that the surfactant compound is 10 000 times more capable of reducing the surface tension than the simple organic molecule!

4.2.3 Structure of an idealised surfactant and its interfacial activity

In simple terms, an idealised surfactant molecule is essentially two chemicals in one; a hydrophilic portion, which tends to keep the molecule in the aqueous phase, and a hydrophobic portion, which will prefer to be expelled from the aqueous phase. There is therefore a dual ('confused') nature to these materials. A wide range of commercially available materials will be discussed below, but for now a common type of material will be exemplified – soap.

Soaps are the most common and greatest volume surfactants in use around the world. They are usually sodium salts of long alkyl chain carboxylic acids, the alkyl chain being from 10–18 carbon atoms long (Figure 4.1).

The long alkyl chain is comparatively bulky when compared to the ionic part of the molecule, which is more frequently described as the headgroup. The simplified view of a surfactant molecule is therefore to picture it as a small matchstick, with an oily tail and a highly polar head.

Consider this molecule in the bulk of an aqueous phase. The polar head will be solvated by polar bonding to the water. However, the oily chain will be bound to the water molecules only by the much weaker dispersion forces. As a consequence of this, the water will become much more structured, as hydrogen bonding on one side is not balanced out by the dispersion forces acting on the other (as occurs for interfacial tension). The result of this imbalance is that an 'iceberg' of structured water is formed around the alkyl chain. This increase in the structure of water is entropically disfavoured, as the system has become more, and not less, ordered.

If the molecule migrates to an oil–water interface, it may orientate itself such that the polar head remains in the aqueous phase, and the oily tail can lose the ordered sheath of water molecules by transferring to the oil phase. This is thermodynamically favoured.

Now consider the surfactant molecule in the bulk of the oil phase. Here, the only forces operating are the weak dispersion type, and both the polar head and the oily tail are unencumbered by solvation effects. But, should the molecule migrate to the oil–water interface, it is energetically favourable for the polar head to be solvated by the water molecules (as salt readily dissolves in water). However, as above, there is an energy barrier due to entropic effects which disfavours the easy passage of the molecule fully into the aqueous phase.

In summary, then, it is thermodynamically favoured for the model surfactant molecule to orientate itself at a water–oil interface such that the polar head is preferentially in the water phase, and the oily tail is preferentially in the oil phase. This is illustrated in Figure 4.2.

It should be noted that the ability of surfactants to migrate to an interfacial region is driven from both sides by the strength of the hydrogen

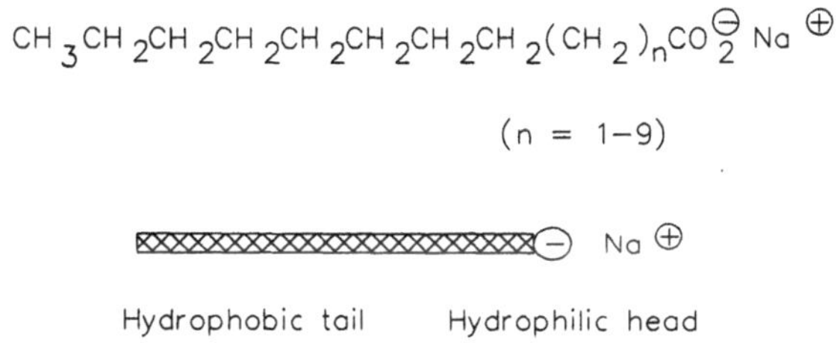

Figure 4.1 Chemical structure of soap, and an idealised surfactant model.

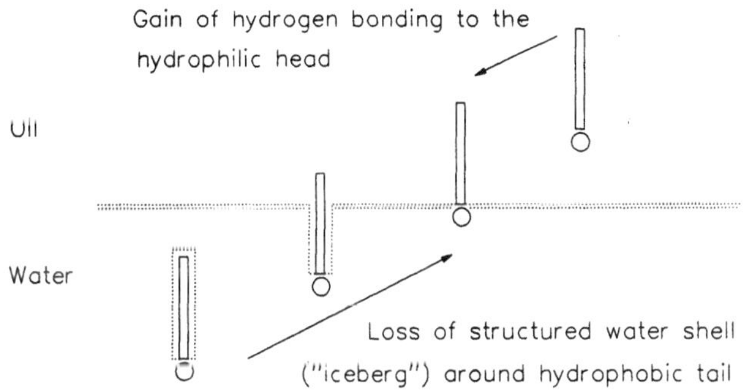

Figure 4.2 Forces leading to the preference of surfactants for an interface.

bonding interactions of the aqueous phase, and is not simply the result of the polar head group preferring the polar environment, and the oily tail preferring the oily phase. The surfactant emulsifier may therefore be described as having a limited solubility in both the aqueous and oil phases. In the case of a simple solute, such as acetone, this is clearly not the case.

4.2.4 Surfactancy and the reduction of interfacial tension

The above section has discussed why it is thermodynamically favourable for surfactants to collect at an oil–water interface. This interface represents the boundary between very polar, densely packed water molecules, and less dense, weakly interacting oil compounds. The presence of the surfactant markedly reduces the difference between these two environments by appearing to be a surface demonstrating properties similar to the phase from which it is viewed. The aqueous environment is bounded by the polar heads of the surfactant, and the oily phase is bounded by the oily tails.

However, seen from the other side, the water appears oily (bounded by the surfactant tails), and the oil appears highly polar (bounded by the polar heads).

The result of this is that the water surface assumes the dispersion characteristics of the oil, and the imbalance of forces between this 'skin' and the true oil layer is dramatically reduced. The interfacial tension is therefore markedly reduced, and the aqueous phase becomes much easier to disperse into the oil. Conversely, the oil surface can assume the polar characteristics of the aqueous phase, leading to ease of dispersion in water.

This argument can be extended to explain why surfactants have no effect on the interfacial tension between oils and air; on both sides of the interface only dispersion forces are in operation, so there is no driving force to concentrate common surfactants at the surface, and no real energy difference to result in a change in interfacial tension. The effect is, of course, finite. The reduction in interfacial tension will continue as more surfactant concentrates at the interface, until a complete monolayer is formed. The interfacial tension will effectively reach a steady value, even though surfactant concentration may continue to build up at the interface.

4.2.5 Micellisation

Once the amount of surfactant available has increased beyond the point at which an interfacial monolayer has been formed, the interfacial tension becomes relatively constant. In addition to this, a number of other solution properties in the aqueous phase also change, such as conductivity and detergency (Figure 4.3).

Although more surfactant molecules cannot adsorb at interfacial sites, once the interface is saturated a further phenomenon occurs, which explains the changes mentioned above. This is called micellisation. This occurs in aqueous solution when the molecules aggregate together and order themselves with their hydrocarbon tails towards each other, and their hydrophilic groups outermost. In effect, a small spherical interface is created inside the bulk of the solution (see Figure 4.4). Even though this orientation is more structured than the monomeric forms of surfactant dissolved in water, and is therefore disfavoured on entropy grounds, it is not as disfavoured as the state in which all the individually dissolved molecules carry around their own 'iceberg' shell of oriented water molecules. This is due to the appearance of the micelle as a collection of

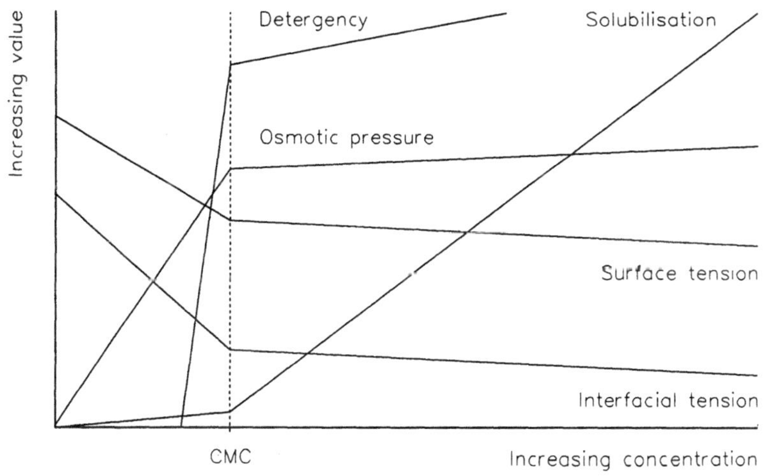

Figure 4.3 Change in aqueous solution characteristics due to micellisation around the critical micelle concentration (CMC).

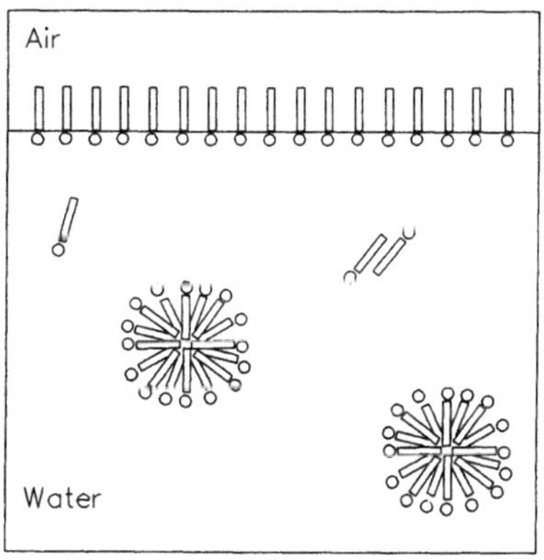

Figure 4.4 Formation of micelles in an aqueous medium after surface saturation.

hydrophilic headgroups arranged in a sphere, allowing hydrogen bonding with the surrounding water readily to take place.

Once again, the case for micellisation of surfactant in the oil phase is less clear cut, as the driving mechanism for aggregation of molecules in the non-polar environment is considerably less.

The concentration of surfactant at which the change in solution properties takes place when micelles form is known as the critical micelle concentration (CMC). The number of surfactant molecules needed for this to happen ranges from 30–100 for ionic molecules, where micellisation is countered by repulsion between the similar charges on the headgroups, to several thousand for non-ionic molecules, where there is no charge repulsion to oppose micellisation. Not surprisingly, this lower resistance to forming a micelle means that non-ionics will form these structures at lower concentration than ionics, typical values being at 0.002% and 0.1%, respectively.

Temperature, pH, other ions and organic solutes also affect the process. However, as for surfactants concentrating at interfaces, it should be remembered that micelles are dynamic entities, with individual members joining and leaving at a very high rate. In addition, micelles need not be simply spherical in shape, and have also been shown to occur as discs, rods, and even in sheet structures.

4.3 Surfactant phenomena and effects

Section 4.2 discussed the driving force for the concentration of surfactants at an interface between a polar environment (usually water) and a second phase (air, oil or solid). That driving force was identified as attributable to the strongly polar molecular interactions of the polar phase (the hydrogen bonding of water).

This section deals with some of the more useful applications of this phenomenon (emulsification, foaming, wetting, solids dispersion, solubilisation, microemulsions and finally, detergency). Section 4.4 introduces more of the chemical structure of surfactants and the processes used for their production, with section 4.5 discussing the practical aspects of using these materials.

4.3.1 Emulsification

An emulsion can be defined as the dispersion together of two immiscible liquids in the presence of a third component, the emulsifier. It should be recognised at the outset that an emulsion is a thermodynamically metastable state, and that even in the presence of the emulsifier, this mixture will only have a finite lifetime. This definition applies to macroemulsions, which are defined as mixtures with particle sizes between about 0.1 to 20 μm. For mixtures with particle sizes of less than 0.1 μm, the term microemulsion is used (see section 4.3.6).

The most common emulsions are either oil in water, or water in oil. For oil in water (O/W), such as milk or bitumen emulsions, the water is the continuous phase. For water in oil (W/O), such as butter or hydraulic fluids, the continuous phase is the oil.

During the emulsification process, when one phase is dispersed in the other, a massive increase in the interfacial area between the two phases takes place. The difference in free energy between the two-phase state and the emulsion is primarily dependent on the need to increase interfacial area in the presence of interfacial tension between the two. Therefore, the ability of the surfactant to lower this interfacial tension is key to the ease of emulsification.

Two further points are worth noting. The first is that there is obviously an entropy term working in favour of dispersing one phase into a multitude of droplets. However, the overall thermodynamics of the process is unfavourable, due to the large amount of energy (large enthalpy) required to generate the extra area. Secondly, several orders of magnitude of energy above that which may nominally be required to effect the emulsification process are needed to produce the emulsion, as the process itself is usually carried out with much agitation, and energy is lost as heat and material motion.

Given that emulsions are only kinetically stable, the role of the surfactant in slowing the breakdown processes is now considered. The breakdown processes are illustrated in Figure 4.5.

4.3.1.1 Creaming or sedimentation. This is dependent upon the density difference between the continuous phase and the dispersed phase. If it is not possible to match the densities closer than those for the chosen system, the addition of thickeners will help to increase the emulsion lifetime, as will the generation of smaller particle sizes during processing. Surfactancy changes have little effect here.

4.3.1.2 Flocculation. This is the process whereby droplets form clusters, and is a reversible phenomenon. This low energy interaction is due to the balance of a weak attractive force (Van der Waals') with a steric repulsion between the two droplets. By choosing a surfactant which can increase the steric repulsion further than the surfactant used for the initial emulsification, flocculation may be prevented altogether. If we consider an O/W system emulsified with an ethoxylated alcohol, then the hydrophilic ethoxylate chain would be on the outside of the oil droplets in the water phase. Increasing the ethoxylate chain length would force the droplets further apart until the Van der Waals' attractive forces fade away. Should the

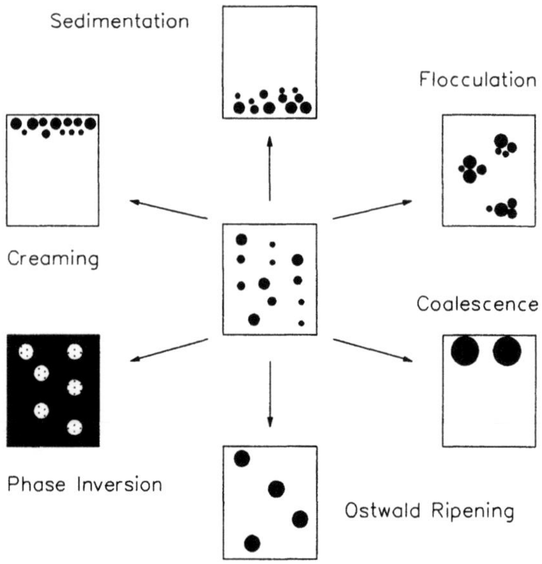

Figure 4.5 Emulsion breakdown mechanisms.

droplets form clusters which cannot be readily redispersed, the emulsion is said to aggregate.

4.3.1.3 Ostwald ripening. As two emulsified liquids are seldom completely immiscible, there can be some transfer of material from one dispersed droplet to another via the continuous phase. The driving force for this transfer of material is partially dependent directly on the interfacial tension, and also on the inverse of the droplet surface area, that is, smaller droplets have a larger internal pressure than large droplets. Ostwald ripening is therefore manifest when larger droplets grow at the expense of smaller ones. A surfactant may be employed to counteract this method of emulsion decay in two ways. The reduction in interfacial tension which will occur by using a surfactant will reduce the driving force for the transfer of material, and the adsorption at the interfacial boundary will lead to a steric hindrance of the material being transferred.

4.3.1.4 Inversion. When one phase is dispersed in another, it is possible to keep adding this first phase until it reaches a critical volume fraction of the emulsion as a whole. At this critical volume, the emulsion may invert, for example change from O/W to W/O. For most emulsions, this is usually manifest by a sudden drop in viscosity of the mixture. Unfortunately, although there are several empirical rules which describe this phenomenon, it is not well understood quantitatively.

4.3.1.5 Coalescence and the Gibbs–Marangoni effect. Coalescence can occur if two droplets come close enough together that the film of the second phase which separates them ruptures. The effectiveness of a surfactant adsorbed at the interface to prevent the collapse of the emulsion can be seen as threefold.

1. It can reduce the overall potential energy of the system by reducing the interfacial tension.
2. If the adsorbed surfactant carries an electric charge, repulsion between approaching droplets is possible.
3. If the molecule adsorbed at the interface is sufficiently bulky, a steric interaction may also prevent coalescence.

Beyond these effects, there are further phenomena which occur when the film continues to thin prior to coalescence. This is the phenomenon of self healing, or the Gibbs–Marangoni effect.

Consider an interface on a thin film, with an equilibrium saturation of surfactant molecules. As the film is stretched (by the second phase molecules moving out from between the two droplets of the first phase) some surfactant is dragged out from the interface. Two things now take place which oppose this disruption of the film.

1. A local increase in interfacial tension takes place, as there is not sufficient surfactant at the interface. The increase in interfacial tension opposes the force stretching and thinning the film. This is known as the Gibbs effect.
2. As surfactant moves back towards the area of lower concentration, in order to re-establish the equilibrium at the interface, molecules of the second phase are dragged back with it, restoring the film thickness. This is known as the Marangoni effect.

4.3.2 Foams

Foam phenomena can be seen as an extension of the attributes associated with emulsions, but replacing the liquid–liquid with a liquid–gas mixture. As for emulsions, a third element (e.g. surfactant) is required if a kinetically stable foam is to be produced, as pure liquids will not allow foams to be produced. As for emulsions, the primary role of the surfactant is the ability to oppose film thinning and ultimate coalescence of the gas phase trapped in the liquid. The Gibbs–Marangoni effects are therefore crucial in this area too, and the driving force for film thinning that this effect opposes is the draining of the film under gravity.

The optimum concentration of surfactant for maximum foaming is usually around the CMC for a number of systems. The onset of micelle formation can be seen as a way of providing a reservoir of surfactant monomers which can aid the Marangoni effect. However, if there is an oversupply of surfactant within the foam, then there is a smaller driving force for the restoration of the thinning film, as the need to transfer surfactant (and associated solvent) to the site of thinning is now less.

In practice, foams tend to be much more difficult to study and to draw conclusions from than emulsions, as most applications involve the simultaneous production and decay of the foam. This rapid turnover makes close study of the surfactant mechanisms for stabilisation and their role in the ultimate destruction of a foam very difficult. The ability to control foam is sometimes very important for particular applications. Some uses (e.g. shampoos) require the generation of a thick and creamy foam for performance or aesthetic reasons. However, should the same formulation be used in a washing machine, undesirable consequences could easily result!

Antifoaming agents generally work through a reverse Marangoni effect. The rationale is that addition of a substance which displaces the surfactant from the interfacial film not only decreases the surfactant content there, but also drags solvent molecules away with it. This may be achieved by displacing the surfactant with a more interfacially active material, but one which interrupts the interfacial packing and stability of the original surfactant film. Polyamides, EO–PO (ethylene oxide–propylene oxide) copolymers and polydimethyl siloxanes are well-known examples of this type of compound.

4.3.3 Wetting

If a drop of liquid is placed on a solid surface and spreads over it, it is said to wet the solid. If it does not wet the surface, the liquid will remain as a drop. For wetting to take place, the liquid must have a surface tension below the critical surface tension (CST) which is a characteristic of the solid. As water has a surface tension of about 72 mN m^{-1}, it will not wet polyethylene terephthalate or polystyrene, which have CSTs of 43 and 32 mN m^{-1}, respectively. However, any surface with a CST above 72 mN m^{-1} (e.g. glass) will wet easily. The addition of a surfactant to water may allow the surface tension to be lowered beyond the CST of a substrate, therefore promoting wetting on a surface normally hard to wet.

Substrates which readily wet with water are usually hydrophilic so that direct polar interactions can allow the wetting to occur. However, with non-polar substrates, water will again reject interaction with the non-polar surface in favour of the stronger hydrogen bonding to itself. Following the addition of a surfactant, rejection of the surfactant hydrophobic chains to the solid interface will occur such that the hydrophilic head is orientated towards the aqueous phase. As a consequence, the solid becomes wettable by water.

In practice, the speed of wetting must also be taken into account. For some applications (e.g. textiles), the processes require a fast wetting of a fibre surface to take place in order to improve cycle times, and hence overall economic efficiency. While it would seem that the best practical surfactant for this application would be the material demonstrating the greatest reduction in interfacial tension, this is not necessarily so. Most interfacial tension measurements are carried out when an equilibrium has been reached for a particular system, but the processing requirements may require a kinetic effect. Consequently, the speed of migration must also be taken into account. For a particular surfactant series, the expectation would be that as the alkyl chain length increased, the surfactancy effect would also increase. However, as the alkyl chain increases together with the overall size of the molecule, the ability to migrate quickly to an interface will fall away. As a consequence of this, there is usually an optimum size of alkyl chain for a given headgroup of surfactant, and on both sides of this optimum, wetting power will fall away due to either the alkyl chain being too short or too long, both effects decreasing wetting power.

4.3.4 *Solid dispersion in liquids*

Dispersions of solids in liquids are usually made by the nucleation of materials from solution, or by grinding up of large particulate solid into finer particles followed by their suspension in a liquid. Whichever method is used for their generation, the important aspects of the resultant mixture is the large surface area of the solid–liquid interface. If the particles are small enough to prevent rapid settling out to either the top or bottom of the liquid, then there are two main mechanisms available for the stabilisation of the resultant slurry; electrostatic and steric. Use of these two methods of stabilisation is normally necessary, as fine particles will tend to attract and collide with each other due to the similar nature of their inherent molecular properties, as they are the same type of solid.

Assuming again that the liquid phase is water, the two types of solid particle that may be dispersed in it are polar and non-polar solids.

Addition of an ionic surfactant to a suspension of a polar solid in water may result in adsorption of the polar headgroup onto the solid surface. The non-polar surfactant tails now present the outward disposition of the solid particles, and hence approach and aggregation of the solid is disfavoured on steric grounds when the alkyl groups begin to overlap.

Addition of an ionic or non-ionic surfactant into a suspension of a non-polar solid in water may result in the adsorption of the non-polar tails onto the solid surface. Stabilisation in this case can be provided via the mutual repulsion of the similarly charged headgroups in the case of ionic surfactants, or by the steric repulsion between the long hydrophilic heads in the case of the nonionic surfactants (usually polyoxyethylene chains).

4.3.5 *Solubilisation*

One consequence of the formation of micelles in aqueous solution is the production of a zone inside the water which is essentially an oily droplet. The presence of hydrophilic heads on the outside of the assembled surfactant molecules ensures that this microscopic oil droplet is a relatively stable structure in water. This structure allows organic materials, which are usually sparingly soluble in water, to give clear solutions, within the ability of the micelles to absorb organics. For example, ethyl benzene can be solubilised at levels of up to one mole of organic compound for each mole of soap in solution. The resultant mixture is clear and stable. Not surprisingly, the ability to solubilise organic compounds depends upon the factors which affect micellar formation.

The effect is not limited to non-polar organics. Micelles formed from surfactants containing polyoxyethylene (POE) units have a core of alkyl chains, and this is surrounded by a polyether layer of chains into which water is essentially excluded due to close steric packing of the units. Materials which are not easily soluble in the alkyl region (ethers and esters) may be absorbed preferentially into this region (usually known as the pallisades layer).

This phenomenon can be used to advantage when pharmaceutical or pesticide formulations are made up. Many of the active materials in such formulations are not water soluble, and yet only a small amount of these usually potent materials are required to give the desired effects. Not only is solubilisation conferred upon the material, but the surfactant may also aid penetration of the material into the host organism, e.g. by promoting wetting. Solubilisation is also beginning to be used to remediate land contaminated with undesirable organic chemicals (e.g. chlorinated biphenyls, PCBs) which may simply not be washed out with water treatments.

4.3.6 *Microemulsions*

Between simple micellar solubilisation and normal macroemulsions, a third state of oil and water mixtures can be defined. This state is called a microemulsion, and is a swollen version of the solubilising micelles in solution. Microemulsions share some of the properties of the other two states. Like macroemulsions, there is less surfactant than that necessary to solubilise the organic phase, but like solubilisation, the mixture is still clear and thermodynamically stable. This combination of properties can be extremely useful when the other states cannot deliver the desired effects.

The mixture of oil, water and surfactant is still as before, but addition of a short or medium chain alcohol as a cosurfactant can aid the formation of the microemulsion. This is due to the hydroxyl headgroup being able to moderate the charges on ionic surfactants, and the steric hindrance of long chain nonionics by packing at the surface of the microsphere and spacing out the primary surfactant.

Applications for microemulsions are based on the properties which allow microemulsions to carry more solubilised organics than a simple swollen micelle, but still to retain their thermodynamic stability.

A microemulsion can be the final form of the product, or can be used to allow physical processing or as a special reaction medium in its own right. One of the earliest uses was in wax polishes, where the small size of the wax particles formed from molten microemulsions not only allowed easy processing, but also imparted a ready shine as the particles were less than the wavelength of visible light.

4.3.7 Detergency

Detergency is discussed as an illustration of the principal properties of surfactants because detergents probably need all the processes associated with surfactants to operate efficiently. Detergents also account for the use of about 70% of all surfactants manufactured in the world!

Detergency is the removal of solid or liquid material from a surface in preparation for some further treatment, or as an end in itself. Domestic detergency involves the cleaning of fabric fibres of polar or non-polar solids and liquids for aesthetic and hygienic effects. Initial replacement of the fibre–air interface with a fibre–water interface will naturally allow ready removal of polar materials from the cloth. However, this leaves non-polar oils and solids which must be removed and separated from the fibre environment and disposed to drain. When considering these processes, it is easy to see how the principal properties of surfactants will be beneficial in this application.

Micellisation will not only provide reservoirs of surfactant monomers, but will also allow solubilisation to occur and micro- and macro-emulsions to form. The surfactant monomers in solution will aid wetting of the fabric and the impurity interface, particularly for fabrics with low critical surface tensions such as polyester. Controlled foaming properties will aid suspension and transport of the dirt.

Although the general properties of surfactants account in a gross way for cleaning power, two further phenomena have been observed for detergency operations with regard to the removal of non-polar oils and solids.

4.3.7.1 The roll-up mechanism. For non-polar oils spread over a relatively non-polar surface such as polyester, dispersion forces act to keep the two adhering substances together. Replacement of the fibre and oil to air interfaces with aqueous interfaces raises the potential energy of the system due to the high fibre–water and oil–water interfacial tensions. Addition of a surfactant to the aqueous phase then allows reduction in the overall potential energy of the system to take place as surfactant adsorbs at the appropriate interfaces.

Should the surfactant be nonionic in composition, it is possible for it slowly to adsorb at the oil–water interface. When this occurs at the edges of the oil which is spread on the fibre, the surfactant may migrate to the fibre–oil interface. It is then possible for the surfactant to form a continuous layer under the oil extending out to the fibre–water region. A reduction in the potential energy of the system as a whole is then possible if the polar surfactant head is absorbed into the aqueous layer rather than the oily layer. This preference for hydrogen bonding with water makes the oil retreat from the fibre. By absorbing more surfactant at the oil–water interface, the potential energy of the system is again reduced.

Overall, the oil will retreat from the fibre and 'roll up' into a ball, which can then be shaken loose from the fibre (see Figure 4.6). Microscopic examination of soiled clothing during the cleaning process with nonionic surfactants has revealed these globules of oil as part of the washing cycle.

4.3.7.2 Particulate dispersion. With a non-polar solid attached to a non-polar fibre, adsorption of the typical anionic present in most detergent formulations onto both surfaces will be such that the hydrophobic tail is anchored, and the charged head will be outermost. Repulsion between the like charges on both the solid and fibre, and the reduction in potential energy between the water and both solids allows the soil to be easily detached from the fibre.

Although this mechanism is difficult to detect or prove, it is quite clear that addition of doubly charged cations such as calcium or magnesium ions have been shown to have a very deleterious effect on detergency, and such ions, present in hard water, usually cause a soap scum of insoluble salts to form.

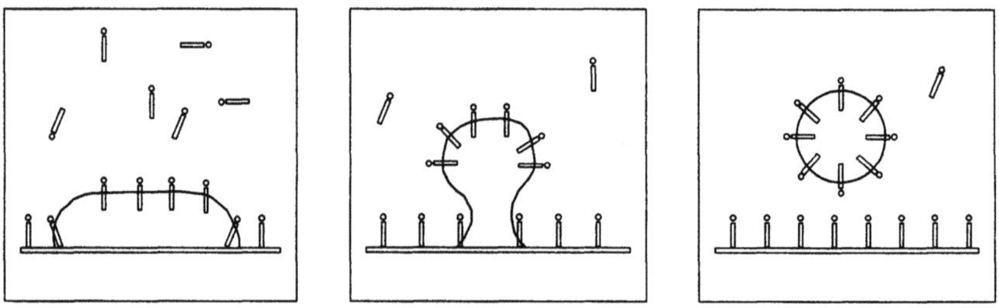

Figure 4.6 The 'roll-up' mechanism of detergency.

4.4 Chemistry of surfactants

In this chapter so far, the chemical nature of a surfactant has been confined to a vague description of a hydrophobic tail and a hydrophilic head, which looks something like a matchstick. The real situation is much more complex, but an understanding of the broad principles of their origins and composition is necessary before useful selection and employment of materials for a target application is possible, particularly in view of the vast range of commercially available products.

The correct balance of hydrophobe and hydrophile are important in order to maximise the desired effect. For example, if we consider common soap, which has an alkali metal salt of a carboxylic acid function as the headgroup, it is possible for the alkyl chain to be too short (less than 8 carbon atoms), which will result in the material being too water soluble to act as a surfactant (it will not be rejected from an interface). However, if the carbon chain is too long (more than 18 carbon atoms), then the material is usually too water insoluble, and may behave much more like an oil or fat itself.

Let us first consider the origin of the hydrophobe in more detail, before going on to the major technologies surrounding the production of hydrophilic functionalities. There are two sources of hydrophobe used to generate commercially available surfactants; plant- and animal-derived oleochemicals, and petrochemicals (see Figure 4.7).

4.4.1 Oleochemicals

Oleochemicals are derived from oil-containing plants (e.g. coconut or palm kernels), and from beef tallow fat (a byproduct of the beef industry). Saponification of these esters is easily carried out to give either the fatty acid feedstocks for further transformation, or alkali metal salts (soaps). The ease of production of soaps has allowed the use of these surfactants in one form or another for almost 2000 years. Some 30% of western Europe and about 70% of all worldwide surfactant usage is accounted for by soaps.

There are three important characteristics to be noted about the use of oleochemical feedstocks for surfactant synthesis. The first is that the alkyl chains built up in nature give only linear materials, and principally even numbers of carbon atoms, due to their biological origin. A small number of odd-numbered alkyl chains are detected, but branching of the alkyl chain is very rare. Secondly, splitting oleochemical esters gives fatty acid materials which are mixtures of isomers, as Table 4.3 shows. Use of mixed isomer materials is usually preferred, not only from an economical point of view (efficient use of all material produced), but also because mixtures almost invariably perform better than single isomer products (a happy coincidence!).

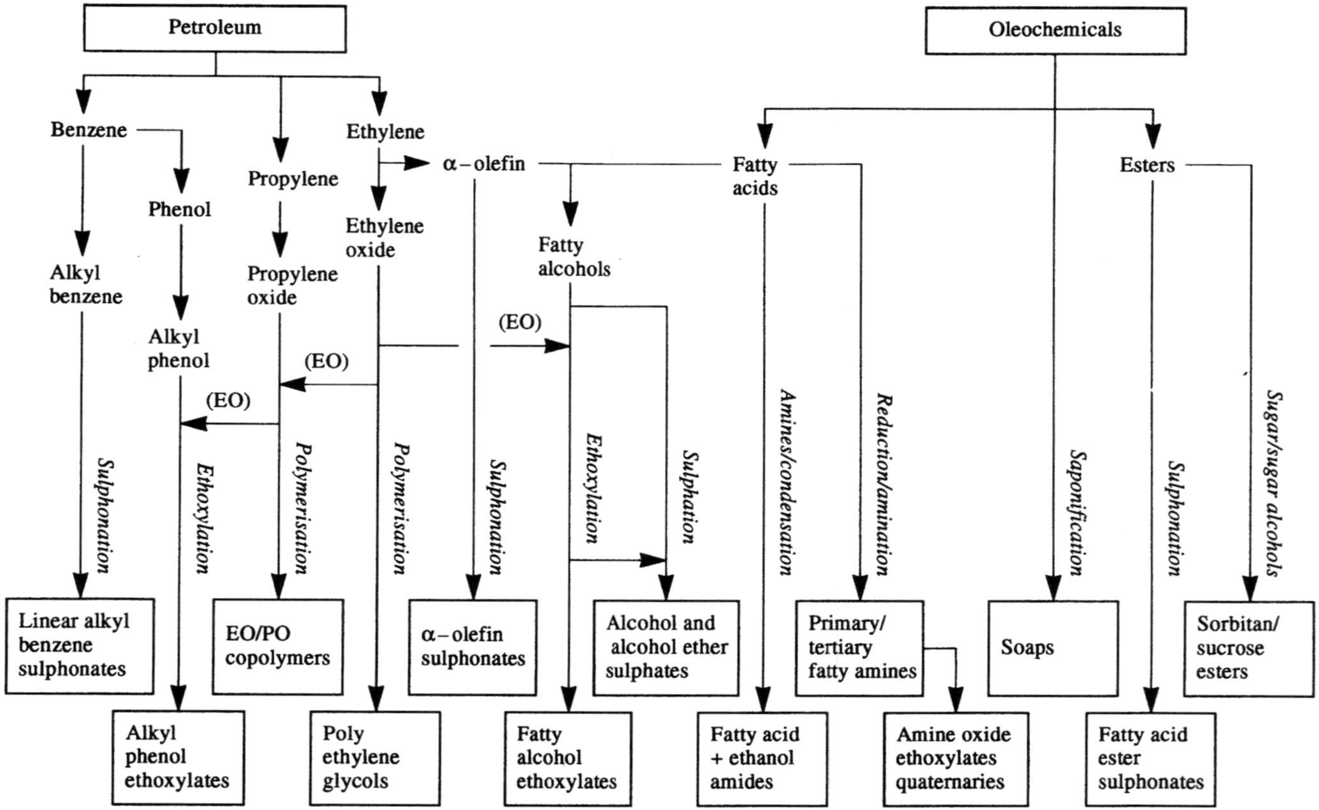

Figure 4.7 Chemical origin of surfactants.

Table 4.3 Composition of oleochemical fatty acids

Alkyl chain length	Fatty acid name		Coconut oil	Tallow fat	Commercial grades	
	Systematic	Common			Stearic	Oleic
C-10	Decanoic	Capric	6	–	–	–
C-12	Dodecanoic	Lauric	57	–	–	–
C-14	Tetradecanoic	Myristic	19	1	3	
C-16	Hexadecanoic	Palmitic	11	26	45	4
C-18	Octadecanoic	Stearic	2	26	52	7
C-18′	9*c*-Octadecanoic	Oleic	5	43	–	71
C-18″	9*c*,12*c*-Octadecadienoic	Linoleic	–	4	–	9

C-18′ = Carbon chain with one double bond. C-18″ = Carbon chain with two double bonds.

The third point to note is that not all fatty alkyl groups derived from oleochemicals are fully saturated. For example, a C-18 fully saturated alkyl group is known as a stearic functionality. With one double bond between the C-9 and C-10 position in a *cis* arrangement, the alkyl group is known as an oleic chain. Linoleic acid is the fatty carboxylic acid with a C-18 chain and two unsaturated linkages. Each different oleochemical source will have its own blend of fatty acid types following the above general rules. Although soaps remain the major tonnage products in this area, chemical transformation of this monoalkyl fatty carboxylic acid gives rise to a number of industrially important materials.

4.4.1.1 Condensation with alcohols or polyols. This gives other ester products, examples being the reaction of myristic acid (C-14 saturated) with isopropanol to form isopropyl myristate for personal care applications, or condensation of fatty acids with sorbitol to form the sorbitan esters which have wide uses in many industrial and personal care areas.

4.4.1.2 Condensation of the fatty acids with primary or secondary amines. This gives rise to amide groups. Reaction with diethanolamine, for example, gives rise to diethanolamides, which are used as foam boosting agents for bubble baths and dishwashing liquids. Reaction with amines derived from reductive amination of glucose gives rise to fatty alkyl glucamides.

4.4.1.3 Reduction of the carboxylate group with hydrogen. This gives fatty alcohols, which may be ethoxylated and sulphated as described below.

4.4.1.4 Reductive animation. This can be carried out in several chemical steps to give fatty amine products. These are more fully described under cationics (section 4.4.4). In summary, therefore, a great deal of chemistry can be carried out on the oleochemical fatty acids, leading to a wide variety of interesting materials.

4.4.2 Petrochemicals

The rise of the petrochemical industry over the last 100 years has provided an enormous range of cheap building blocks for industry. These building blocks can be put together in such a way that surfactant hydrophobes can be produced easily on a large scale. Ethylene can be polymerised to give α-olefins, or the appropriate fatty alcohols, depending on the technology used. Olefins can be joined to either benzene or phenol in an alkylation reaction, and even paraffins can be oxidised to fatty acids. In this way, hydrophobes are built up which have the same basic properties as the oleochemical fatty groups, that is, they can be built up to give the same amount of water insolubility as the C-10–C-18 fatty oleochemical groups. This is done by adjusting the hydrophobe length to suit the surfactant type. For α-olefins, which may be reacted to give the sulphonate via reaction of sulphur trioxide directly with the olefin double bond, the chain length needs to be C-10–C-18, which is also true for the fatty alcohols. However, for an alkyl benzene or an alkyl phenol, the alkyl group need only be C-6–C-9 to achieve the same effect. The main technologies for the hydrophilisation of these feedstocks are ethoxylation or sulphation, described in section 4.4.3. Again, it should be remembered that even the simple descriptions of the types of petrochemical hydrophobe given above do not give a complete picture.

Mixtures of chemical entities are again the rule, rather than the exception. For example, consider the chemistry that takes place during the production of a linear alkyl benzene hydrophobe. Alkylation of benzene with a linear olefin will generally result in the production of all different isomers of the secondary alkyl benzene. This is in spite of even starting with a single pure olefin, as rearrangement to a more randomised mixture can take place under the conditions of the reaction itself.

These petrochemical processes give products which are complementary to the oleochemical types, and use of material from all of these production processes depends upon the final combination of effects required, how best to match chemical structure to them and the cost of each potential solution. For oleochemicals, the hydrophobic part is defined by which oil or fat is sourced to provide the feedstock, and the chemistry is more concentrated around the transformation of the head carboxylate group prior to its final derivatisation. For petrochemicals, the chemical transformation element is evident over the whole production pathway.

4.4.3 Major hydrophilising technologies

Two technologies have become common in order to convert hydrophobic materials into water-soluble surfactant products-sulphation and ethoxylation.

4.4.3.1 Sulphation or sulphonation. Reaction of sulphur trioxide with a number of different substrates will form a sulphate or sulphonate group in its acid form, which may revert easily to starting materials when the sulphating medium is removed. In order to prevent this happening, sulphated products are neutralised with base, for example sodium hydroxide, and this then prevents their reversion from taking place. If the reaction is carried out on feedstock containing hydroxyl functions, a sulphate group ($-OSO_3Na$) is formed. If the reaction is carried out on an alkyl benzene, α-olefin or an alkyl methyl ester, the sulphur trioxide bonds directly to a carbon, forming a sulphonate group ($-SO_3Na$). The reaction is usually very exothermic, and as sulphur trioxide is easily generated from sulphur, a large range of hydrophobes can be easily converted to families of surfactant molecules at low cost.

These groups of materials are structurally similar to soap, in that the head group is an alkali metal ionic form of an oxy acid. However, there is a crucial difference with regard to their performance characteristics which has allowed this technology to grow in importance over the years. The detergency property of soaps falls off markedly when calcium or magnesium doubly charged cations are present in wash water, as these cations precipitate the soaps into an insoluble scum. Sulphates or sulphonates are markedly less sensitive to concentrations of these ions, and these have been the standard basic ingredients of detergent formulations. As these compounds have this performance advantage in the major tonnage outlet for surfactants, it is not surprising that this technology has become so important.

4.4.3.2 Ethoxylation. Reaction of sulphur trioxide with a hydrophobe is a simple way of producing a step change in hydrophilic character. However, if different degrees of hydrophilisation are required in order to try to match product structure to a needed performance, ethoxylation offers a much more controlled way of doing this.

Ethylene oxide (EO) is a highly reactive molecule, as it is a three-membered ring under considerable strain. Opening of the ring under acid or basic conditions is highly favoured as it releases this strain energy. If the ethylene oxide is reacted with an alcohol, the product from the reaction is also an alcohol, and so further reaction to give poly(oxyethylene) chains is easily achieved (Figure 4.8).

As the poly(oxyethylene) chain is built up, the hydrophobic starting

$$R{-}OH \xrightarrow{(EO)} R{-}OCH_2CH_2OH \xrightarrow{(EO)} R{-}(OCH_2CH_2)_n{-}OH$$

Figure 4.8 Ethoxylation of an alcohol functionality.

material becomes more and more water soluble. It is therefore a relatively simple matter to control the ratio of ethylene oxide to feedstock to obtain the degree of solubility of surfactancy required. As the ethylene oxide reagent itself is used on a large tonnage scale for other business areas, the material is an economically viable feedstock for surfactant production.

One important point should be noted about this process. Addition of a certain number of moles of EO to a material containing a reactive hydrogen site (e.g. −OH, −NH, −COOH) does not give a single compound. As the addition takes place, for example between a fatty alcohol and two moles of EO, a spread of molecules is formed, as the hydroxyl end of one homologue will react similarly to the hydroxyl end of another homologue. A spread of isomers will build up which will range from the starting alcohol itself, with no EO added to it, through molecules with 1, 2, 3, 4 and 5 EO, up to substantial quantities of material with six ethoxylate groups attached. Although ethoxylation is an imprecise way of generating a required molecular species, it appears that it is more often the case than not that this spread of products is actually more useful in most applications than a single homologue would be, and this is probably an example of the synergistic effects discussed in section 4.5.2.

Reactions can also be carried out by polymerising propylene oxide, but this is much less common, as propylene oxide tends to add hydrophobicity rather than hydrophilicity.

4.4.4 Cationics

As described above, 70% of all surfactant production is accounted for by a few product families. These include anionics (soap, sulphates and sulphonates) and nonionics (principally ethoxylates). Cleaning is the major use for these materials.

It is also possible to create families of cationic compounds which are derived essentially from oleochemical fatty acids via reductive amination processes requiring several steps. These nitrogen-based materials display an ability to bond particularly strongly with some substrates, a feature known as substantivity. Many materials in nature display a surface negative charge, especially when wet, and a positively charged surfactant will bind strongly to this material, imparting a hydrophobic effect. If the surface is textile fibres, the hydrophobic effect allows the fibres to slide over one another, and to untangle easily. This offsets the usual matting of the material, and the fabric becomes soft to the touch. In the same way as fabrics are softened by this mechanism, hair can also be conditioned.

Cationics also display interesting biocidal properties which have been related to their ability to disrupt bacterial cell walls, a property not displayed by anionics or nonionics.

4.4.5 A full range of functionality

The surfactant families given in Table 4.4 are a first introduction to the main chemical processes used by industry. An exhaustive list is not offered and the reader is referred to more detailed texts for this information and to the literature of surfactant-producing companies for an apparently bewildering array of commercially available materials. However, the table does give a good preliminary idea of some of the major elements which are combined to give surfactant molecules.

In addition to the product families shown in Figure 4.7, a few more types of compound have been illustrated. Phosphates have been added and extended anionics are possible, as any reaction which can be carried out on a fatty alcohol can also be carried out on the appropriate ethoxylate. Reaction of an ethoxylate with chloroacetic acid gives the ether carboxylate, while reaction of the same reagent with a tertiary amine gives a betaine structure (two charge centres in the same molecule).

Table 4.4 Selection of hydrophobes and hydrophiles

Main hydrophobes:		
Oleochemical	Linear alkyl (acids, alcohols, esters)	
Petrochemical	Linear alkyl benzene	
	Alkyl phenol	
	α-olefin	
Main hydrophiles:		
Anionic	– COONa	Soaps
	– SO_3Na	Sulphonates
	– OSO_3Na	Sulphates
	– OPO_3Na	Phosphates
	– $CH(SO_3Na)CO_2CH_3$	Ester sulphonates
Extended anionic	– $(EO)_nSO_3Na$	Ether sulphate
	– $(EO)_nPO_3Na$	Ether phosphate
	– $(EO)_nCH_2\ CO_2Na$	Ether carboxylate
Nonionic	– $(EO)_nH$	Ethoxylates
	– $CON(CH_2CH_2OH)_2$	Ethanolamides
	– $N(CH_3)_2{-}O$	Amine oxides
Cationic	– $N^+(R)_3X^-$	Quaternaries
	– $N^+(CH_3)_2CH_2CO_2^-$	Betaines

Each combination of hydrophilic groups adds to the possible range of uses that these molecules can display. For example, an ether sulphate displays a higher degree of hard water tolerance and lower skin irritation when compared to the parent alcohol sulphate. Many other product families have not been mentioned due to the nature of this article, but most will be found in the Further reading section.

4.4.6 Understanding surfactant chemistry

The major theme that runs through this short introduction to the chemistry of surfactants is that all of the commercially available products are mixtures. Although this is usually an advantage in terms of performance, it is something of a handicap when a simple picture of structure and effect is required.

One of the major themes for the pharmaceutical industry is the idea of designing high purity and highly specific molecules, to act at a particular site to create one specific end effect for the patient. This concept is one of key and lock, and the search consists in trying to find a specific key to turn in the lock and alleviate the illness without creating other complications (turning other locks).

Understanding surfactant effects is usually approached in terms of simplified models, and this simplification process can mean that much of the real problem becomes lost in the modelling process. This is not always a bad thing, trying to extend current understanding with carefully worked out theories, but it is essential that a good grasp of product composition should be a part of this process.

Consider the likely chemical composition of a fatty alcohol which has been reacted with ethylene oxide. The product displays three main compositional elements.

1. The alkyl group will be composed of a mixture of alkyl chains which will depend upon the fat or oil from which it was sourced.
2. The ethoxylation process will give a spread of ethoxylate chains for each of the individual alkyl chains, creating a matrix of ethoxylate products.
3. If the amount of ethoxylation is low, the free alcohol itself (with no ethylene oxide added to it) will be a large part of the product. This is because the alcohol hydroxyl group reacts slower than the ethoxylate hydroxyl, with the result that ethoxylates continue growing in preference to the alcohol reacting.

Surfactancy therefore cannot be seen as lock and key. It is a teamwork of molecules and their end effects. To achieve the best results, a single key will not give the answer. The lock is usually a combination lock.

4.5 Formulating with surfactants

This chapter has so far introduced three concepts; that surfactant molecules are essentially driven to an interface by forces arising from a polar medium, that this phenomenon can lead to a series of useful effects, and that commercial grades of surfactants, although similar in structure to the idealised molecules originally used to explain these effects, are in fact gross mixtures of many chemical entities. Given the complexity of available products, some helpful guidelines are required in order to demonstrate that practical formulating is not as random as section 4.4 may suggest.

4.5.1 The hydrophile–lipophile balance (HLB)

Surfactants have been widely studied in liquid–liquid emulsions, as the production and destruction of emulsions are fundamental to many commercial processes and their longevity has allowed detailed studies to be carried out. Consequently, a wealth of data has built up with time. In an attempt to convert this knowledge into a practical guide, Griffin proposed a system in 1949 which sought to classify the surfactant nature of a particular product. This he called the HLB of the surfactant (the hydrophile–lipophile balance). Lipophile here means essentially the same as the term hydrophobe. The proposal was to regard the emulsification of an oil into an aqueous system as requiring a certain balance of hydrophobe and hydrophile in order to allow the formulation of one in the other. That is, only materials with the right balance will enable the formulation to be achieved efficiently.

4.5.1.1 The HLB scale. Griffin defined the HLB scale as having numerical values between 1 and 20. Surfactants which are hydrophilic would lie towards the top end of the scale, and hydrophobic materials would lie towards the bottom end. The HLB of an ethoxylated nonionic surfactant could be calculated using the equation:

$$\text{HLB} = \text{molar \% EO groups}/5$$

The idea was extended by Rideal and Davies, and attempts were made to place numerical values on each functional part of the surfactant molecule and to link these together with the following equation:

$$\text{HLB} = 7 + \text{sum (hydrophilic numbers)} - \text{sum (hydrophobic numbers)}$$

Some typical numbers for various groups are listed in Table 4.5.

For example, let us take cetyl alcohol plus 10 ethylene oxide units. The measured HLB value for this surfactant is 12.9.

Table 4.5 Typical HLB numbers for common functional groups

Hydrophilic		Hydrophobic	
$-SO_4Na$	38.7	$-CH_2-$	0.48
$-CO_2Na$	21.1	$-CH_3$	0.48
$-CO_2H$	2.1	$-CH_2CH(CH_3)-$	0.15
$-OH$	1.9		
$-CH_2CH_2O-$	0.33		

- From Griffin, the percentage of the molecule composed of hydrophilic groups is 66%. The HLB is (66/5) = 13.2.
- From Rideal and Davies, the HLB = 7 plus 10 EO groups, plus one OH group = 7 + 3.3 + 1.9 = 12.2

The above example illustrates the usefulness of such simple rules. It does not give anything other than a rough guide, however, as it makes no allowance for changes in the conditions of use, the nature of the oil and water phases, or the presence of other ingredients.

4.5.1.2 The required HLB. It has been shown that it is possible to generate a numerical value which gives an approximate idea of the character of the surfactant being considered. However, the other essential element in the use of the HLB concept is the idea that for each emulsification carried out between a particular oil and an aqueous system, a certain balance of hydrophobe and hydrophile is required to make this work. For example, the required HLB for the generation of an oil-in-water emulsion of isopropyl myristate (IPM) is 11.

In order to find an HLB suitable for the duty required, the following procedure can be adopted.

- Make up small samples of a mixture of oil and water and add into each sample a set amount of surfactant of known HLB. Agitate the samples, and observe them for the ease of formation of an emulsion. By using different surfactant combinations of known HLB, it is possible that a particular part of this range will be obviously better than other parts.
- If all of the samples emulsify satisfactorily, repeat the experiments but use a lower level of surfactant until an optimum is reached.
- If none of the samples emulsifies well, repeat the experiment but increase the level of emulsifier in each.

A table of typical blends of Span (sorbitan esters) and Tween (polysorbate products) which give unitary HLB values are included in Table 4.6.

As a general rule of thumb, the HLBs for particular applications are as shown in Table 4.7.

Table 4.6 Surfactant combinations giving unitary HLB numbers

HLB	Composition			
2	Span 80	8%	Span 85	92%
4	Span 80	88%	Span 85	12%
6	Span 80	83%	Tween 80	17%
8	Span 80	65%	Tween 80	35%
10	Span 80	46%	Tween 80	54%
12	Span 80	28%	Tween 80	72%
14	Span 80	9%	Tween 80	91%
16	Tween 20	60%	Tween 80	40%

(Span and Tween are registered trademarks of ICI surfactants)

Table 4.7 Typical HLB numbers for different applications

HLB range	Use
4–6	Water-in-oil (W/O) emulsifiers
7–9	Wetting agents
8–18	Oil-in-water (O/W) emulsifiers
10–18	Solubilisers
13–15	Detergents

4.5.2 Synergy

While the above is a good starting point, the possibility of achieving a better end effect by the blending of two surfactants should be considered. This is for a number of reasons, such as:

- Ease of matching the required HLB by blending.
- A denser packing may be achieved at the interface between dissimilar alkyl and headgroup functionalities.
- The higher packing will result in an even lower interfacial tension.
- Electrostatic repulsion between similar headgroups (e.g. sulphates) may be mitigated by the use of nonionics (e.g. ethoxylates).

An example of this effect is the synergism that exists between the Span- and Tween-type products. Spans are low HLB materials formed by the condensation of fatty acids and sorbitol. Tweens are produced by ethoxylation of unreacted hydroxyl groups in the Spans, giving high HLB materials. Blends of these two materials can be seen to produce two additive effects. The Span is oil soluble and will approach the interface from the oil side. The Tweens are water soluble and approach the interface from the water side. Packing at the interface is therefore increased by the two surfactants essentially keeping the bulk of their structures on different sides of the interface.

4.5.3 Physical and chemical compatibility

For a successful formulated product, it is likely that the correct blend of surfactants and other chemicals will be required. It is not enough to arrive at a correct HLB, as many other considerations must be taken into account.

4.5.3.1 Viscosity. The final product may be needed with a particular viscosity. Hard surface cleaners and disinfectants may be clear liquids, creams, gells or suspensions, depending upon the consumer preference for performance or perception. However, it is not enough to manufacture the best product for the outlet, if after a month on the supermarket shelf the clear solution clouds, the cream splits into two layers, the gells collapse into a thin liquid, or the solid either falls out of solution or sets the mixture solid. Viscosity over time is a very problematical area, and although accelerated ageing tests can be carried out, either by storing at temperatures above normal (40°C), or by centrifugation, there is yet no easy way of predicting emulsion or suspension lifetime.

4.5.3.2 pH Stability. If the product must necessarily be of a low pH, materials such as soaps would be of little use, as they would be present in their protonated forms, rather than as their alkali salts. Compare the hydrophilicity of the alkali form with the protonated form in Table 4.5 (21 versus 2). However, if the product is strongly alkali itself, there would be little sense in adding an ester surfactant, as this would break down to give a fatty acid salt (a soap) and a non-surfactant polyol.

4.5.3.3 Temperature of use. Anionics such as sodium lauryl sulphate tend to crystallise out of solution below certain temperatures, as the alkyl groups reach or go below their paraffinic melting points. This general loss of utility for anionics is known as the Krafft point. In contrast, nonionics such as ethoxylates tend to lose their water of hydration (the ordered hydrogen bonding of water neighbours which gives them their aqueous solubility) at higher temperatures, and so can separate out of aqueous solution as a result of this. The temperature is generally referred to as the cloud point, as a clear solution of surfactant clouds when the solubility is lost upon warming. Cooling down will return the clear solution. Surfactancy effects seem to be at a maximum at or around the cloud point, falling away on both sides of this. A complete understanding of this observation has yet to be advanced.

4.5.3.4 Chemical compatibility. The easiest way to illustrate this is to refer to the most well known of all surfactant systems; the washing and conditioning of clothing. The major surfactant types used in washing are the anionics. However, after washing, softening of the garment is commonly carried out by the addition of a cationic. These two operations are usually carried out in a two-step process, rather than both together, for the simple reason that blending the two together will neutralise the effects needed from each! This is one of the most obvious examples where there is a need to consider the formulation as part of a total process, rather than as an isolated exercise in its own right.

4.5.4 Cost

Once the primary and secondary effects have been clearly identified and the physical and chemical compatibilities have been established, the next consideration when choosing a surfactant is cost. While there are a large number of readily available, high tonnage products at relatively cheap prices, it is not, however, always the best advice to eliminate higher cost surfactants at too early a stage in the screening process. Obviously, if a simple sulphate or alkoxylate is required and this performs as wanted, then that is all that needs to be considered. However, if by use of the higher cost surfactant, a further advantage can also be gained in processing or in final product performance, this may justify its use for that application.

4.5.5 Other considerations

If the primary screening exercise is complete and some technically successful candidates have been identified, this is not the end of the process. Depending upon the final use, other considerations may need to be met. Detergents may need to fulfil certain biodegradability criteria. A skin care formulation must not corrode or sensitise the treated area. The more specialised the end effect required, the more likely it is that commodity surfactants will not be adequate performers, or that a special grade may be required to suit the application.

4.6 Some final thoughts

In section 4.1, the intention was outlined to provide the reader with a first impression of the main principles of surfactancy. What has been adumbrated has been a necessarily simplified snapshot of a very complex field. For example, surfactants have been described as essentially medium molecular weight species, with a hydrophobic portion containing 10–18 carbons, and an appropriate hydrophilic head. However, many polymeric surfactant types are known, with molecular weights of 3000–5000 being common. It is clearly not sufficient to describe such molecules in terms of a simple matchstick model! In addition, the speed with which such large molecules migrate to an interface will dictate how they may be used to advantage for desired effects.

Similarly, while Figure 4.7 seeks to described common technologies and product types, each field of surfactant application will use other product types which will be common to that area. Sulphosuccinates are known to be mild surfactants, used in the personal care area, and alkyl polysaccharides are used as adjuvants in agrochemical formulations. The reader may wish to determine for themselves where these products would fit into the scheme of Figure 4.7.

Accordingly, the texts listed in the following section have been selected as books which take on themes of surfactancy in an easily readable format, or in a way in which the themes can provide a good understanding of particular essential aspects. These provide some of the broader background which it has not been possible to include here, and which has in turn led to the necessary qualifications and generalisations of this short introduction.

Further reading

Clint, J.H. (1992) *Surfactant Aggregation*, Blackie, Glasgow.
Falbe, J. (1987) *Surfactants in Consumer Products*, Springer-Verlag, Heidelberg.
Meyers, D. (1992) *Surfactant Science and Technology* (2nd edn), VCH, New York.
Porter, M.R. (1991) *Handbook of Surfactants*, Blackie, Glasgow.
Schick, M.J. (1987) *Nonionic Surfactants, Physical Chemistry*, Vol. 23, Surfactant Science Series, Dekker, New York.
Tadros, Th.F. (1984) *Surfactants*, Academic Press, London.

5 Biodegradation of surfactants

G.F. WHITE

5.1 Introduction

5.1.1 Background

The purpose of this chapter is to outline the principles and pathways involved in the biodegradation of synthetic surfactants. Biodegradation may be defined concisely as a biologically catalysed reduction in complexity of chemicals.[1] Although higher organisms metabolize surfactants,[2,3] we are concerned here with biodegradation by microorganisms, of which the most versatile and important contributors to surfactant biodegradation are the bacteria.

Restricted availability of animal fats for soap making during the 1914–18 and 1939–45 World Wars led to the production of alternative synthetic surfactants based on petrochemical feedstocks. Subsequent diversification within the synthetic surfactant industry spawned a range of surfactants with properties suitable for the more varied applications evident in the accompanying chapters. The growth in the use of synthetic surfactants in detergent formulations faltered in the 1960s when significant pollution episodes became apparent.[4,5] Investigations in Europe and the USA enabled the detergent industry to trace the cause of the resistance of some surfactants, especially highly branched alkylbenzene sulphonates, to biodegradation by bacteria. This period saw the beginning of much work on the biodegradation of surfactants which continues to this day, so that synthetic surfactants are probably the most thoroughly studied group of compounds in terms of their biodegradability by bacteria. Necessarily, therefore, this account can only attempt to distil an essence of the extensive literature that exists on biodegradation of surfactants. More detailed accounts and compilations of data can be found in Cain,[6,7] Karsa and Porter,[8] Swisher,[5] White and Russell[9,10] and White.[11]

What we know about surfactant biodegradation is largely a reflection of the environmental pollution background against which the work has developed. Thus most work has focused on the biodegradation of spent/waste surfactants in relatively dilute solution (e.g. effluent streams, river water) using environmental isolates from activated sewage sludge, river water and sediments, or soils. Moreover, the perspective usually taken has been that of seeking to achieve maximum biodegradation (extent and rates) in order to minimize environmental impact. In contrast, in the present context we are concerned not so much with the biodegradation of spent or waste surfactant formulations (ultimately desirable though that may be), but rather with minimizing the undesirable biodeterioration of surfactants before they have served their intended purpose. Unfortunately, there are very few published studies on the microbial processes and mechanisms involved in the biodeterioration of surfactants in formulations, so our metabolic and mechanistic knowledge must, at least for now, be based on studies of the biodegradation process in microbes in the environmental context.

5.1.2 Surfactants and bacterial nutrition

To provide a framework in which to describe the how and why of surfactant degradation in microbes, we need first to appreciate some aspects of environmental microbiology.

Although there is an anthropocentric distinction between the desirable biodegradation of waste surfactant and the undesirable spoilage of surfactant formulations, from a microbial perspective the difference all but vanishes because in both situations opportunistic microorganisms are exploiting potential sources of nutrition. Microbes do not collude to spoil our formulations, any more than they are altruistic in clearing up our waste! Bacterial life is characterized by periods of rapid growth of the population by cell division when nutrients are plentiful, interspersed with periods of starvation. Successful proliferation of bacteria means that nutrients are depleted rapidly so that for much of the time bacteria exist in low nutrient conditions. Consequently microorganisms have evolved several adaptations to survive such widespread privation. A central requirement for the survival of microbial cells is that they must be able to generate a minimum metabolic energy in order to maintain their structural and functional integrity. The vast majority of microorganisms are chemo-organotrophs, i.e. they obtain their energy by breaking down organic nutrients (catabolism) and oxidizing the reduced carbon/hydrogen that is present. Consequently for many microorganisms, survival depends on scavenging for reduced organic carbon compounds in their environment and organisms which develop catabolic diversity enabling them to utilize a wider range of chemicals as potential nutrients will have an advantage.[12] This brings us back to synthetic surfactants.

In recent times on the evolutionary timescale, microbial survival has been assisted by human mobilization of organic carbon in fossil deposits. Hitherto, this organic carbon has been of relatively restricted accessibility to microorganisms because of the localized nature of such deposits. However, the exploitation of fossil carbon for feedstocks in the chemical industry has mobilized this resource into a much wider environment. For example, many synthetic surfactants derive their hydrophobic chains from paraffin fractions in crude oil via intermediates such as olefins and alcohols.[13] In addition, agriculture provides animal and vegetable oils such as tallow, coconut oil and palm-kernel oil, the lipids which have been traditional starting materials for the soap industry and are now also used as raw materials for the production of 'synthetic' (i.e. non-soap) surfactants.

The key step in production of most surfactants is covalent attachment of the hydrophobic alkyl chains found in petrochemical and agricultural lipid feedstocks, to a hydrophilic group, commonly for example the $-SO_3^-$ group either as a sulphonate or a sulphate ester. This clearly alters the properties of the molecule to human advantage in the intended application, but it also potentially occludes the molecule from the normal catabolic pathways which would otherwise have been involved in its biodegradation by bacteria. Thus, the biodegradation of surfactants needs to be viewed in terms of how microorganisms exploit their catabolic versatility to meet the challenge of accessing the carbon in these synthetic compounds.

Broadly speaking, microorganisms can use two general strategies to gain access to the reduced carbon in the hydrophobic chains of surfactants:[10,14] either by separating the hydrophilic group from the hydrophobic group to leave an alkyl chain with a readily metabolizable functional group such as $-CH_2OH$ or $-COOH$, or by direct attack at the ω-position of the intact molecule with the hydrophilic group still in place. In some surfactants the hydrophilic moiety is itself comprised of significant amounts of polymeric carbon, for example in the form of polyethylene glycols. In such cases a third strategy presents itself, namely progressive destruction of the hydrophilic group by successive removal of glycol units. All three strategies (separation of hydrophile from hydrophobe, or the erosion of either of these groups from the ends of the molecule) will result in the loss of the surfactant properties of the molecule. This phase of the process, which is easily assessed (for example by loss of foaming, increase in surface tension, or specific dye-based assays for the parent surfactant), is termed primary biodegradation. Subsequent bacterial metabolism of the products of primary biodegradation leads ultimately, at least for the more readily degradable surfactants, to the production of CO_2, H_2O and compounds which are the 'normal' metabolic intermediates of bacterial cells. This phase is referred to as ultimate biodegradation. There is evidence that for some industrial compounds, the products of primary biodegradation are more toxic to aquatic and other organisms than are the parent compounds (e.g. alkylphenols from alkylphenol ethoxylates, see section 5.3.4). Thus in the environmental context, it is important to assess the fates of all parts of the surfactant molecule, i.e. both primary and ultimate biodegradation. However in terms of microbial spoilage of surfactant-containing formulations, the primary steps are particularly important. In succeeding sections, both aspects will be addressed but with particular emphasis on primary biodegradation.

5.2 Biodegradation of anionic surfactants

5.2.1 Alkyl sulphates

Alkyl sulphate (alcohol sulphate) surfactants contain ester bonds between long chain alcohols and sulphuric acid. Primary biodegradation of these compounds is achieved in bacteria by enzymic hydrolysis of the sulphate group to liberate the parent alcohol[15–17] which then undergoes oxidation via the alkanal to the fatty carboxylic acid (Figure 5.1). Fatty acids either undergo elongation to C_{16} and C_{18} homologues and incorporation into phospholipids, or they enter central metabolic pathways via β-oxidation to acetyl-coenzyme A, thus achieving ultimate biodegradation. β-Oxidation is an ubiquitous catabolic process in living cells which leads to the degradation of carboxylated aliphatic chains by sequential removal, from the carboxyl terminus, of 2-carbon acetyl fragments in the form of acetyl-coenzyme A.[18] Secondary alkyl sulphates such as decyl-4-sulphate follow a similar biodegradation pathway, except that the liberated secondary alcohols loop out of the pathway to undergo C—C bond scission[19] at the substituted carbon atom via a dione intermediate (Figure 5.1). The resulting alkyl chain fragments are alkanals and alkanoic acids which then re-enter the central metabolic pathways to yield acetyl-CoA.

Alkylsulphatase enzymes catalysing the hydrolysis of alkyl sulphates in *Pseudomonas* spp. have been purified and well characterized in terms of their substrate specificity and kinetics and the mechanisms of action (for reviews see Dodgson *et al.*;[20] Dodgson and White;[21] Cain, 1987;[7] White and Russell[10]). Bacteria which contain alkylsulphatases are commonly found in natural environments.[22–24] Naturally-occurring alkyl sulphates are equally ubiquitous;[20] examples include the sulphated lipids and chlorosulpholipids present in freshwater algae,[25,26] and bile salt sulphates in the gastrointestinal tract of animals.[27] These compounds may be natural substrates for bacterial alkylsulphatases. Thus, evolution of the biodegradative capacity towards alkyl sulphate surfactants is probably not a recent response to synthetic compounds, but dates from much earlier encounters with natural analogues. Some enteric microorganisms develop a resistance to high concentrations of surfactant through mechanisms not involving biodegradation.[28] Presumably this results from exposure in the gut to natural bile-salt surfactants, microbial biodegradation of which might compromise the microorganism's symbiotic relationship with the host.

Most of the work described in the foregoing paragraphs relates to

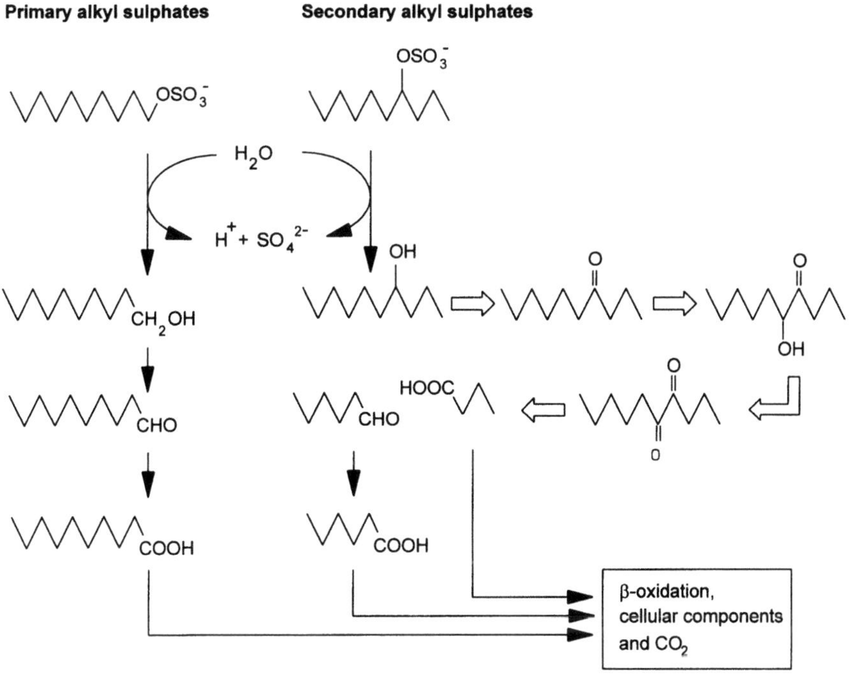

Figure 5.1 Metabolic pathways for bacterial biodegradation of primary and secondary alkyl sulphates. Open arrows show the extra reactions involved in C—C bond cleavage for secondary alcohols. Adapted from White and Russell.[10]

aerobic systems. Alkyl sulphates are also readily degradable in anaerobic waste treatment plants,[29,30] and several non-fermentative denitrifying bacteria have been isolated from river sediment which can grow anaerobically on sodium dodecyl sulphate.[31] Extracts of cells grown aerobically or anaerobically contained alkylsulphatase activity, implying that the hydrolytic mechanism operates under both conditions. Sulphatase-mediated hydrolysis also occurs in the anaerobic degradation of naturally-occurring sulphated bile-salt surfactants by intestinal bacteria.[32,33]

5.2.2 Alkyl phosphates

Alkyl phosphates (alcohol phosphates) are analogous to the alkyl sulphates in so far as they are alcoholic esters of a mineral oxyacid, except of course that the tribasic nature of phosphoric acid permits both monoalkyl and dialkyl esters ($ROPO_3^{2-}$ and $(RO)_2PO_2^-$) to serve as anionic surfactants. Every living cell produces a plethora of phosphate esters such as phospholipids, DNA, RNA, ATP and many other intermediary metabolites. Phosphatase enzymes which are important for the turnover of these compounds are equally ubiquitous. In contrast, industrial production of alkyl phosphates occurs on a relatively small scale. The consequent lack of interest in the biodegradability, together with natural and ubiquitous occurrence of many phosphatases, has led to the tacit assumption that these enzymes act also on the synthetic long chain alkyl phosphates. Such data as is available indicates that linear primary alkyl phosphates and ethoxylate phosphates[34,35] are readily biodegradable. Moreover the diminution of phosphate ester linkages accompanying the loss of foaming and increase in surface tension during biodegradation suggests that phosphate ester hydrolysis may indeed be the initiating step (cf. alklyl sulphates, section 5.2.1).

5.2.3 Dialkyl sulphosuccinates

Like the alkyl sulphates and phosphates, dialkyl sulphosuccinate (DASS) surfactants also contain ester bonds, but in this case the acid components are the carboxylic acid groups of succinic acid rather than mineral acid. Not surprisingly, hydrolytic scission of the alcohol from ester linkage is again the mechanism[36] that allows bacteria to gain access to the long chain alcohol. Removal of the first alkyl group yields monoalkyl sulphosuccinate (MONAS) and depending upon which alcohol is removed first, both α- and β-isomers may accumulate transiently (Figure 5.2). Removal of the alkyl chain from either α- or β-MONAS completes the primary biodegradation process, yielding sulphosuccinate. To achieve ultimate biodegradation, this hydrophilic residue containing a $-SO_3^-$ unit still in organic linkage with the succinate moiety must be further biodegraded. Recent work has shown that sulphosuccinate is catabolized by a *Pseudomonas* sp. under aerobic conditions via a monooxygenation pathway that first introduces an oxygen atom from O_2 at the sulphonated carbon to produce a hydroxyl group. The resulting structure (2-sulphomalate, see Figure 5.2) is equivalent to the bisulphite adduct of a carbonyl group and undergoes spontaneous decomposition to eliminate bisulphite and form the common intermediary metabolite, oxaloacetate.[37] Under anaerobic conditions, mixed cultures accomplish complete primary biodegradation of DASS but there was no catabolism of the sulphosuccinate group.[36] This is consistent with hydrolytic removal of alkyl chains and subsequent inability to attack the sulphosuccinate in the absence of O_2.

As with the alkylsulphatases, esterase reactions involving natural analogues[38] can be readily identified in lipid metabolism and so the capacity to attack DASS is likely to be widely distributed. Monooxygenation of sulphosuccinate on the other hand, may be a more specialized capability.

In alkyl sulphates, alkyl phosphates and dialkyl sulphosuccinates, a long chain alcohol is linked to an acidic hydrophilic group by an ester bond.

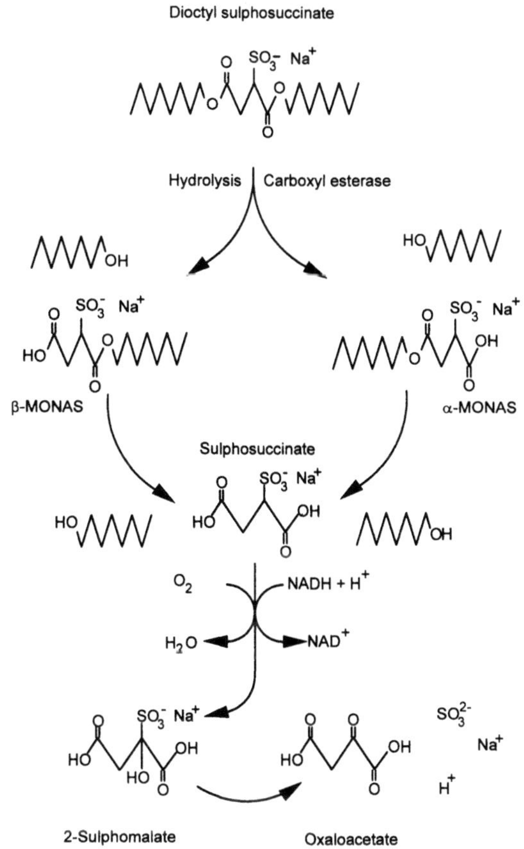

Figure 5.2 Metabolic pathway for bacterial biodegradation of dioctylsulphosuccinate.

Consequently a simple hydrolytic mechanism is all that is needed to access the alkyl chain. This has advantages for bacteria because enzymic hydrolysis proceeds without the need for any energetically expensive cofactors to be consumed.

5.2.4 *Alkylethoxy sulphates*

Alkylethoxy sulphate (alcohol ethoxylate sulphate) surfactants may be considered as alkyl sulphate analogues in which a polyethylene glycol moiety has been inserted between an alkyl chain and the sulphate group, and the hydrocarbon chain is attached via an ether bond to polyethylene glycol (e.g. sodium dodecyltriethoxy sulphate, Figure 5.3). This structure provides several points of attack for primary biodegradation in bacteria, all of which are known to occur.

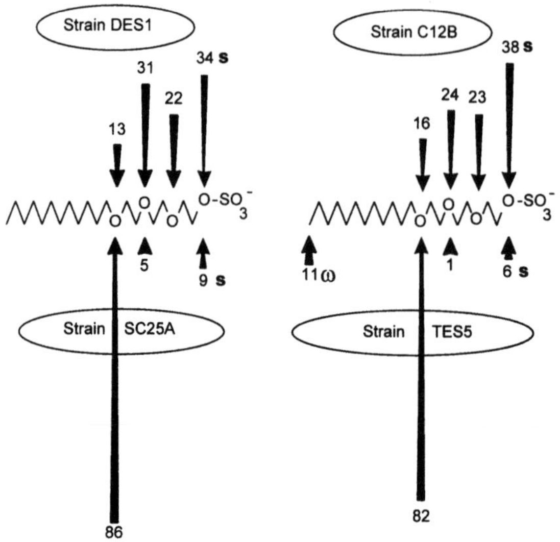

Figure 5.3 Relative contribution of different routes to the primary biodegradation of dodecyltriethoxy sulphate in four strains of *Pseudomonas*. Arrows marked **s** and ω indicate sulphatase-mediated hydrolysis and ω/β-oxidation, respectively. Other arrows indicate ether fission. Lengths of arrows are proportional to the percentages of surfactant (quoted at the base of each arrow) which undergo primary biodegradation at the indicated site. Adapted from White and Russell.[9]

First, all *Pseudomonas* spp. so far isolated by enrichment culture on this surfactant as growth substrate, are capable of breaking the alkyl–glycol ether bond (sometimes called central ether cleavage) in order to gain access to the alkyl chain (Figure 5.3). Thus, separation of hydrophobic from hydrophilic groups is an important route of primary biodegradation for this type of surfactant. In some bacteria,[39] the responsible 'etherase' enzyme is very specific for this bond (e.g. strains SC25A and TES5, Figure 5.3), whereas in other isolates the enzyme(s) are relatively non-specific in so far as they will also attack other ether bonds within the polyethylene glycol (stains DES1 and C12B).

Secondly, alkylethoxy sulphate surfactants are also substrates for enzymic hydrolysis of the sulphate group (Figure 5.3), cf. alkyl sulphates, yielding alcohol ethoxylates (Figure 5.4). Thirdly, direct attack on the alkyl chain of the intact alkylethoxy sulphate surfactant by ω/β-oxidation has also been demonstrated for strain TES5 (Figure 5.3). ω-Oxidation involves hydroxylation of the terminal methyl group of alkyl chain by mono-oxygenase which inserts oxygen from O_2. The resulting hydroxylic group is subsequently oxidized via aldehyde to carboxylic acid which opens the chain to sequential removal of C_2 units by the ubiquitous β-oxidation pathway for catabolism of fatty acid chains. This relatively minor route in alkyl ethoxysulphates assumes much greater importance in the biodegradation of some sulphonated surfactants (see below).

Ether-bond scission of sodium dodecyltriethoxy sulphate by *Pseudomonas* spp., produces metabolites from the hydrophilic end of the molecule[40,41] in which the anionic SO_3^- group remains attached to ethylene glycol residues (Figure 5.4). These metabolites constitute a homologous series of sulphated glycols containing mono-, di- or tri-ethylene glycol units, and they are susceptible to oxidation to the corresponding carboxylic

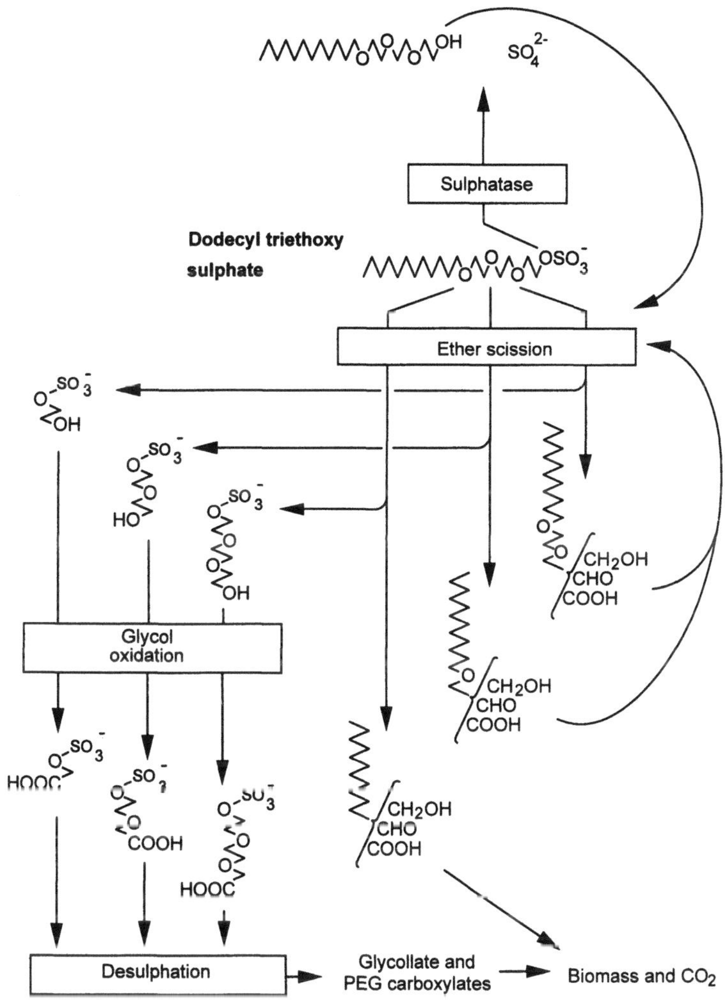

Figure 5.4 Metabolic pathways for bacterial biodegradation of dodecyltriethoxy sulphate in *Pseudomonas* spp.

acids. Not all these metabolites are formed in single organisms, but in mixed environmental cultures, all three glycol sulphates and the three corresponding carboxylates have been detected.[41] These compounds clearly contain nutritionally useful organic carbon (and sulphur). In sewage populations the glycol sulphates are fully mineralized to inorganic sulphate but the mechanism(s) of biodegradation have yet to be established. These sulphate esters are probably not degraded by the surfactant-degrading alkylsulphatases (section 5.2.1) because these enzymes are specific for long chain ($>C_5$) surface-active analogues.[10] Bacterial metabolism of some short chain (C_3–C_7) alkyl sulphates is initiated by hydrolase enzymes[42,43] but others, e.g. monomethyl sulphate[44,45] and isopropyl sulphate,[46,47] involve oxygenation reactions in the initial biodegradation steps.

The main hydrophobic coproducts of etherase and sulphatase actions on dodecyltriethoxy sulphate (Figure 5.4) are the nonionic surfactants mono-, di-, and tri-ethylene glycol dodecyl ethers (see section 5.3.2), together with dodecanol and the four corresponding oxidation products in which the terminal hydroxyls are converted to carboxylic acids.[48] The corresponding long chain alkanals have also been detected but there are no clear indications as to whether the alcohols or the alkanals are the primary cleavage products. These products may serve as substrates in further rounds of ether cleavage[48] until the alkyl chain is made available.

5.2.5 Alkane sulphonates

Alkyl sulphates are formed by the sulphation of an alcohol with SO_3 or close relatives such as oleum or chlorosulphonic acid,[49] a process which is readily reversible by hydrolysis. Alkane sulphonates on the other hand are produced by sulpho-oxidation of paraffins with SO_2 plus O_2 to give secondary isomers, or addition of bisulphite to olefins in the presence of O_2 to yield primary isomers.[50] These processes are less easily reversed so that alkane sulphonates are much more stable to hydrolysis than are sulphate esters;[51] in the presence of hot dilute acids, the latter hydrolyse in minutes or hours but sulphonates remain stable for days. Not surprisingly then, primary biodegradation of alkane sulphonates in bacteria does not occur by simple hydrolysis. Early work[52] suggesting that bacterial desulphonation was occurring to yield sulphite and a long chain alkanal was later confirmed[53,54] by the demonstration that the desulphonation in cell-free bacterial extracts was dependent on the presence of molecular oxygen and NAD(P)H. The postulated mechanism (Figure 5.5) involves the operation of a monooxygenase enzyme to introduce an oxygen atom from O_2 at the sulphonated carbon; the hydroxylated product is equivalent to an aldehyde–bisulphite adduct which spontaneously loses sulphite to yield the alkanal. Secondary alkane sulphonates are also readily biodegradable[5] probably by formation and subsequent hydrolysis of an analogous keto–

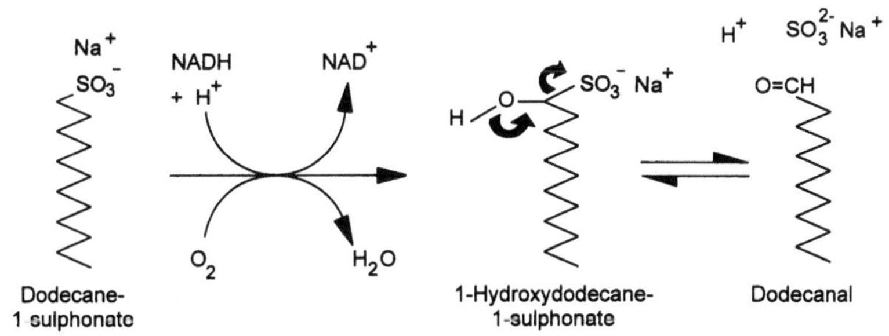

Figure 5.5 Metabolic pathway for bacterial biodegradation of primary alkane-1-sulphonate by hydrophile/hydrophobe separation. Reproduced with permission from White and Russell.[9]

bisulphite adduct.[54] These pathways are analogous to the catabolism of the secondary sulphonate, sulphosuccinate (section 5.2.3).

In addition to the oxidative hydrophile–hydrophobe separation route, there is also evidence to implicate the ω/β-oxidation pathway in the biodegradation of alkane sulphonates.[55] This is based on studies of the biodegradation of dodecane sulphonate in mixed cultures in which disappearance of the parent surfactant preceded the appearance of any inorganic sulphate. No direct information is available to determine how far along the alkyl chain β-oxidation can progress, but by analogy with the bacterial ω/β-oxidation of other surfactants (see section 5.2.7), and with ω/β-oxidation of alkane sulphonates in mammalian systems[56] the process is unlikely to progress closer than 4–5 carbon atoms from the sulphonate group. In view of the requirement for molecular O_2 in both desulphonation and ω/β-oxidation pathways, it is not surprising that alkane sulphonates are resistant to anaerobic biodegradation.[30]

Thus for alkane sulphonates in which hydrophile/hydrophobe separation by hydrolysis is more difficult than for surfactants containing ester linkages (e.g. alkyl sulphates, alkyl phosphates, dialkyl sulphosuccinates or alkylethoxy sulphates), biodegradation mechanisms become more complex and place significantly greater demands on cellular resources such as energetically expensive cofactors. In those surfactants for which alkyl chain separation becomes even more difficult, ω/β-oxidation appears to be the only route yet observed for bacteria to assimilate carbon for growth. This is the case for the linear alkylbenzene sulphonates (LAS) such as 2-(*p*-sulphophenyl) dodecane (section 5.2.7).

Primary biodegradation of hydroxyalkane and alkene (α-olefin) sulphonates is rapid and complete[57–59] but the mechanisms are unknown. By analogy with the structurally similar primary alkane sulphonates, both the hydroxylation/sulphite-elimination pathway and the ω/β-oxidation pathway may operate. It is also possible that other mechanisms involving the hydroxyl or olefinic groups already present, may exist.

5.2.6 *Fatty acid ester sulphonates*

α-Sulphofatty acid methyl esters are known to undergo both primary and ultimate biodegradation in mixed environmental cultures[55,60,61] but little is known about the enzymic mechanisms involved. During investigations on the biodegradation of α-sulphofatty acid methyl esters in a model sewage treatment plant, Steber and Wierich[62] detected short chain sulphonated intermediates that had retained the ester methyl group. These compounds, analogous to sulphosuccinate, were presumably formed by ω/β-oxidation of the aliphatic chain. Separation of the sulphonate group must therefore occur at a late stage, presumably by mechanisms similar to that for sulphosuccinate. Fatty acid ester sulphonates are resistant to biodegradation under anaerobic conditions,[62] in keeping with the requirement for O_2 in ω-oxidation.

5.2.7 *Linear alkylbenzene sulphonates*

The chemical stability of the aromatic carbon–sulphur bond, compared with, for example sulphate esters, makes linear alkylbenzene sulphonates (LAS) relatively resistant to biodegradation by hydrophile removal. Under these circumstances, the alternative strategy of direct attack on the alkyl chain without removal of the hydrophilic group becomes the easiest way by which bacteria can access the alkyl chain. Bacteria achieve this feat by a combination of ω-oxidation and β-oxidation. As we have seen, β-oxidation is a universal pathway for the degradation of carboxylated aliphatic chains by sequential removal, from the carboxyl terminus, of 2-carbon acetyl fragments in the form of acetyl-coenzyme A.[18] In order to biodegrade the alkyl chain in a surfactant like LAS by β-oxidation, it is first necessary to introduce a carboxylic acid functional group at the end of the chain. The capacity for terminal oxidation to yield long chain carboxylic acids is widely distributed in the microbial world, and is important for example in initiating the metabolism of alkanes.[63] When a linear aliphatic chain is already substituted at one end, terminal oxidation at the other end is referred to as ω-oxidation. This is the usual case for surfactants because one terminus is often occupied by the hydrophilic group. Insertion of an oxygen atom at the ω-carbon is achieved by a hydroxylation reaction (usually catalysed by a monooxygenase, Figure 5.6) analogous to that used to labilize the sulphonate group in alkane sulphonates (Figure 5.5). The resulting primary hydroxyl group is oxidized subsequently via alkanal to carboxylic acid which then is the starting point for β-oxidation.

At first sight, ω/β-oxidation seems the most direct route for bacteria to assimilate carbon from the alkyl chain. However, the initiating monooxygenation is NADH dependent and thus requires an initial investment of metabolic energy. Consequently, the initiation of alkyl-chain biodegradation through ω-oxidation is energetically more demanding than is hydrolytic

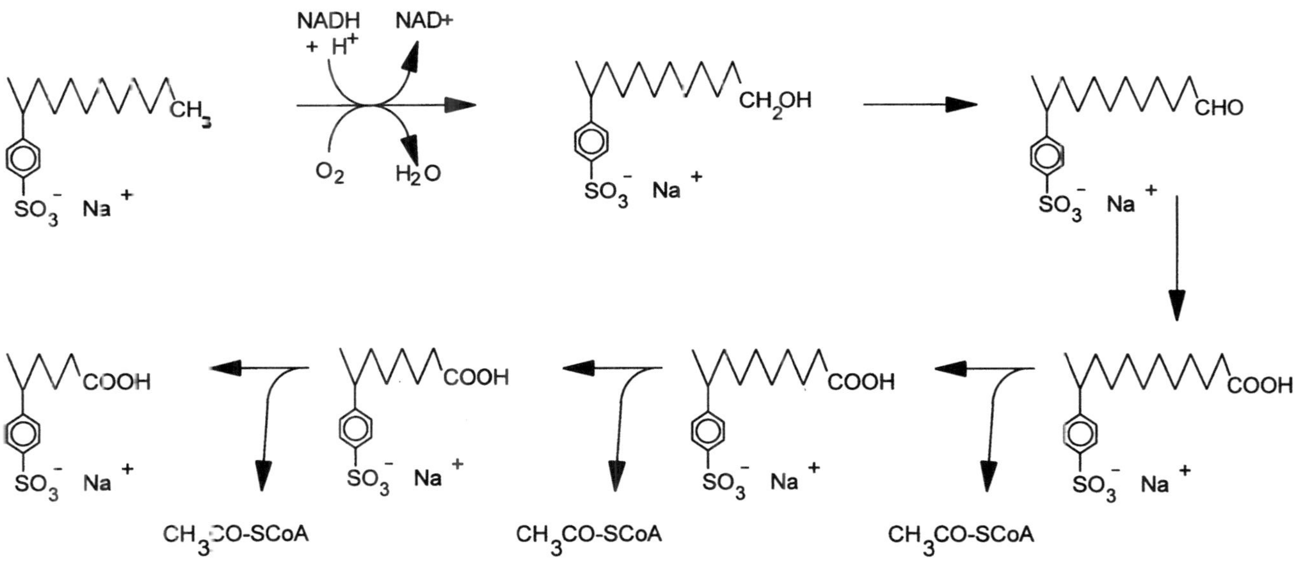

Figure 5.6 Metabolic pathways for bacterial biodegradation of linear alkylbenzene sulphonate. The upper row of reactions achieved ω-oxidation of the alkyl chain and the lower row shows sequential release of C_2 units as acetyl coenzyme A via β-oxidation. Adapted from White and Russell.[9]

separation of hydrophile from hydrophobe. Therefore, ω-oxidation is not normally observed where facile separation of hydrophile is possible, e.g. in microbial biodegradation of alkyl sulphates. However, ω/β-oxidation does occur for those compounds in which separation of hydrophile is more difficult, for example the alkane sulphonates and alkyl ethoxysulphates, and it is the only route so far observed for assimilation of carbon from LAS.

The β-oxidative enzymes which are used in surfactant biodegradation seem to be those of the normal fatty acid oxidation pathway[7] which requires an unsubstituted CH_2 group at the β-position (two carbons away from the carboxyl group) and at least one proton on the α-carbon. This pathway can cope with single methyl branches on the α-carbon, but substitution at the β-carbon or *gem*-dimethyl branching anywhere in the chain prevents further progress in oxidation of the alkyl chain. Those alkylbenzene sulphonates produced and marketed in significant quantities during the 1950s were based on tetrapropylene alkylbenzene feedstocks which produced quaternary substitution in the alkyl chains with *gem*-dimethyl branching. Such highly branched alkyl chains are particularly resistant to β-oxidation which led to environmental persistence of these surfactants, and eventual prohibition of their inclusion in domestic detergent formulations.

β-Oxidation proceeds until the site of oxidation approaches the aromatic ring, usually to within 4–6 alkyl carbon atoms.[7] Further metabolism of the resulting short chain sulphophenyl alkanoates such as 5-(*p*-sulphophenyl) hexanoate then involves desulphonation and ring cleavage, usually[64–66] but apparently not always[67] by other microorganisms constituting a consortium.

Biodegradation of the alkyl chain in linear alkylbenzene sulphonates by ω/β-oxidation yields short chain sulphophenyl alkanoates. Ultimate biodegradation of these intermediates may be viewed as two distinct processes, namely desulphonation and ring opening. The current balance of opinion is that these events occur in the order given,[68–71] but it must be remembered that this conclusion is based on studies not with sulphophenyl alkanoates but with structural analogues. Thus toluene sulphonic acid is metabolized first by monooxygenation at the methyl group to produce 4-sulphobenzoate which in turn undergoes a dioxygenation reaction in which both oxygen atoms of O_2 are introduced on adjacent carbon atoms to destabilize the aromatic nucleus (Figure 5.7). The resulting sulphonodihydrodiol intermediate is the equivalent of the bisulphite adduct of an α-hydroxycyclic ketone which is sufficiently unstable that it breaks down spontaneously with the elimination of sulphite and an electronic rearrangement to produce 3,4-dihydroxybenzoic acid (protocatechuate). Subsequent biodegradation of 3,4-dihydroxybenzoic acid via the so-called *meta*-pathway is well established from studies on the biodegradation of

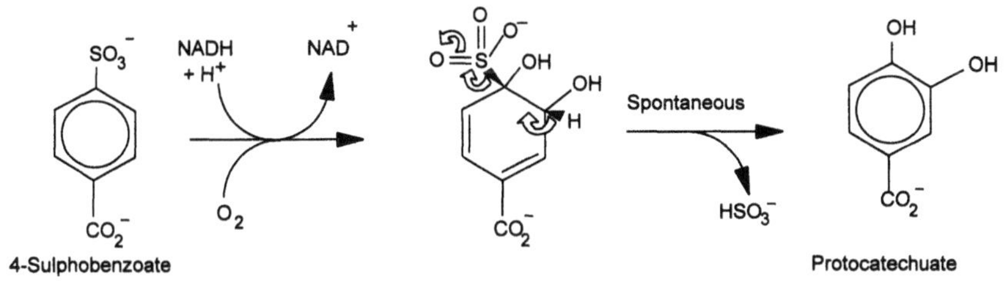

Figure 5.7 Desulphonation of 4-sulphobenzoate by a dioxygenase. Reproduced with permission from White and Russell.[10]

aromatic compounds in general.[72,73] Thus, removal of the highly polar sulphonate group before ring opening enables competent organisms to biodegrade the catechol products using the widely distributed ring-cleavage enzymes which bacteria have evolved to assimilate carbon from the numerous aromatic compounds that occur in nature.

Bacteria which attack the alkyl chain in LAS, do so because they are carbon-limited and they need to acquire further supplies of carbon for energy and growth. Recently Kertesz *et al.*[74] have shown that when a mixed culture derived from a sewage treatment plant was grown with abundant carbon and nitrogen but under conditions of sulphur limitation, primary biodegradation occurred by direct desulphonation of intact LAS. Thus environmental conditions can significantly affect biodegradation patterns.

The requirement for O_2 in ω/β-oxidation and in the desulphonation of arylsulphonates precludes biodegradation under anaerobic conditions[30,75–77] cf. the alkane sulphonates.

5.2.8 Fatty acid alkanolamide sulphates

The sulphate esters of stearic and oleic fatty monoethanolamides, in which the ethanolamine OH is sulphated (e.g. Figure 5.8a), undergo very rapid and complete primary biodegradation (loss of surfactant property) but in fact there is scant evidence on which to base any mechanistic proposals. In the presence of sewage mixed microorganisms, significant amounts of carbon (25–50%) remain as undegraded intermediates. No inorganic sulphate was detectable,[5] showing that primary biodegradation was not achieved by sulphatase action (cf. alkyl sulphates, section 5.2.1). Amide hydrolysis seems improbable because this would liberate most of the carbon (>80%) in the molecule as readily degradable fatty acids (cf. ethoxylated fatty amide nonionics, section 5.3.6). Of the remaining possible sites of primary attack (deamination of the ethanolamine units and ω/β-oxidation of the acyl chain), the latter emerges as the most likely mechanism, because the CO_2 yields were consistent with degradation of

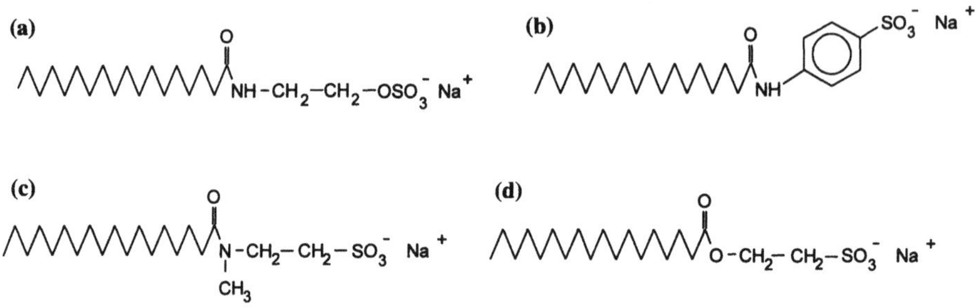

Figure 5.8 Examples of anionic surfactants based on fatty acid amides and esters. (a) Stearic acid monoethanolamide sulphate ester, (b) arylsulphamido stearic acid, (c) stearic amide of *N*-methyl taurine, (d) stearyl isethionate.

the fatty acyl chain to within a few carbons of the amide bond which is typical of that observed in other ω/*β*-oxidations of surfactants (section 5.2.5).

5.2.9 Sulphonated esters and amides of fatty acids

Sulphoaryl amides of fatty acids (arylsulphamido carboxylic acids; fatty acyl sulphanilates, Figure 5.8b) are degraded by *Escherichia coli* in accord with the distance principle[78] which states[5] that the greater the distance between the sulphonate group and the free terminus of the alkyl chain, the faster the biodegradation. This principle was derived from data for LAS and other ω/*β*-oxidized surfactants, thus implying that the arylsulphamido carboxylic acids also suffer the same fate. Disinclination to hydrophile–hydrophobe separation (here by central fission of the amide bond) may be caused by the presence of the aryl nucleus and this is also consistent with parallel observations for alkylphenol ethoxylates (section 5.3.4). Biodegradation data for the closely related sulphoalkyl amides of fatty acids (alkylsulphamido carboxylic acids, Igepon T, Figure 5.8c) and sulphoalkyl esters of fatty acids (Igepon A, Figure 5.8d), are equally scanty but nevertheless indicate rapid and extensive biodegradation in mixed cultures. To account for this, Sawyer and colleagues[79,80] have speculated that an initial hydrolytic attack liberates free fatty acids which are then rapidly assimilated by central metabolic pathways present in all bacteria (cf. nonionic fatty alkanolamides, section 5.3.6). Unfortunately there is no direct experimental evidence with pure cultures to support this claim, and in fact it conflicts with available data for the analogous fatty acid alkanolamide sulphates (section 5.2.8) and for the sulphoaryl amides of fatty acids just described.

Whatever the initiating pathways, it seems likely that biodegradation of sulphoalkyl esters and amides of fatty acids will lead to the formation of isethionate and (methyl-substituted) taurines. Taurine is produced naturally in abundance in animal tissues where it serves many functions including conjugation with toxic compounds as a preliminary to excretion. Consequently large amounts are deposited in the environment where bacteria have evolved to exploit what, for them, is a nutritionally useful source of carbon, nitrogen and sulphur. Extracts of a taurine-degrading species of *Agrobacterium* isolated from river mud[81] degraded taurine with the consumption of O_2 and the release, first, of ammonia, then sulphite (Figure 5.9) which was subsequently converted to sulphate.[81] The occurrence of the implied oxidative deamination prior to desulphonation was later confirmed by the detection of a taurine dehydrogenase which catalysed the conversion of taurine to sulphoacetaldehyde.[82,83] Desulphonation of sulphoacetaldehyde to acetate and sulphite[84] was catalysed by a specific sulpholyase enzyme[85] which required thiamine pyrophosphate and magnesium ions for activity.[86] Although the structurally related isethionate ($HO–CH_2–CH_2·SO_3^-$) served as a carbon source for growth of the same organism (a Gram-negative rod from sewage), the sulphoacetaldehyde sulpholyase (Figure 5.9) was inactive on this compound which required prior conversion to sulphoacetaldehyde through catalysis by a specific dehydrogenase before desulphonation by the sulpholyase could occur. In *Pseudomonas aeruginosa*, conversion of taurine to sulphoacetaldehyde is accomplished by transamination with pyruvate rather than oxidative deamination,[87] before sulpholyase-catalysed desulphonation.[88] Other *Pseudomonas* and *Achromobacter* species also use aminotransferase enzymes to achieve taurine deamination via this route.[89,90]

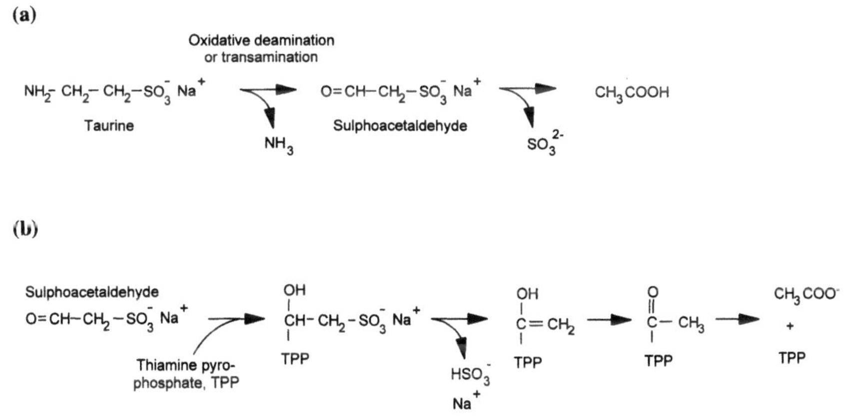

Figure 5.9 Bacterial biodegradation of taurine showing (a) overall pathway; the sequential release of ammonia then sulphite, and (b) the thiamine pyrophosphate dependent desulphonation of sulphoacetaldehyde.

5.3 Biodegradation of nonionic surfactants

5.3.1 Polyethylene glycol (ethoxylate) chains as bacterial nutrients

Anionic groups such as $-SO_3^-$ are highly polar and heavily solvated in aqueous solution, thus making these groups highly hydrophilic. In contrast, sufficient hydrophilicity in ethoxylate nonionic surfactants is achieved only by combination of many units (typically 5–25) of uncharged ethylene glycol units in the form of a polyethylene glycol (PEG) moiety. Thus the hydrophilic group in ethoxylated surfactants contains abundant carbon, often more than in the hydrophobic alkyl chain. These moieties are therefore potential sources of carbon for bacterial growth. Consequently the two strategies of bacterial attack on surfactants considered so far (hydrophile separation and ω/β-oxidation) are now supplemented with a third,[11] namely destruction of the hydrophilic group by successive removal of glycol units from the terminus. Evidence for this contribution to biodegradation is available for a number of nonionic surfactants as will now emerge.

5.3.2 Linear alcohol ethoxylates

A major route for biodegradation of linear alcohol ethoxylates is via a hydrophile/hydrophobe separation pathway, in this case a scission at the central ether link, evidence for which has been collated in earlier reviews[4,6] and more recently by Swisher.[5] From studies with laboratory die-away tests, Patterson *et al.*[91] first proposed that the cleavage sites specifically included the alkyl–ether (central ether) bond and subsequent studies, also with mixed cultures but in other laboratories[92–96] have confirmed this view. Radiotracer experiments[97] with a pure bacterium *Pseudomonas* sp. SC25A able to grow on alcohol ethoxylates, showed that [1-^{14}C]dodecyl decaethoxylate was degraded with a very rapid production of $^{14}CO_2$ and simultaneous liberation of decaethoxylate (Figure 5.10). These observations strongly suggested scission of the central ether bond to liberate the decyl chain (as dodecanol, dodecanal or dodecanoic acid) from the decaethoxylate, followed by oxidation of the decyl chain to acetyl-coenzyme A, thence to $^{14}CO_2$ via β-oxidation and the tricarboxylic acid cycle, respectively.

In addition to the central ether cleavage pathway, there is plenty of evidence demonstrating the occurrence of the ω/β-oxidation route in the biodegradation of alcohol ethoxylates. Thus TLC, IR and NMR analyses[91] have demonstrated the conversion of alcohol ethoxylates to metabolites containing carboxylic acids and ethoxylated material, via pathway(s) additional to, but independent of the central ether cleavage route. By

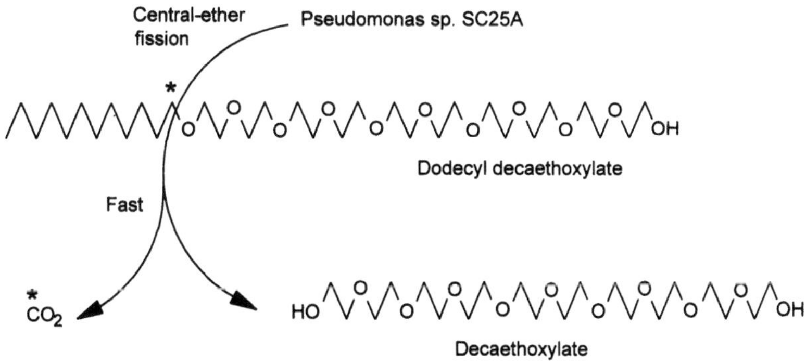

Figure 5.10 Primary biodegradation of dodecyl decaethoxylate via central ether cleavage in a *Pseudomonas* sp.; an example of hydrophile/hydrophobe separation in nonionics. The asterisk indicates the carbon radiolabelled with ^{14}C. Reproduced with permission from White.[11]

combining measurements of rates of $^{14}CO_2$ production from alkyl-labelled and ethoxylate-labelled octadecyl ethoxylate, with IR and NMR analysis of metabolites, Nooi *et al.*[98] also concluded that biodegradation by mixed microbial populations from activated sewage sludge begins with ω/β-oxidation of the alkyl chain of the intact surfactant. The dominance of these two distinct primary biodegradation pathways (ether cleavage and ω/β-oxidation) has been confirmed more recently for biodegradation of linear octadecyl ethoxylate.[96]

Alcaligenes faecalis var. *denitrificans* grew anaerobically on linear alkyl ethoxylates under denitrifying conditions.[99] Although little is known about this process, at least some degree of anaerobic degradation of these nonionics is not surprising because their PEG cousins succumb readily under anaerobic conditions (section 5.3.5).

5.3.3 Branched alcohol ethoxylates

Oxo alcohols produced by Zeigler synthesis are mixtures of linear and 2-methyl-branched primary alcohols. Alcohol ethoxylates derived from them thus contain a proportion of molecules with α-methyl branches close to the central ether bond. Swisher[5] has collated information from several sources which indicates that biodegradation of oxo alcohol ethoxylates by mixed bacterial cultures yields mixtures of metabolites, mainly PEGs plus some carboxylated PEGs. Collectively these results were consistent with most of the surfactant (probably the linear components) being biodegraded via central fission to give PEGs, and the remainder (the branched isomers) undergoing ω/β oxidation. Confirmation that C-2 alkyl-branched ethoxylates are inhibited from central cleavage came from studies on biodegradation of 2-ethyl[3-^{14}C]decyl decaethoxylate in mixed bacterial cultures.[97] No

PEGs were formed, but radioactive intermediates were detected, probably arising from ω/β-oxidation of alkyl chain. These results are consistent with earlier findings[100] that extensive (95%) primary biodegradation of linear secondary alcohol ethoxylates in sewage treatment tests was accompanied by the accumulation of compounds of the type $(HOOC)_2CH-(OCH_2CH_2)_nOH$ arising from ω/β-oxidation of the two alkyl chains attached to the ether-linked carbon atom. Evidently central cleavage was prevented by substitution at the ether-linked carbon. The pattern was thus emerging that branching of the alkyl chain in the vicinity of the central ether bond hindered central ether cleavage, thus allowing the alternative mechanism of ω/β-oxidation to play a more significant role.

Later work showed that biodegradation of the oxo alcohol ethoxylates (linear and branched mixture) in mixed cultures[101,102] also yielded metabolites which were partly ω/β-oxidized, partly ethoxylate shortened (17–25%) but also 6–7% of ethoxylate shortened molecules with the alkyl chain still intact (Figure 5.11a). It appears that, by restricting central scission, the presence of light branching in the alkyl chain makes ethoxylate shortening, as well as ω/β-oxidation, more significant.

The trend towards ethoxylate shortening without oxidation of the alkyl chain would be expected to become more pronounced with increased

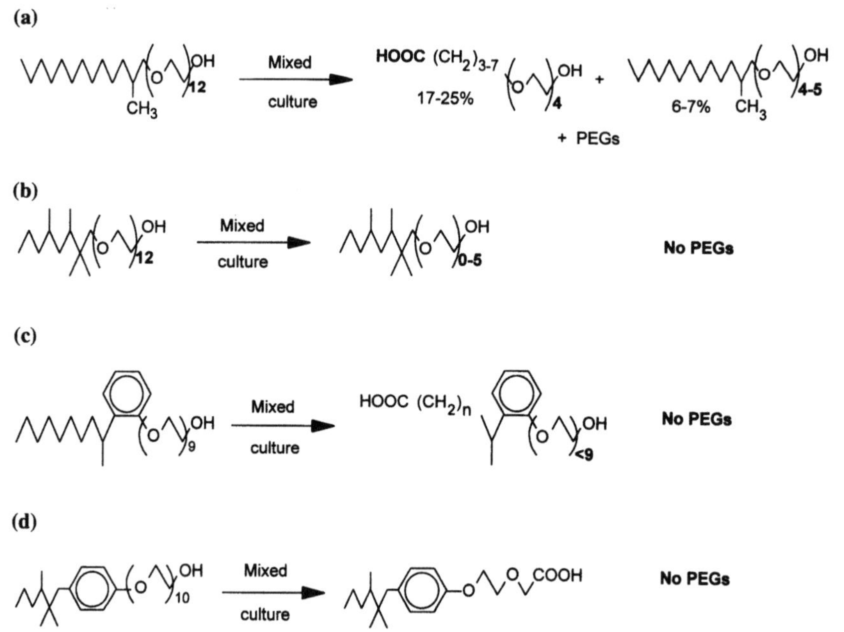

Figure 5.11 Evidence for the occurrence of ethoxylate shortening during the biodegradation of alcohol ethoxylates and alkylphenol ethoxylates. Reproduced with permission from White.[10] (a) Linear oxo alcohol ethoxylates; (b) tetrapropylene ethoxylates; (c) linear alkylphenol ethoxylates; (d) branched alkylphenol ethoxylates.

branching of the alkyl chain which would inhibit not only central ether cleavage but also ω/β-oxidation. The highly branched tetrapropylene alkyl chains are renowned for their recalcitrance in surfactants such as the now notorious tetrapropylene benzene sulphonates which produced numerous environmental pollution problems during the 1950s and 1960s. Use of these groups as hydrophobes in ethoxylate nonionics does indeed produce a marked shift in the biodegradation pattern.[102] Thus biodegradation of tetrapropylene ethoxylate was incomplete even after 7 weeks, PEGs were almost undetectable and the only detectable metabolites were a homologous series of shorter ethoxylates with the alkyl chain intact which could only have arisen from progressive ethoxylate shortening (Figure 5.11b).

5.3.4 Alkylphenol ethoxylates

Biodegradation of linear (primary or secondary) alkylphenol ethoxylates in mixed culture invariably fails to produce PEGs from central scission.[103] The presence of the aryl nucleus probably provides sufficient steric hindrance to prevent central scission; in effect the presence of the aromatic nucleus is equivalent to branching adjacent to the central ether bond. Instead, ω/β-oxidation and ethoxylate shortening combine to produce the metabolites shown in Figure 5.11c. If in addition to the presence of an aryl nucleus, the alkyl chain is made highly branched or shortened, ω/β-oxidation of the hydrophobe would also be precluded so that in mixed culture the dominant biodegradation mechanism expected is ethoxylate shortening (Figure 5.11d). Positive identification of nonylphenol and its mono- and di-ethoxylates in the environment[104,105] and nonylphenol diethoxylate as major accumulating intermediates in the biodegradation of nonylphenol decaethoxylate,[106–108] together with the detection of several lower ethoxylate homologues during biodegradation of 4-*n*-butylphenol hexaethoxylate,[109] clearly demonstrate that sequential degradation of the PEG chain is indeed the major biodegradative route. Environmental accumulation of these biodegradation products is giving cause for concern because not only are they toxic, but they also accumulate in aquatic organisms.[105] Moreover, they are suspected of contributing to 'hormone pollution'[110] because alkylphenols have been shown to cause changes in gender characteristics in developing male trout at very low concentrations.

To summarize, three routes of attack are available in principle by which bateria may gain access to utilizable carbon in ethoxylate surfactants, viz. central ether scission, ω/β-oxidation and ethoxylate shortening. All three operate in the biodegradation of linear alkyl ethoxylates, thus enabling these compounds to undergo rapid primary, and complete ultimate, biodegradation. Introduction of alkyl-chain branching or an aryl group next to the central ether bond restricts central scission, so that biodegradation at the alkyl and ethoxylate extremities of the surfactants becomes relatively more significant. Extensive branching of the alkyl chain further precludes ω/β-degradation of the alkyl chain, leaving ethoxylate shortening as the dominant pathway. Ethoxylate shortening is probably operative for all types but is more readily apparent for those surfactants in which the other routes are restricted.

5.3.5 *Polyglycols*

We have already encountered polyethylene glycols (PEGs) as hydrophilic components in alcohol ethoxylates, and in free form as the products of central ether cleavage of those surfactants. PEGs and the analogous polypropylene glycols (PPGs) are industrially important in their own right in, for example, pharmaceuticals, emolients and lubricants.

There is ample evidence that PEGs are readily degradable aerobically by soil and aquatic bacteria (for reviews, see Cox;[111] Kawai[112]), and depolymerization of low molecular weight PEGs[113–115] and higher polymers[116,117] in pure culture is well documented. The occurrence of anaerobic degradation is also firmly established,[99,118,119] and for both aerobic[116] and anaerobic[119] microorganisms, extracellular depolymerizing enzymes have been reported which break down high molecular weight PEGs to smaller oligomers more easily assimilated into the cells. On the other hand, intracellular enzymes are usually involved in the assimilation and degradation of low molecular weights PEGs and their metabolites.

Cain[6] has identified four possible mechanistic strategies that microorganisms might use to depolymerize PEGs: (i) oxygenative cleavage by monooxygenases, (ii) oxidation of the carbon atom α to the ether bond followed by hydrolysis of the resulting hemiacetal or ester, (iii) direct hydrolysis of the C–O–C ether bond, (iv) carbon–oxygen lyase-mediated cleavage. Of these, the commonest mechanisms for *exo*-cleavage of PEGs (i.e. sequential removal of terminal glycol units) which is emerging, involves the enzymic formation, then spontaneous hydrolysis, of a hemiacetal intermediate. However, there is considerable variety in the routes to formation of the hemiacetal. For example, in the anaerobic *Pelobacterium venetianus* an intracellular diol dehydratase removes H_2O from the terminal glycol to yield a vinyl intermediate which is then rehydrated in the reverse orientation to produce a hemiacetal (Figure 5.12a) carrying a terminal methyl group. The hemiacetal is easily hydrolysed to produce acetaldehyde and a PEG now shortened by one glycol unit, which can undergo further rounds of *exo*-cleavage.[118,120,121] A similar non-oxidative mechanism has been proposed for an obligately aerobic strain of *Acinetobacter*.[122] In a *Pseudomonas* sp. capable of aerobic growth on a range of PEGs, an intracellular PEG-dehydrogenase is believed to extract a hydrogen atom from each methylene in the terminal glycol unit to produce an enol which then hydrates to the hemiacetal

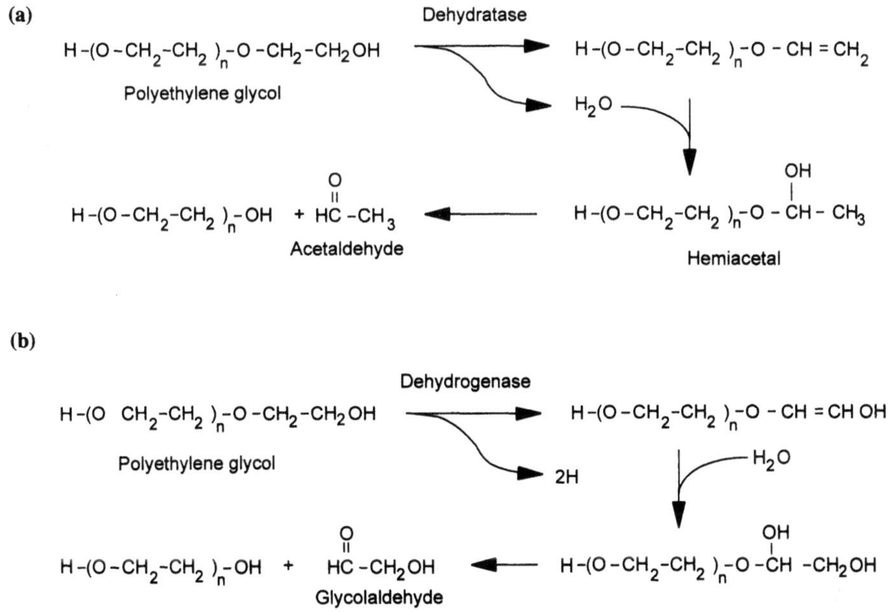

Figure 5.12 Metabolic pathways for the biodegradation of polyethylene glycols (PEGs). Both mechanisms involve formation of an olefinic group in the terminal glycol unit which then hydrates to introduce OH at the ether-linked carbon to yield an unstable hemiacetal. The mechanisms differ in the route to unsaturation of the terminal glycol which involves either (a) dehydration (non-oxidative mechanism) or (b) dehydrogenation (oxidative mechanism).

(Figure 5.12b). The terminal HO group of the PEG is unaffected by this mechanism and its retention would lead to production of glycolaldehyde during spontaneous hydrolysis of the hemiacetal, although there is no direct evidence for this.[123] A *Flavobacterium* sp. growing under aerobic conditions, also employed an intracellular dehydrogenase enzyme, but in this case acting on terminally oxidized PEG-carboxylates (see below) via an unknown mechanism to produce glyoxylate (OHC–COOH) as the putative reaction product.[112,124]

In the environment and in laboratory cultures, PEGs are frequently oxidized by microbial dehydrogenase enzymes which convert a terminal C–OH group to COOH via CHO to yield the so-called 'PEG carboxylates'. In several instances it appears that oxidation of the terminal hydroxyl of the PEG to carboxyl is a prerequisite to *exo*-cleavage of the glycol unit.[113,117,125,126]

Biodegradation of polypropylene glycol (PPG) has been shown to occur and, with a single notable exception,[99] appears to be mediated by a separate system from those described for PEG.[113,127,128] For example, PPG was degraded to 80% in 8 days by a *Corynebacterium* species which could not utilize PEGs.[129] Further studies involving this isolate acting on commercial dipropylene glycol as a model compound, were complicated by the existence of several positional and optical isomers arising from variation in the position of the methyl groups.[130] Nevertheless the results indicated that isomers containing secondary alcohols were oxidized preferentially compared with primary alcohols, that these particular PPG-degrading enzymes were not extracellular and that the mechanisms for PPG- and PEG-degradation were probably similar, i.e. oxidation of the terminal alcoholic group prior to cleavage of an *exo*-ether bond. The *Corynebacterium* also grew on PEG/PPG/PEG block copolymers but only those with an excess of PPG. Similarly Cain[131] reports that PEG degraders will grow on PEG in block copolymers as long as the PEG:PPG ratio is high (at least 3:1 and preferably 5:1 or 7:1).

5.3.6 *Fatty acid alkanolamides and ethoxylated fatty amines*

Fatty acid alkanolamides such as fatty acid monoethanolamide (the unsulphated form of the structure in shown Figure 5.8a), diethanolamide and monoisopropanolamide, are known to undergo extensive and rapid biodegradation in mixed environmental or sewage cultures.[58,132,133] Recently, studies with closed bottle tests to assess oxygen uptake as a measure of biodegradation, have been used to propose hydrophile/hydrophobe separation mechanisms for primary biodegradation of both ethoxylated fatty amines and fatty alkanolamides.[134] In the proposed mechanisms, fatty amides undergo amide bond hydrolysis to liberate the free fatty acid which is then rapidly oxidized. For the ethoxylated fatty amines, the speculative mechanism is that C–N bond cleavage yields a long chain alkanal which is oxidized to fatty acid before further catabolism. Although these mechanisms are plausible, the data do not preclude contributions from, for example, ω/β-oxidation. More work with pure cultures is needed to establish with certainty the mechanisms of biodegradation of these fatty amine and amide surfactants, and indeed of the related anionic surfactants as well (sections 5.2.8 and 5.2.9).

5.4 Biodegradation of cationic surfactants

Cationic surfactants particularly the quaternary ammonium chlorides (QAC, Figure 5.13), possess potent germicidal properties and they are used in a variety of biocides, sanitizers and disinfectants. As a consequence, these compounds might be expected to be extremely recalcitrant to microbial degradation, and indeed there is plenty of evidence to show that biodegradation of cationics can be impeded by their toxicity to the degrading microflora (see e.g. Games *et al.*[135]). However, many environmental bacteria appear to be much less susceptible to cationic surfactant

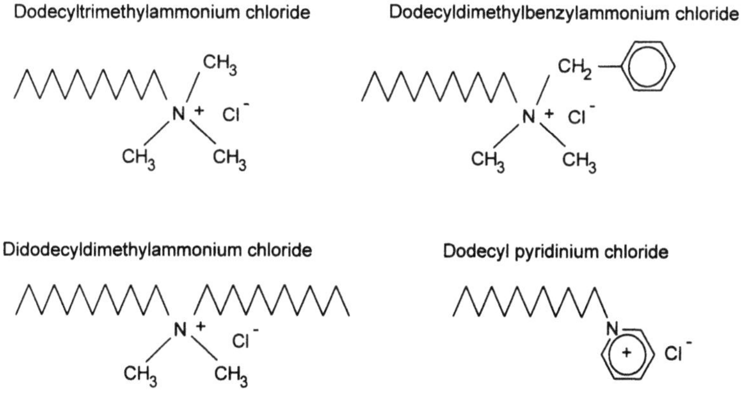

Figure 5.13 Examples of some common quarternary ammonium chloride cationic surfactants.

toxicity than are the target pathogenic bacteria such as *Streptococcus*.[136] There is no doubt that, given the right conditions, these compounds are metabolized by microorganisms but an extensive review[137] has revealed a confusing complexity of data with, for example, reported biodegradabilities ranging between 0 and 100%. Although such factors as the test system used and methods of analysis will contribute to this variability, two features stand out as major determinants of the environmental fate of cationics, viz. adsorption and acclimation of microorganisms.

5.4.1 Adsorption

The utility of cationics as fabric softeners and hair conditioners depends upon their adsorptive capacity to surfaces. QACs are strongly sorbed to a wide range of materials and form complexes with anionic substances. Because both toxicity and biodegradation rates are concentration-dependent, and biodegradation and adsorption reduce solution concentrations, the interactions between biodegradation and adsorption are likely to be very significant for cationics.[138] Depletion of cationics by adsorption may be commensurate with losses from true biodegradation and it is important to distinguish their relative contributions to the apparent 'primary biodegradation' as measured by degrees in solution concentration.[5] Moreover adsorption of cationics may itself affect biodegradation rates either positively or negatively. For example, if solution concentrations exceed bactericidal toxicity thresholds, adsorption may alleviate toxication by removal of surfactant from solution. Indeed it is well established that cationic surfactants which are recalcitrant at high concentrations (in the order of 10^1–10^2 ppm) are readily degraded at lower concentrations (see e.g. Dean-Raymond and Alexander,[139] Gerike *et al.*[140]). There is also a general failure to observe significant biodegradation of QACs in low biomass systems in which detoxication by adsorption to biomass is necessarily minimal. On the other hand, at low concentrations of QAC, biodegradation rates increase linearly with concentration[137] so that if biodegradation occurs predominantly in solution, then adsorption reduces bioavailability of surfactant and lowers biodegradation rates. Evidently the extent of adsorption of cationics and the magnitude and direction of its effects on biodegradation need to be assessed for individual systems. This is an important point to appreciate in the present context because in surfactant–product formulation the nature and extent of surfaces available for adsorption and the concentration of cationics will usually be very different from those found in wastewater treatment systems and receiving waters, to which previously published work applies.

5.4.2 Acclimation

Acclimation (or 'adaptation') is defined operationally as an increase in the biodegradation rate of a chemical as a result of prior exposure of the bacterial population to the compounds. Several factors may contribute to this phenomenon[14] including induction of enzymes, genetic changes (mutations) producing new metabolic capabilities, and, most commonly, growth of that part of the microbial population competent in the biodegradation process. Challenging unexposed microbial populations with high concentrations of toxic QACs is likely to decimate the bacterial population beyond the point where it can recover and accomplish biodegradation; many authors have observed the recalcitrance of QACs at high concentrations.[137,139,140] However at lower doses, a higher survival rate of potential degraders will allow them to function, reduce the toxic threat and increase the numbers of competent bacterial cells through growth. Subsequent additions at the same concentration will be dealt with even more effectively, and additions at higher concentration may also be accommodated. In this way the dose level may be progressively raised, and the microbial population declared to be 'acclimated' to the surfactants. Thus for example, when dodecyltrimethylammonium chloride was dosed at 0.1 ppm into either a continuous flow river microcosm[138] or into a batch model lake ecosystem[141] over a period of about 2 weeks, there were very significant increases in the rate of biodegradation of surfactant. Similarly dodecyldimethylbenzylammonium chloride disappeared in river water tests within a few days at 5 ppm but there was little degradation at 30 ppm. Moreover, repeated addition in increasing doses, with each dose delayed until the preceding one had degraded, enabled high concentrations to be degraded easily.[142] The effects of acclimation can be expected to be most profound for alkylpyridinium chloride and dialkyldimethyl- and alkyldimethylbenzyl-ammonium chlorides which are generally more resistant than the alkyl trimethyl compounds.[137]

5.4.3 *Biodegradability and metabolic pathways*

Several studies have shown that rates of primary biodegradation of QACs depend largely on the balance between long chain alkyl and methyl groups attached to nitrogen.[5,136,139,142–144] Thus QACs with a single long alkyl chain (monoalkyltrimethylammonium chloride) undergo primary biodegradation faster than the dialkyldimethyl analogues. Substitution of a methyl group with benzyl in the latter type slightly retards biodegradation, and the alkylpyridinium chlorides in which nitrogen and three of its bonds are actually part of an aromatic ring are slower still. The most resistant group are the trialkylmonomethyl compounds.

The length of the alkyl chains also influences primary biodegradation rates, which increase with increasing chain length up to about C_{10-12} but then decrease for higher homologues. This pattern contrasts markedly with the anionics for which the distance principle (longer chains enabling ever faster biodegradation, see section 5.2.9) applies. The decrease in biodegradability at longer chain lengths is probably attributable to the bacteriostatic/bactericidal properties of cationics which increase with increasing chain length. Thus as chain length is increased to around C_{10-12} toxic effects begin to outweigh the distance principle. These observations imply operation of an ω/β-oxidation of the long alkyl chain, thus supporting more direct evidence obtained using an isolated *Xanthomonas* species.[139] This organism degraded decyltrimethylammonium chloride with the formation of 9–carboxynonyltrimethylammonium ion (the product of ω-oxidation of the parent surfactant) and 7-carboxyheptyltrimethylammonium ion (the product of a first round of β-oxidation). There was also evidence for the formation of 5-carboxypentyl-, 3-carboxypropyl- and carboxymethyltrimethyl ammonium ions which correspond to further rounds of acetyl-group removal by β-oxidation. The presence of these shorter compounds in small amounts is consistent with the slowing of β-oxidation as the site of oxidation approaches the hydrophilic group, as seen in other surfactants.

Other metabolic pathways must also contribute because in acclimated sewage treatment units all the carbon in the long alkyl chains and in the methyl groups in dialkyldimethylammonium chloride were equally accessible to microorganisms.[145] Intermediates that were formed, although unidentified, were short-lived and converted to CO_2. Demethylations clearly occur and Cain[4] has accumulated several pieces of circumstantial evidence that support the possibility, but whether they precede or follow ω/β-oxidation is unknown.

5.5 Relevance to biodeterioration

Almost all the pathways described in the foregoing sections relate to studies with bacteria from natural (e.g. rivers, lakes) or manmade (e.g. wastewater treatment plants) environments. Chemical and physical conditions in these environments are likely to be quite different from those present in a surfactant-containing formulation in terms of, for example, physical form and properties (viscosity, fluidity), concentrations of surfactant and other organic and inorganic components, and the microflora with which the sample may come into contact. So how valid is it to extrapolate from biodegradation of surfactants in environmental situations to biodeterioration in surfactant formulations?

First, members of the genus *Pseudomonas* pervade the ranks of those known surfactant degraders which have been studied mechanistically. *Pseudomonas* species are also extremely versatile and ubiquitously distributed in water supplies, in the air and on solid surfaces. They are thus likely contaminants in surfactant preparations, and indeed have been specifically identified in documented cases.[146,147]

Secondly, although high concentrations of surfactant may affect microbial cell physiology, they are unlikely to divert metabolic pathways in competent organisms.

Thirdly, in most pathways, the surfactant undergoes an initial biochemical step(s) which destroys its amphipathic properties (primary biodegradation), followed by a series of catabolic steps which further degrade the molecule to components which are normally present in the cells and which serve as sources of carbon and energy for growth (ultimate biodegradation). Although these latter stages involving growth and replication of the cells may be influenced by environmental conditions such as concentrations of surfactant and of other nutrients (e.g. nitrogen, phosphorus), the primary biodegradation step is less likely to be affected. Thus notwithstanding possible differences in ultimate biodegradation, the primary steps of biodegradation in the environment and of biodeterioration in the formulation, are likely to be comparable.

Finally the organisms so far studied as biodegraders of surfactants (which include members of the genera *Arthrobacter*, *Achromobacter*, *Alcaligenes*, *Bacillus*, *Citrobacter*, *Corynebacterium*, *Flavobacterium*, *Klebsiella*, *Micrococcus*, *Mycobacterium*, *Nocardia*, *Pseudomonas*, *Sphaerotilus*, *Streptococcus* and *Vibrio*,[4] have been isolated by enrichment culture on relatively high concentrations of surfactants (typically around 0.1–1%). Such concentrations are necessary to provide sufficient growth and they create a selection pressure for those organisms that cannot only withstand the lytic effects of surfactants on cells but can also utilize them as nutrients. These are characteristics that might be expected of microbial contaminants in surfactant formulations. Moreover, the experimental studies so far carried out on the mechanisms of biodegradation and the associated enzymology have required work with high culture densities of bacterial cells which in turn can only be achieved by growth in high concentrations of surfactants. As Cain[110] has pointed out, this approach may be criticized as having little relevance to true environmental conditions where typical concentrations are in the ppm range (i.e. ~0.0001%), but in the present context, it is fortuitously appropriate because it approaches the high concentrations of surfactant usually incorporated in formulations.

References

1. Alexander, M. (1994) *Biodegradation and Bioremediation*, Academic Press, New York.
2. Gloxhuber, C. (1980) Anionic surfactants – biochemistry, toxicology, dermatology. *Surfactant Science Series*, Vol. 10, Marcel Dekker, New York.
3. Biedlingmaier, S. and Schmidt, A. (1983) Alkylsulfonic acids and some *S*-containing detergents as sulfur sources for growth of *Chlorella fusca*. *Arch. Microbiol.*, **136**, 124–130.
4. Cain, R.B. (1977) Surfactant biodegradation in waste waters. In *Treatment of Industrial Effluents*, Callely, A.G., Forster, C.F. and Stafford, D.A. (eds)), Hodder and Stoughton, London, pp. 283–327.
5. Swisher, R.D. (1987) Surfactant biodegradation. *Surfactant Science Series*, Vol. 18, Marcel Dekker, New York.
6. Cain, R.B. (1981) Microbial degradation of surfactants and builder components. In *Microbial Degradation of Xenobiotics and Recalcitrant Compounds, FEMS Symposium No. 12*, Leisinger, T., Hütter, R., Cook, A.M. and Neüsch, J. (eds), Academic Press, London, pp. 325–370.
7. Cain, R.B. (1987) Biodegradation of anionic surfactants. *Biochem. Soc. Trans.*, **15**, 7S–22S.
8. Karsa, D.R. and Porter, M. (eds.) (1995) *Biodegradability of Surfactants.*. Blackie A & P, Glasgow.
9. White, G.F. and Russell, N.J. (1992) Biodegradation of anionic surfactants. In *Surfactants in Lipid Chemistry; Recent Synthetic, Physical and Biodegradative Studies*, Tyman, J.H.P. (ed.) Royal Society of Chemistry, Cambridge, pp. 99–122.
10. White, G.F. and Russell, N.J. (1993) Biodegradation of anionic surfactants and related compounds. In *Biochemistry of Microbial Degradation*, Ratledge, C. (ed.) Kluwer, Dordrecht, pp. 143–177.
11. White, G.F. (1993) Bacterial biodegradation of ethoxylated surfactants. *Pesticide Sci.*, **37**, 159–166.
12. Poindexter, J.S. (1987) Bacterial responses to nutrient limitation. In *Society for Microbiology Symposium 41*, Fletcher, M., Gray, T.R.G. and Jones, J.G. (eds), Cambridge University Press, Cambridge, pp. 283–317.
13. Linfield, W.M. (1976) Anionic surfactants. *Surfactant Science Series*, Vol. **7**, Parts 1 and 2. Marcel Dekker, New York.
14. White, G.F. and Russell, N.J. (1995) What is biodegradation? In *Biodegradability of Surfactants*, Karsa, D.R. and Porter, M.R. (eds), Blackie A & P, Glasgow, pp. 28–64.
15. Payne, W.J. (1963) Pure culture studies of the degradation of detergent compounds. *Biotechnol. Bioeng.*, **5**, 355–365.
16. Payne, W.J. and Feisal, V.E. (1963) Bacterial utilization of dodecyl sulfate and dodecylbenzene sulfonate. *Appl. Microbiol.*, **11**, 339–344.
17. Thomas, O.R.T. and White, G.F. (1989) Metabolic pathway for the biodegradation of sodium dodecyl sulfate by *Pseudomonas* sp. C12B. *Biotechnol. Appl. Biochem.*, **11**, 318–327.

18. Ratledge, C. (1993) Biodegradation of oils, fats and fatty acids. In *Biochemistry of Microbial Degradation*, Ratledge, C. (ed.), Kluwer Academic, Dordrecht, pp. 89–141.
19. Lijmbach, G.W.M. and Brinkhuis, E. (1973) Microbial degradation of secondary *n*-alkyl sulfates and secondary alcohols. *Antonie Van Leeuwenhoek*, **39**, 415–423.
20. Dodgson, K.S., White, G.F. and Fitzgerald, J.W. (1982) *Sulfatases of Microbial Origin*, CRC Press, Boca Raton.
21. Dodgson, K.S. and White, G.F. (1983) Some microbial enzymes involved in the biodegradation of sulphate surfactants. *Top. Enzyme Ferment. Biotechnol.*, **7**, 90–155.
22. White, G.F., Russell, N.J. and Day, M.J. (1985) A survey of sodium dodecyl sulphate (SDS)-resistance and alkylsulphatase production in bacteria from clean and polluted river sites. *Environ. Pollution*, **A37**, 1–11.
23. Anderson, D.J., Day, M.J., Russell, N.J. *et al.* (1988) Temporal and geographical distributions of epilithic sodium dodecyl sulfate (SDS)-degrading bacteria in a polluted South Wales river. *Appl. Environ. Microbiol.*, **54**, 555–560.
24. White, G.F., Anderson, D.J., Day, M.J. *et al.* (1989) Distribution of planktonic bacteria capable of degrading sodium dodecyl sulphate (SDS) in a polluted South Wales river. *Environ. Pollution*, **57**, 103–115.
25. Haines, T.H. (1973) Halogen- and sulfur-containing lipids of *Ochromonas*. *Ann. Rev. Microbiol.*, **27**, 403–411.
26. Mercer, E.J. and Davies, C.L. (1979) Distribution of chlorosulpholipids in algae. *Phytochemistry*, **18**, 457–462.
27. Haslewood, G.A.D. (1967) *Bile Salts*, Methuen, London.
28. Kramer, V.C., Calabrese, D.M. and Nickerson, K.W. (1980) Growth of *Enterobacter cloacae* in the presence of 25% sodium dodecyl sulfate. *Appl. Environ. Microbiol.*, **40**, 973–976.
29. Maurer, E.W., Cordon, T.C., Weil, J.K. *et al.* (1965) The effect of tallow-based surfactants on anaerobic digestion. *J. Am. Oil Chem. Soc.*, **42**, 189–192.
30. Bruce, A.M., Swanwick, J.D. and Ownsworth, R.A. (1966) Synthetic detergents and sludge digestion: Some recent observations. *J. Proc. Inst. Sewage Purification*, 427–447.
31. Dodgson, K.S., White, G.F., Massey, J.A. *et al.* (1984) Utilization of sodium dodecyl sulphate by denitrifying bacteria under anaerobic conditions. *FEMS Microbiol. Lett.*, **24**, 53–56.
32. Huijghebaert, S.M. and Eyssen, H.J. (1982) Specificity of bile salt sulfatase activity from *Clostridium* sp. strain S1. *Appl. Environ. Microbiol.*, **44**, 1030–1034.
33. Huijghebaert, S.M., Mertens, J.A. and Eyssen, H.J. (1982) Isolation of a bile salt sulfatase-producing *Clostridium* strain from rat intestinal microflora. *Appl. Environ. Microbiol.*, **43**, 185–192.
34. Kassem, T.M. (1982) Synthesis of new surfactants from alkylated phenols. *Tenside*, **19**, 213–215.
35. Langowska, I. (1982) Utilization of organic phosphates by selected bacteria and algae. *Water Res.*, **16**, 161–167.
36. Hales, S.G. (1993) Biodegradation of the anionic surfactant dialkyl sulphosuccinate. *Environ. Chem. Toxicol.*, **12**, 1821–1828.
37. Quick, A., Russell, N.J., Hales, S.G. *et al.* (1994) Biodegradation of sulphosuccinate: direct desulphonation of a secondary sulphonate. *Microbiology*, **140**, 2991–2993.
38. Harwood, J.L. and Russell, N.J. (1984) *Lipids in Plants and Microbes*, George Allen and Unwin, London.
39. Hales, S.G., White, G.F., Dodgson, K.S. *et al.* (1986) A comparative study of the biodegradation of the surfactant sodium dodecyltriethoxy sulphate by four detergent-degrading bacteria. *J. Gen. Microbiol.*, **132**, 953–961.
40. Hales, S.G., Dodgson, K.S., White, G.F. *et al.* (1982) Initial stages in the biodegradation of the surfactant sodium dodecyltriethoxy sulfate by *Pseudomonas* sp. strain DES1. *App. Environ. Microbiol.*, **44**, 790–800.
41. Griffiths, E.T., Hales, S.G., Russell, N.J. *et al.* (1986) Metabolite production during the biodegradation of the surfactant sodium dodecyltriethoxy sulphate under mixed-culture die-away conditions. *J. Gen. Microbiol.*, **132**, 963–972.
42. White, G.F. and Matts, P.J. (1992) Biodegradation of short-chain alkyl sulphates by a coryneform species. *Biodegradation*, **3**, 83–91.

43. Matts, P.J., White, G.F. and Payne, W.J. (1994) Purification and characterization of the short-chain alkylsulphatase of *Coryneform* B1a. *Biochem. J.*, **304**, 937–943.
44. Davies, I., White, G.F. and Payne, W.J. (1990) Oxygen-dependent desulphation of monomethyl sulphate by *Agrobacterium* sp. M3C. *Biodegradation*, **1**, 229–241.
45. Higgins, T.P., Snape, J.R. and White, G.F. (1993) Comparison of pathways for biodegradation of monomethyl sulphate in *Agrobacterium* and *Hyphomicrobium* species. *J. Gen. Microbiol.*, **139**, 2915–2920.
46. Crescenzi, A.M.V., Dodgson, K.S. and White, G.F. (1984) Purification and some properties of the D-lactate-2-sulphatase of *Pseudomonas syringae* GG. *Biochem. J.*, **223**, 487–494.
47. Crescenzi, A.M.V., Dodgson, K.S., White, G.F. *et al.* (1985) Initial oxidation and subsequent desulphation of propan-2-yl sulphate by *Pseudomonas syringae* GG. *J. Gen. Microbiol.*, **131**, 469–477.
48. Griffiths, E.T., Hales, S.G., Russell, N.J. *et al.* (1987) Identification of hydrophobic metabolites formed during biodegradation of alkyl ethoxylate and alkyl ethoxy sulphate surfactants by *Pseudomonas* sp. DES1. *Biotechnol. Appl. Biochem.*, **9**, 217–229.
49. Shore, S. and Berger, D.R. (1976) Alcohol and alcohol ether sulfates. In *Surfactant Science Series, Vol. 7, Part 1, Anionic Surfactants*, Linfield, W.M. (ed.), Marcel Dekker, Basel and New York, pp. 135–217.
50. Bluestein, C. and Bluestein, B.R. (1976) Petroleum sulfonates. In *Anionic Surfactants, Part II*, Linfield, W.M. (ed.), Marcel Dekker, New York, pp. 315–343.
51. Wagner, F.C. and Reid, E.E. (1931) The stability of the carbon–sulphur bond in some aliphatic sulphonic acids. *J. Am. Chem. Soc.*, **53**, 3407–3413.
52. Cardini, G., Catelani, D., Sorlini, D. *et al.* (1966) La degradazione microbica dei detergenti di sintesi. *Annal. Microbiolog. Enzimolog.*, **16**, 217–223.
53. Thysse, G.J.E. and Wanders, T.H. (1972) Degradation of *n*-alkane-1-sulfonates by *Pseudomonas*. *Antonie van Leeuwenhoek*, **38**, 53–63.
54. Thysse, G.J.E. and Wanders, T.H. (1974) Initial steps in the degradation of *n*-alkane-1-sulfonates by *Pseudomonas*. *Antonie van Leeuwenhoek*, **40**, 25–37.
55. Cordon, T.C., Maurer, E.W. and Stirton, A.J. (1970) Course of biodegradation of anionic detergents by analyses for carbon, methylene blue active substance and sulfate ion. *J. Am. Oil Chem. Soc.*, **47**, 203–206.
56. Taylor, A.J., Olavesen, A.H., Black, J.G. *et al.* (1978) The metabolism of the surfactants dodecyl sulfonate and hexadecyl sulfonate in the rat. *Toxicol. Appl. Pharmacol.*, **45**, 105–117.
57. Oba, K., Mori, A. and Tomiyama, S. (1968) Biochemical studies of *n*-α-olefin sulfonates. I. *Yukagaku*, **17**, 517–520.
58. Sekiguchi, H., Miura, K., Oba, K. *et al.* (1975) Biodegradation of α-olefin sulfonates (AOS) and other surfactants. *Yukagaku*, **24**, 145–148.
59. Kravetz, L., Chung, H. and Rapean, J.C. (1982) Ultimate biodegradation studies of α-olefin sulfonates. *J. Am. Oil Chem. Soc.*, **59**, 206–210.
60. Maurer, E.W., Weil, J.K. and Linfield, W.M. (1977) The biodegradation of esters of α-sulfo fatty acids. *J. Am. Oil Chem. Soc.*, **54**, 582–584.
61. Lotzsch, K., Neufahrt, A. and Taeuber, G. (1979) Comparative study of the biodegradation of secondary alkanesulfonates using ^{14}C-labelled preparations. *Tenside Surfact. Deterg.*, **16**, 150–155.
62. Steber, J. and Wierich, P. (1989) The environmental fate of fatty acid α-sulfomethyl esters. *Tenside Surfact. Deterg*, **26**, 406–411.
63. Morgan, P. and Watkinson, R.J. (1993) Biodegradation of components of petroleum. In *Biochemistry of Microbial Degradation*, Ratledge, C. (ed.), Kluwer Academic, Dordrecht, pp. 1–31.
64. Hrsak, D., Bosnjak, M. and Johanides, V. (1982) Enrichment of linear alkylbenzene-sulphonate (LAS)-degrading bacteria in continuous culture. *J. Appl. Bacteriol.*, **53**, 413–422.
65. Sigoillot, J.C. and Nguyen, M.H. (1990) Isolation and characterisation of surfactant degrading bacteria in a marine environment. *FEMS Microbiol. Ecol.*, **73**, 59–68.
66. Jimenez, L., Breen, A., Thomas, N. *et al.* (1991) Mineralization of linear alkylbenzene sulfonate by a four-member aerobic bacterial consortium. *Appl. Environ. Microbiol.*, **57**, 1566–1569.

67. Willetts, A.J. and Cain, R.B. (1972) Microbial metabolism of alkylbenzene sulfonates. Bacterial metabolism of undecylbenzene-*p*-sulphonate and dodecylbenzene-*p*-sulphonate. *Biochem. J.*, **129**, 389–402.
68. Locher, H.H., Leisinger, T. and Cook, A.M. (1989) Degradation of *p*-toluenesulphonic acid via side chain oxidation, desulphonation and *meta* ring cleavage in *Pseudomonas (Comamonas) testosteroni* T–2. *J. Gen. Microbiol.*, **135**, 1969–1978.
69. Locher, H.H., Leisinger, T. and Cook, A.M. (1991) 4-Sulphobenzoate 3,4-dioxygenase. *Biochem. J.*, **274**, 833–842.
70. Thurnheer, T., Zurrer, D., Hoglinger, O. *et al.* (1990) Initial steps in the degradation of benzene sulfonic acid, 4-toluene sulfonic acid, and orthanilic acid in *Alcaligenes* sp. strain O–1. *Biodegradation*, **1**, 55–64.
71. Cook, A.M. and Leisinger, T. (1991) Desulfonation of aromatic compounds. In *Proceedings of the International Symposium on Environmental Biotechnology*, Royal Flemish Society of Engineers, Ostend, pp. 115–122.
72. Hopper, D.J. (1991) Aspects of the aerobic degradation of aromatics by micro-organisms. In *Biodegradation: Natural and Synthetic Materials*, Betts, W.B. (ed.), Springer-Verlag, London, pp. 1–14.
73. Smith, M.R. (1993) The physiology of aromatic hydrocarbon degrading bacteria. In *Biochemistry of Microbial Degradation*, Ratledge, C. (ed.), Kluwer Academic, Dordrecht, pp. 347–378.
74. Kertesz, M.A., Kolbener, P., Stockinger, H. *et al.* (1994) Desulfonation of linear alkylbenzenesulfonate surfactants and related compounds by bacteria. *Appl. Environ. Microbiol.*, **60**, 2296–2303.
75. Swanwick, J.D., Bruce, A.M. and Vandyke, K.G. (1968) Inhibition of sludge digestion by synthetic detergents. *Water Pollution Control*, **67**, 91–99.
76. Wood, A.A., Claydon, M.B. and Finch, J. (1970) Synthetic detergents: some problems. *Water Pollution Control*, **69**, 675–683.
77. McEvoy, J. and Giger, W. (1986) Determination of linear alkylbenzenesulfonates in sewage-sludge by high-resolution gas-chromatography mass-spectrometry. *Environ. Sci. Technol.*, **20**, 376–383.
78. Kölbel, H., Kurzendorfer, P. and Zahiruddin, M. (1976) Constitution and properties of surfactants. IV. Influence of structure on the biodegradation of anionic surfactants. *Tenside Surfact. Deterg.*, **1**, 7–18.
79. Sawyer, C.N., Bogan, R.H. and Simpson, J.R. (1956) Biochemical behaviour of synthetic detergents. *Ind. Eng. Chem.*, **48**, 236–240.
80. Ryckman, D.W. and Sawyer, C.N. (1957) Chemical structure and biological oxidizability of surfactants. *Purdue Conf.*, **12**, 270–284.
81. Ikeda, K., Yamada, H. and Tanaka, S. (1963) Bacterial degradation of taurine. *J. Biochem. (Tokyo)*, **54**, 312–316.
82. Kondo, H., Anada, H., Ohsawa, K. *et al.* (1971) Formation of sulfoacetaldehyde from taurine in bacterial extracts. *J. Biochem. (Tokyo)*, **69**, 621–623.
83. Kondo, H., Kagotani, K., Oshima, M. *et al.* (1973) Purification and some properties of taurine dehydrogenase from a bacterium. *J. Biochem. (Tokyo)*, **73**, 1268–1278.
84. Kondo, H. and Ishimoto, M. (1972) Enzymatic formation of sulfite and acetate from sulfoacetaldehyde, a degradation product of taurine. *J. Biochem. (Tokyo)*, **72**, 487–489.
85. Kondo, H. and Ishimoto, M. (1975) Purification and properties of sulfoacetaldehyde sulfolyase, a thiamine pyrophosphate-dependent enzyme forming sulfite and acetate. *J. Biochem. (Tokyo)*, **78**, 317–325.
86. Kondo, H. and Ishimoto, M. (1974) Requirement for thiamine pyrophosphate and magnesium for sulfoacetaldehyde sulfo-lyase activity. *J. Biochem. (Tokyo)*, **76**, 229–231.
87. Shimamoto, G. and Berk, R.S. (1979) Catabolism of taurine in *Pseudomonas aeruginosa*. *Biochim. Biophys. Acta*, **569**, 287–292.
88. Shimamoto, G. and Berk, R.S. (1980) Taurine catabolism. II. Biochemical and genetic evidence for sulfoacetaldehyde sulfolyase involvement. *Biochim. Biophys. Acta*, **632**, 121–130.
89. Toyama, S., Miyasato, K., Yasuda, M. *et al.* (1973) Occurrence of taurine-pyruvate aminotransferase in bacterial extract. *Agric. Biol. Chem.*, **37**, 2939–2941.

90. Toyama, S., Misono, H. and Soda, K. (1978) Properties of taurine: α-ketoglutarate aminotransferase in *Achromobacter superficialis*. Inactivation and reactivation of enzyme. *Biochim. Biophys. Acta*, **523**, 75–81.
91. Patterson, S.J., Scott, C.C. and Tucker, K.B.E. (1970) Nonionic detergent degradation: III. Initial mechanism of the degradation. *J. Am. Oil Chem. Soc.*, **47**, 37–41.
92. Tobin, R.S., Onuska, F.I., Brownlee, B.G. *et al.* (1976) The application of an ether cleavage technique to a study of the biodegradation of a linear alcohol ethoxylate nonionic surfactant. *Water Res.*, **10**, 529–535.
93. Ichikawa, Y., Kitamoto, Y. and Hosoi, N. (1978) Degradation of polyethylene glycol dodecyl ethers by a Pseudomonad isolated from activated sludge. *J. Ferment. Technol.*, **56**, 403–409.
94. Kravetz, L., Chung, H., Guin, K.F. *et al.* (1982) Ultimate biodegradation of an alcohol ethoxylate and a nonylphenol ethoxylate under realistic conditions. Part 1. *Household and Personal Products Industries*, March 1982 and April 1982, 46–72 and 62–70.
95. Vashon, R.D. and Schwab, B.S. (1982) Mineralization of linear alcohol ethoxylates and linear alcohol ethoxy sulfates at trace concentrations in estuarine water. *Environ. Sci. Technol.*, **16**, 433–436.
96. Steber, J. and Wierich, P. (1985) Metabolites and biodegradation pathways of fatty alcohol ethoxylates in microbial biocenoses of sewage treatment plants. *Appl. Environ. Microbiol.*, **49**, 530–537.
97. Watson, G.K. and Jones, N. (1979) The microbial degradation of nonionic surfactants. *Soc. Gen. Microbiol. Quart.*, **6**, 78–79.
98. Nooi, J.R., Testa, M.C. and Willemse, S. (1970) Biodegradation mechanisms of fatty alcohol nonionics. Experiments with some ^{14}C-labelled stearyl alcohol/EO condensates. *Tenside Surfact. Deterg.*, **7**, 61–65.
99. Grant, M. and Payne, W.J. (1983) Anaerobic growth of *Alcaligenes faecalis* var. *denitrificans* at the expense of ether glycols and non-ionic detergents. *Biotechnol. Bioeng.*, **25**, 627–630.
100. Wickbold, R. (1966) Analysis for nonionic surfactants in water and wastewater. *Vom Wasser*, **33**, 229–241.
101. Schoberl, P. and Bock, K.J. (1980) Surfactant degradation and its metabolites. *Tenside Surfact. Deterg.*, **17**, 262–266.
102. Schoberl, P. (1981) Comparative investigations on the microbial metabolism of a nonylphenol and an oxoalcohol ethoxylate. *Tenside*, **18**, 64–72.
103. Osburn, Q.W. and Benedict, J.H. (1966) Polyethoxylated alkyl phenols: relationship of structure to biodegradation mechanism. *J. Am. Oil Chem. Soc.*, **43**, 141–146.
104. Giger, W., Brunnen, P.H. and Schaffner, C. (1984) 4-Nonylphenol in sewage sludge: accumulation of toxic metabolites from nonionic surfactants. *Science*, **225**, 623–625.
105. Ahel, M., McEvoy, J. and Giger, W. (1993) Bioaccumulation of the lipophilic metabolites of nonionic surfactants in freshwater organisms. *Environ. Pollution*, **79**, 243–248.
106. Rudling, L. (1972) Biodegradability of nonionic surfactants: a progress report. *Report B134 of Swedish Water and Air Pollution Research Laboratory*, Stockholm.
107. Stephanou, E. and Giger, W. (1982) Persistent organic chemicals in sewage effluents. 2. Quantitative determination of nonylphenol ethoxylates by glass capillary gas chromatography. *Environ. Sci. Technol.*, **16**, 800–805.
108. Maki, H., Masuda, N., Fujiwara, Y. *et al.* (1994) Degradation of alkylphenol ethoxylates by *Pseudomonas* sp. strain TR01. *Appl. Environ. Microbiol.*, **60**, 2265–2271.
109. Baggi, G., Beretta, L., Galli, E. *et al.* (1978) Biodegradation of alkylphenol ethoxylates. In *The Oil Industry and Microbial Ecosystems*, Chester, K.W.A. and Somerville, H.J. (eds), Heyden, London, pp. 129–136.
110. Cain, R.B. (1994) Biodegradation of detergents. *Curr. Opinion Biotechnol.*, **5**, 266–274.
111. Cox, D.P. (1978) The biodegradation of polyethylene glycols. *Adv. Appl. Microbiol.*, **23**, 173–194.
112. Kawai, F. (1987) The biochemistry of degradation of polyethers. *CRC Crit. Rev. Biotechnol.*, **6**, 273–307.
113. Watson, G.K. and Jones, N. (1977) The biodegradation of polyethylene glycols by sewage bacteria. *Water Res.*, **11**, 95–100.

114. Jenkins, L.D.L. and Cook, K.A. (1979) Microbial degradation of polyethylene glycols. *J. Appl. Bacteriol.*, **47**, 75–85.
115. Wagener, S. and Schink, B. (1988) Fermentative degradation of nonionic surfactants and polyethylene glycol by enrichment cultures and by pure cultures of homoacetogenic and propionate-forming bacteria. *Appl. Environ. Microbiol.*, **54**, 561–565.
116. Haines, J.R. and Alexander, M. (1975) Microbial degradation of polyethylene glycols. *Appl. Microbiol.*, **29**, 621–625.
117. Obradors, N. and Aguilar, J. (1991) Efficient biodegradation of high-molecular-weight polyethylene glycols by pure cultures of *Pseudomonas stutzeri*. *Appl. Environ. Microbiol.*, **57**, 2383–2388.
118. Schink, B. and Stieb, M. (1983) Fermentative degradation of polyethylene glycol by a strictly anaerobic, Gram negative, non-spore-forming bacterium, *Pelobacter venetianus* sp. nov. *Appl. Environ. Microbiol.*, **45**, 1905–1913.
119. Dwyer, D.F. and Tiedje, J.M. (1986) Metabolism of polyethylene glycol by two anaerobic bacteria, *Desulfovibrio desulfuricans* and a *Bacteroides* sp. *Appl. Environ. Microbiol.*, **52**, 852–856.
120. Strass, A. and Schink, B. (1986) Fermentation of polyethylene glycol via acetaldehyde in *Pelobacter venetianus*. *Appl. Microbiol. Biotechnol.*, **25**, 37–42.
121. Frings, J., Schramm, E. and Schink, B. (1992) Enzymes involved in anaerobic polyethylene glycol degradation by *Pelobacter venetianus* and *Bacteroides* strain PG1. *Appl. Environ. Microbiol.*, **58**, 2164–2167.
122. Pearce, B.A. and Heydeman, M.T. (1980) Metabolism of di(ethylene glycol) [2-(2′-hydroxyethoxy)ethanol] and other short poly(ethylene glycol)s by Gram-negative bacteria. *J. Gen. Microbiol.*, **118**, 21–27.
123. Thelu, J., Medina, L. and Pelmont, J. (1980) Oxidation of poly(oxethylene) oligomers by an inducible enzyme from *Pseudomonas* P400. *FEMS Microbiol. Lett.*, **8**, 187–190.
124. Kawai, F. and Yamanaka, H. (1986) Biodegradation of polyethylene glycol by symbiotic mixed culture (obligate mutualism). *Archiv. Microbiol.*, **146**, 125–129.
125. Kawai, F., Kimura, T., Fukaya, T. *et al.* (1978) Bacterial oxidation of polyethylene glycol. *Appl. Environ. Microbiol.*, **35**, 679–684.
126. Kawai, F., Kimura, T., Tani, Y. *et al.* (1978) Purification and characterization of polyethylene glycol dehydrogenase involved in the bacterial metabolism of polyethylene glycol. *Appl. Environ. Microbiol.*, **40**, 701–705.
127. Kawai, F., Fukaya, M., Tani, Y. *et al.* (1977) Identification of polyethylene glycols (PEGs)-assimilable bacteria and culture characteristics of PEG 6000 degradation in a mixed culture. *J. Ferment. Technol.*, **55**, 429–437.
128. Ogata, K., Kawai, F., Fukaya, M. *et al.* (1975) Isolation of polyethylene glycol-assimilable bacteria. *J. Ferment. Technol.*, **53**, 557–761.
129. Kawai, F., Hanada, K., Tani, Y. *et al.* (1977) Bacterial degradation of water-insoluble polymer (polypropylene glycol). *J. Ferment. Technol.*, **55**, 89–96.
130. Kawai, F., Okamoto, T. and Suzuki, T. (1985) Aerobic degradation of polypropylene glycol by *Corynebacterium* sp. *J. Ferment. Technol.*, **63**, 239–244.
131. Cain, R.B. (1992) Microbial degradation of synthetic polymers. In *Microbial Control of Pollution, Society of Microbiology Symposium 48*, Fry, J.C., Gadd, G.M., Herbert, R.A., *et al.* (eds), Cambridge University Press, Cambridge, pp. 293–338.
132. Knaggs, E.A. (1964) Alkylolamides in soft detergents. *Soap Chemical and Specialties*, **40**, 79–82.
133. Garrison, L.J. and Matson, R.D. (1964) A comparison of Warburg respirometry and die-away studies of the degradability of select non-ionic surfactants. *J. Amer. Oil Chem. Soc.*, **41**, 799–804.
134. van Ginkel, C.G., Stroo, C.A. and Kroon, A.G.M. (1993) Biodegradability of fatty amines and amides and the non-toxicity of their biodegradation products. *Tenside Surfact. Deterg.*, **30**, 213–216.
135. Games, L.M., King, J.E. and Larson, R.J. (1982) Fate and distribution of a quaternary ammonium surfactant, octadecyltrimethyl ammonium chloride (OTAC), in wastewater treatment. *Environ. Sci. Technol.*, **16**, 483–488.
136. Baleux, B. and Caumette, P. (1977) Biodégradation de quelques agents de surface cationiques. *Water Res.*, **11**, 833–841.

137. Boethling, R.S. (1984) Environmental fate and toxicity in wastewater treatment of quaternary ammonium surfactants. *Water Res.*, **18**, 1061–1076.
138. Shimp, R.J. (1989) Adaptation to a quaternary ammonium surfactant in aquatic sediment microcosms. *Environ. Toxicol. Chem.*, **8**, 201–208.
139. Dean-Raymond, D. and Alexander, M. (1977) Bacterial metabolism of quaternary ammonium compounds. *Appl. Environ. Microbiol.*, **33**, 1037–1044.
140. Gerike, P., Fischer, W.K. and Jasiak, W. (1978) Surfactant quaternary ammonium salts in aerobic sewage digestion. *Water Res.*, **12**, 1117–1122.
141. Ventullo, R.M. and Larson, R.J. (1986) Adaptation of aquatic microbial communities to quaternary ammonium compounds. *Appl. Environ. Microbiol.*, **51**, 356–361.
142. Ruiz Cruz, J. (1979) Contamination of natural watercourses by synthetic detergents. XVI. Influence of surfactant concentration and acclimation on the biodegradation of cationic detergents in river water. *Grasa Aceites*, **30**, 293–299.
143. Ruiz Cruz, J. and Dobarganes Garcia, M.C. (1979) Pollution of natural waters by synthetic detergents. XV. Relation between structure and biodegradation of cationic surfactants in river water. *Grasa Aceites*, **30**, 67–74.
144. Larson, R.J. and Vashon, R.D. (1983) Adsorption and biodegradation of cationic surfactants in laboratory and environmental systems. *Develop. Indust. Microbiol.*, **24**, 425–434.
145. Sullivan, D.E. (1983) Biodegradation of cationic surfactant in activated sludge. *Water Res.*, **17**, 1145–1151.
146. Williams, O.B. and Rees, H.B., Jr. (1949) Bacterial utilization of anionic surfactants. *J. Bacteriol.*, **58**, 823–824.
147. Flawn, P.C., Malcolm, S.A. and Woodroffe, R.C.S. (1973) Assessment of the preservative capacity of shampoos. *J. Soc. Cosmetic Chem.*, **24**, 229–238.

6 Preservation of agrochemicals

D.A. KNOWLES

6.1 Introduction

Agrochemical products have been used widely for many years to increase the yield and improve the quality of food and fibre crops all over the world. The agrochemical industry has become a major business producing products with a total world sales value estimated in 1993 at about US$25 billion, and it plays an important part in the economies of most countries. The agrochemical business represents a significant opportunity for surfactants and other essential formulation additives and adjuvants in spray applications. Although the agrochemical industry has reached maturity in North America, Western Europe and Japan, there is still considerable scope for new more environmentally friendly formulations. Developing areas, especially the Asia–Pacific region, will show increasing need for agrochemical products.

Changes in the population of the world and increasing urbanisation and industrialisation of communities are placing a great demand on the efficient use of available land for agriculture. For example, the United Nations have forecasted that if present trends continue, the population of the world will increase from about 5 billion now to about 10 billion by the year 2040, and the fastest rate of growth will be in the less developed areas, particularly the Asia–Pacific region.[1] There will, therefore, be an increasing need for agrochemical products as an important input in the management of food and fibre crops to improve their yield and quality.

The ability to protect growing crops from weeds, pests and diseases has been known since ancient times in the Old World of the Middle East, Asia and China. However, the greatest improvements in crop protection efficiency and productivity in terms of crop yield and quality have occurred mainly in the West and within the last century. Simple emulsifiable oils and soaps have been used as agricultural sprays for many years. The modern era of weed control can be said to have started in the 1940s with the development of the phenoxy acid herbicides such as 2,4-D acid. Since the Second World War, particularly in the 1960s, 1970s and 1980s, many new synthetic pesticides have been introduced to combat a very wide range of weeds, pests and diseases. A great deal of research and development has been carried out by all the major agrochemical companies to produce formulations which can be applied easily to crops and which will optimise the activity of the pesticide.[2]

Although the last few decades have seen remarkable development in new agrochemical active ingredients and formulations, most companies are now reviewing their product strategies and government regulatory authorities are introducing controls and legislation which are leading to the introduction of safer and more environmentally friendly active ingredients and formulations. There is also a need to reduce the total amount of active ingredients applied per hectare. The increasingly high cost of the development of new products is causing the industry to consolidate by mergers of companies or joint ventures between companies.

The wide variety of agrochemical formulations which is available requires a range of different formulation additives to produce safe and usable products. Two of the most important formulation additives are surface active agents, or surfactants, and preservatives. Surfactants have been obtained from natural products by extraction or modification for thousands of years. Many examples of surfactants are well known, like soaps for cleaning, greases and tallows for waterproofing and glue, egg white and natural gums as dispersing and emulsifying agents.

Synthetic surfactants, which have been specially synthesised in order to obtain surface active effects, represent a relatively modern development which may be said to have evolved from the 'sulphonated oils' of the last century. The period between the two World Wars was a very active phase in the development of sulphated and sulphonated anionic surfactants with long hydrocarbon chains.

Since the Second World War, the development of surfactants entered a more specialised phase with the introduction of amphipathic molecules for specific applications. Non-ionic surfactants became available in which the hydrophilic part of the molecule was based on condensed chains of ethylene oxide. A wide range of surfactant properties can be achieved by varying the ethylene oxide chain length. This development has led to a better understanding of the colloid and surface chemistry principles involved in the fundamental functional properties of wetting, dispersing, emulsification and solubilisation in the formulation of pesticides. As a result of all this work, it is now possible for surfactant suppliers to prepare 'tailor-made' surfactants to suit particular functions.[3–5]

For nearly all formulations the most important formulation additive is the surfactant in terms of preparation and production. The surfactant often determines the maximum concentration of the formulation that can be achieved, the particle or droplet size, long term stability and sometimes even the biological activity of the formulation.

Preservatives are also an important additive to formulations to prevent biodegradation during preparation and storage, particularly where the formulations are aqueous based and contain carbohydrates, or where the products are exposed to the atmosphere after application, as in the case of baits and pellets.

This chapter reviews the current state-of-the-art with conventional agrochemical formulations and the trend towards safer and more convenient formulations. It also describes the major surfactants used in these systems and the reasons for their choice. The chapter also reviews the major types of spoilage microorganisms which are encountered in agrochemical formulations, the preservatives which are used to prevent microbial spoilage during preparation, storage and use, and testing protocols for quality control. Finally regulatory constraints on the use of formulation additives are discussed.

6.2 Agrochemical formulations

Since the 1940s there has been a rapid development of synthetic organic chemical pesticides for crop protection and pest control by all the major international chemical companies. Farmers and growers in all the main agricultural areas of the world rely substantially upon crop protection chemicals to help them meet the ever increasing demand for food and other materials such as natural fibres. The consumer continues to seek higher quality and more variety of produce. In post-war years, the chemical industry has endeavoured to satisfy these demands by the continuous development and introduction of novel crop protection chemicals into the international marketplace. Today, there is an effective herbicide, insecticide or fungicide to combat almost every significant problem faced by the modern farmer and grower.

This development has led to a need for a wide range of product formulations to accommodate the variety of physical and chemical properties of the pesticide active ingredients. For example, water-soluble active ingredients may be marketed as aqueous solutions or powder formulations, whereas oily liquid active ingredients are sold as hydrocarbon-based emulsifiable concentrates. Active ingredients which have very low solubility in either water or hydrocarbon oils, may be formulated as suspensions, powders or water-dispersible granules.[6]

In recent years, pressure from the consumer has highlighted a need for products and formulations which are safer and more convenient to use and which will be more effective at much lower application rates. By far the most important method of application is by spraying, usually with water but occasionally with oils as the principal carrier. Formulations are also made for direct application to the soil or for treating seeds before planting, and for protecting stored crops from various pests and diseases (fungi, insects or rodents) which in some countries could destroy as much as 30–40% of the harvest.

Pesticidal active ingredients or toxicants encompass a broad range of chemicals each with its unique chemical and physical properties and mode of action. The main categories of pesticide are herbicides, insecticides, fungicides, plant growth regulators, molluscicides and rodenticides. A great deal of research work has been carried out into understanding the modes of action and physiological effects of active ingredients and the influence of formulation type on the biological performance of the pesticide.[7] The successful use of any active ingredient depends on its correct formulation into a preparation which can be applied for crop protection with safety to those applying the material, to animal life and to the environment in general.

The earliest pesticide formulations were based on simple dusts, powders, aqueous solutions and mineral oil-in-water emulsions. In the post-war period, particularly during the period 1970 onwards, there has been a rapid development of more sophisticated formulations based on the availability of more powerful surfactants and a much better understanding of the principles of colloid and surface chemistry.

The main objectives of formulation can be summarised as follows:

> To provide the user with a convenient, safe product which will not deteriorate over a period of time, and to obtain the maximum activity inherent in the active ingredient.

Usually it is the physicochemical properties of the active ingredient which are the main indicators of the choice of the specific formulation type, but other factors which need to be taken into account are:

- biological activity and mode of action
- method of application
- safety in use
- formulation costs
- market preference.

Once these parameters have been determined, proper selection can be made of the final formulation and the use of inert ingredients including surfactants, preservatives and other additives.

The most common formulations are still soluble concentrates for water-soluble chemicals, emulsifiable concentrates for oil-soluble chemicals and wettable powders and suspension concentrates for insoluble solids. Granules and seed treatments for direct application have also been produced for many years. In recent years the number of formulation types has increased enormously to meet the needs of operator and environmental safety or to improve the activity and persistence of the active ingredient by controlled release. An international coding system was, therefore, devised by GIFAP in 1984 (GIFAP; International Group of National Associations of Manufacturers of Agrochemical Products, based in Brussels, Belgium).

Table 6.1 Major types of pesticide formulations

Formulation	GIFAP Code
Granules	GR
Wettable powder	WP
Solution concentrate	SL
Emulsifiable concentrate	EC
Suspension concentrate	SC
Seed treatments	DS, WS, LS, FS
O/W emulsions	EW
Suspoemulsions	SE
Microemulsions	ME
Water-dispersible granules	WG
Microcapsules	CS

The major types of formulation with their international codes are shown in Table 6.1.

The most important formulations are those which are made for dilution into water in a spray tank. In these cases the choice of formulation additives is very important to ensure that the product mixes and dilutes easily. Sometimes products may be mixed together in the spray tank or may be mixed with spray adjuvants to enhance biological activity. Products such as granules or seed treatments are usually applied undiluted to the soil or to the seed respectively. A few products are formulated to be diluted and sprayed in oils, and there are many minor formulations such as baits, pellets, smokes and aerosols for special purposes.

6.2.1 *Conventional formulations*

6.2.1.1 Wettable powders (WP). Wettable powder formulations of pesticides have been known for many years and are made from solid active ingredients which are suitable for fine grinding through a hammer or pin-type mill or a fluid energy microniser. The powders contain dry surfactants as powder wetting and dispersing agents and inert carriers or fillers.

Wettable powders frequently contain more than 50% active ingredient and the upper limit is usually determined by the amount of inert material such as silica required to prevent the active ingredient particles fusing together during processing in the dry grinding mills. This is influenced by the melting point of the active ingredient, but an inert filler is also needed to prevent the formulated product from caking or aggregating during storage.

Wettable powders have a high proportion of particles less than 5 μm and all the particles should pass through a 44 μm screen. Ideally, the amount of surfactants should be sufficient to allow the spray droplets to wet and spread over the target surface, but the particles should not be easily washed off by rain.

Powder formulations contain a wetting agent to lower the interfacial tension between the solid particles and water and ensure that the powder wets and mixes with water in the spray tank easily. A dispersing agent is also necessary to prevent the particles in the spray tank from flocculating or aggregating together. This ensures that the particles remain suspended in water during the spraying operation.

The types of wetting agents commonly used are:

- sodium dodecylbenzene sulphonate
- sodium lauryl sulphate
- dioctyl sulphosuccinate
- fatty alcohol ethoxylates
- nonyl phenol ethoxylates.

Examples of dispersing agents are:

- sodium lignosulphonates
- naphthalene sulphonic acid–formaldehyde condensates.

A typical wettable powder formulation is shown below:

	% By weight
Active ingredient	25–80
Wetting agent	1–3
Dispersing agent	2–5
Inert filler/carrier	to 100

Wettable powders can also be made from liquid pesticides by using absorbent fillers such as diatomaceous earth or high surface area synthetic silica. However, in this case the active ingredient concentration is usually limited to 40%.

Many pesticides, especially herbicides and fungicides are formulated as wettable powders. However, due to their 'low tech.' image arising from their dustiness, which creates hazards on handling, they are now being superseded by suspension concentrates or water-dispersible granules.

6.2.1.2 Solution concentrates (SL). The simplest of all formulations to make is the solution concentrate, an aqueous solution of the active ingredient which merely requires dilution in the spray tank. The number of pesticides which can be formulated in this way is limited by solubility and hydrolytic stability. Some solution concentrate formulations contain a surfactant, usually to assist wetting onto the leaf surface. These wetting agents are generally of the nonionic type, such as nonylphenol condensed with 8–9 mols of ethylene oxide.

Solution concentrate formulations are usually very stable and, therefore, present few storage problems. Some problems do occur occasionally, such as precipitation during dilution and corrosion of metal containers or spray applicators. A typical solution concentrate formulation is shown below:

	% By weight
Active ingredient	20–50
Wetting agent	3–10
Antifreeze	5–10
Water Water-miscible solvent }	to 100

Nonylphenol or tallow amine ethoxylates are often used as tank mix wetters for solution concentrate formulations. Alternatively, the wetting agent may be built into the formulation to ensure that the correct rate of wetting agent is applied to optimise biological activity. This is often the case, for example, with paraquat and glyphosate formulations. A considerable amount of work is being carried out on new surfactant wetting agents for glyphosate formulations.[8] In some cases preservatives may be necessary to prevent mould growth or bacterial spoilage during long term storage.

6.2.1.3 Emulsifiable concentrates (EC). Emulsifiable concentrate formulations have been very popular for many years and represent the biggest volume of all pesticide formulations in terms of consumption.

Emulsifiable concentrates are made from oily active ingredients or from low melting, waxy solid active ingredients which are soluble in non-polar hydrocarbon solvents, such as xylene, C-9–C-10 solvents, solvent naphtha, odourless kerosene or other proprietory hydrocarbon solvents. Surfactant emulsifiers are added to these formulations to ensure spontaneous emulsification with good emulsion stability properties in the spray tank. Careful selection of a 'balanced pair' emulsifier blend is frequently necessary to ensure emulsion dilution stability is maintained over widely differing climatic conditions and degrees of water hardness. Emulsion droplets of 0.1–5 μm are produced.

The formulation of emulsifiable concentrates has been greatly facilitated by the commercial development over the last 20 years of nonionic emulsifying agents in which the hydrophilic portion of the molecule consists of a polyethylene oxide chain. The nonionic surfactant which is commonly used is a nonyl phenol hydrophobic chain condensed with 12 or more moles of ethylene oxide. The other component of the balanced pair is generally an anionic surfactant such as the oil-soluble calcium salt of dodecylbenzene sulphonic acid.

The total concentration of the emulsifier blend is usually 5–10% of the formulation. There are no definite rules to determine the ratio of anionic to nonionic surfactant in the mixed emulsifiers, but guidance can be obtained from the HLB system. HLB stands for hydrophile–lipophile balance and the higher the HLB the more hydrophilic (water soluble) is the surfactant. The HLB range 8–18 will normally provide good oil-in-water emulsions. The optimum ratio of anionic and nonionic surfactants is determined experimentally to give spontaneous emulsification in water, and to give a stable emulsion with very little creaming and no coalescence.

Emulsifiable concentrates are limited in the number of active ingredients for which they are suitable. Many pesticides are not soluble enough to be supplied economically in this form. However, it may be possible to boost the solubility of the active ingredient by the addition of a more polar solvent without increasing the risk of crystallisation in the spray tank.

A typical emulsifiable concentrate formulation is shown below:

	% By weight
Active ingredient	20–70
Emulsifier blend	5–10
Solvent Cosolvent }	to 100

The presence of solvents and emulsifiers in emulsion concentrate formulations can sometimes give enhanced biological efficacy compared with other formulations. Many insecticides, e.g. organophosphorous compounds and pyrethroids, are oil-soluble active ingredients and are readily formulated as emulsifiable concentrates, and a few active ingredients need to be formulated with solvents for optimum biological activity.

Health, safety and environmental pressures on the use of petroleum-based solvents generally are influencing a move away from these solvent-based formulations. However, it seems unlikely that solvents can be replaced entirely for some products, and safer high flash point solvents are being introduced along with new ideas for packaging to reduce physical contact between the product and the operator.

6.2.1.4 Suspension concentrates (SC). Suspension concentrate technology has been widely applied to the formulation of many solid crystalline pesticides since the late 1960s and early 1970s. Pesticide particles may be suspended in an oil phase, but it is much more usual for suspension concentrates to be dispersions in water. Considerable attention has been given in recent years to the production of aqueous suspension concentrates by fine grinding processes such as bead milling. The use of surfactants as wetting and dispersing agents has also led to a great deal of research on the colloidal and surface chemistry aspects of dispersion and stabilisation of solid/liquid dispersions.[9]

Suspension concentrate formulations offer many advantages such as ease of handling and application, safety to the operator and environment and economy. They also enable water-soluble wetting agents to be built into the formulation to give enhanced biological activity.

In most cases, suspension concentrates are made by dispersing the active ingredient powder in an aqueous solution of a wetting and/or dispersing agent, followed by a wet grinding process in a bead mill to give a particle size distribution in the range 0.1–5 μm. The wetting/dispersing agent aids the wetting of the powder into water and the breaking of aggregates, agglomerates and single crystals into smaller particles. In addition, the surfactant which becomes adsorbed onto the freshly formed particle surface should prevent reaggregation of the small particles and should ensure colloidal stability of the dispersion. Typical wetting/dispersing agents used in suspension concentrate formulations are:

- lignosulphonates
- naphthalene sulphonic acid–formaldehyde condensates
- nonyl phenol ethoxylates
- fatty alcohol ethoxylates
- tristyrylphenol ethoxylate phosphate ester
- ethylene oxide/propylene oxide copolymers

Also in development are polymeric surfactants which show considerable potential for the stabilisation of suspension concentrates for long term storage.[10] A typical suspension concentrate formulation is shown below:

	% By weight
Active ingredient	20–50
Wetting/dispersing agent	2–5
Propylene glycol antifreeze	5–10
Anti-settling agent	0.2–2
Water	to 100

The antisettling agent is added to prevent separation of particles during long term storage. These agents are often cellulose derivatives, natural gums or other types of polysaccharides, such as xanthum gum, and they are generally susceptible to microbial attack. For this reason, preservatives are sometimes added to suspension concentrate formulations to prevent degradation of the antisettling agent so that long term stability of the product is not impaired.

Many crystalline solid active ingredients are now available as suspension concentrates. However, there is increasing pressure, especially in Europe and the USA, to enforce stringent pack rinsing and disposal regulations, which may have a serious impact on the future of suspension concentrates and their packaging.

6.2.1.5 Seed treatments (DS, WS, LS, FS). Although most pesticide formulations are applied by spraying onto crops or weeds, a significant amount of pesticide products are applied directly onto seeds prior to planting into the soil. It is estimated that the market value of seed treatment formulations in 1993 was about US$750 million representing about 3% of the total market for agrochemical products. Approximately 50% of seed treatment formulations are applied to cereal seeds.

Products for seed treatment fall into four categories:

- powder for dry seed treatment – DS
- water slurriable powder for seed treatment – WS
- non-aqueous solution for seed treatment – LS
- flowable suspension for seed treatment – FS

Of these formulations only water slurriable powders and flowable suspensions use appreciable amounts of surfactants. The technology for producing flowable suspensions is similar to that for producing suspension concentrates, and therefore preservatives are needed to prevent degradation of the antisettling thickeners. Surfactants used are similar to those used for suspension concentrate formulations.

Because seed treatments are applied directly to the seed, there is very little loss of active ingredient. Seed treatments are, therefore, seen as a very efficient means of targeting pesticide to crops and are regarded as an environmentally safe way of applying pesticides.

6.2.2 *New generation formulations*

Over the last few years there has been increasing pressure from government and regulatory authorities to develop formulations which have less impact on the environment generally.[11] The main issues which are being addressed are:

- safety in manufacture and use
- convenience for the user
- ease of pack disposal or re-use
- reduction of the amount of pesticide applied
- reduction of waste and effluent of all kinds.

The likely trends over the next few years in the development of pesticide formulations are likely to be:

- To use safer solvents or to eliminate solvents wherever possible and use aqueous emulsions.
- To replace wettable powders by aqueous suspension concentrates or water-dispersible granules.
- To develop multi-active ingredient formulations with built-in bio-enhancing surfactant wetters.

- To control release rate and targeting of pesticides by encapsulation techniques.
- To develop more effective spray adjuvants to enhance biological activity and reduce pesticide dosage.

These complex requirements are being met by technical advances in surfactants and other formulation additives, particularly blends of surfactants, more powerful dispersing agents and a better understanding of the principles of colloid and surface chemistry and rheology.[12] There is also a need for better spray application systems to reduce pesticide waste, and a greater understanding of the physiological effects of spray droplets on leaf surfaces.

It is recognised that some components of formulations are less desirable than others and that some formulation types exhibit more operator exposure problems than others. Examples of these are aromatic solvents in emulsifiable concentrates which are coming under scrutiny due to their flammability, toxicity both alone and in the formulation, and package disposal problems.

Wettable powders also cause problems with toxic dust being created during the loading operation to a spray tank. The ideal product would seem to be one which is solvent free, gives no operator exposure hazard, has the maximum biological activity at the lowest dose level and produces no pack disposal problem. A dry powder or water-dispersible granule in a water-soluble sachet which can be added directly to a spray tank goes a long way towards meeting these requirements, and considerable work is being carried out on this option by all the major agrochemical companies. However, it will never be possible to formulate all active ingredients this way and so other options are being evaluated extensively, along with ideas for packaging and close-coupled spray application systems. Aqueous-based formulations will be a necessary and safe alternative to water-dispersible granule formulations and these options include (in addition to suspension concentrates which have been already discussed):

- suspoemulsions
- emulsions or concentrated emulsions
- microemulsions
- microencapsulation.

Other possibilities involving specialised packaging will be gels and effervescent tablets.

6.2.2.1 Water-dispersible granules (WG). Water-dispersible granules, or dry flowables, as they are sometimes known, are a relatively new type of formulation and are being developed as safer and more commercially attractive alternatives to wettable powders and suspension concentrates.

They are becoming more popular because of their convenience in packaging and use, being free-flowing granules which should disperse quickly when added to water in the spray tank. They therefore represent a technological improvement on wettable powders and imitate liquids in their handling characteristics.

Water-dispersible granules are complex formulations because they can be formulated using various processing techniques, but in each case the resultant product must redisperse in the spray tank to give the same particle size distribution as the original powder or suspension from which it is made. This requires careful choice of both the surfactant additives and the process of granulation which is usually one of the following techniques:[13]

- pan granulation
- mixing agglomeration
- extrusion granulation
- fluid bed granulation
- spray drying.

Several factors, such as the physicochemical properties of the active ingredient and additives, need to be considered when deciding upon which process to use. These factors and the various processing techniques used to make water-dispersible granules determine the main properties of the final product in terms of granule shape and size, degree of dustiness and ease of dispersion into water. The dispersion time in water is a very important property, and to ensure that no problems occur in the spray tank it is usually necessary for all the granules to disperse completely within two minutes at varying degrees of water temperature and hardness.

Water-dispersible granules usually contain a wetting agent and a dispersing agent in the same way as a wettable powder or a suspension concentrate. They may also contain a water-soluble salt to act as a disintegrant in the spray tank. The remainder of the formulation is usually a water-soluble or a water-dispersible filler. A typical water-dispersible granule formulation is shown below:

	% By weight
Active ingredient	50–90
Wetting agent	1–5
Dispersing agent	5–20
Disintegrating agent	0–15
Soluble or insoluble filler	to 100

Wetting and dispersing agents commonly used in water-dispersible granules are often similar to those used in wettable powder and suspension concentrate formulations. Because they are essentially dry solids there is usually no need to add a preservative to these formulations.

6.2.2.2 Suspoemulsions (SE). Mixed formulations are becoming more popular due to their convenience, because they ensure that the farmer applies the correct amount of each component pesticide and because they enable problems of tank mix incompatibility to be overcome. If one active ingredient is a solid and the other is a liquid, it is necessary to produce a suspoemulsion formulation, which consists of three phases:

- liquid oil droplets
- solid dispersed particles
- continuous phase, usually water.

Suspoemulsions can, therefore, be considered to be mixtures of suspension concentrates with oil-in-water (O/W) emulsions with added surfactants to prevent flocculation and thickeners to prevent separation of the dispersed phases. Surfactants used as dispersing agents for the solid phase are similar to those already mentioned for suspension concentrates. Emulsifiers for the oily liquid phase will be discussed in section 6.2.2.3 on emulsions (EW). As these formulations are aqueous based and generally thickened with polysaccharides, it is necessary to add a preservative to prevent degradation of the thickener. Some problems of heteroflocculation between the solid particles and the oil droplets can occur and extensive storage testing of these formulations is necessary.[14]

6.2.2.3 O/W Emulsions (EW). Emulsions are now receiving considerable attention because of the need to produce stable emulsions in

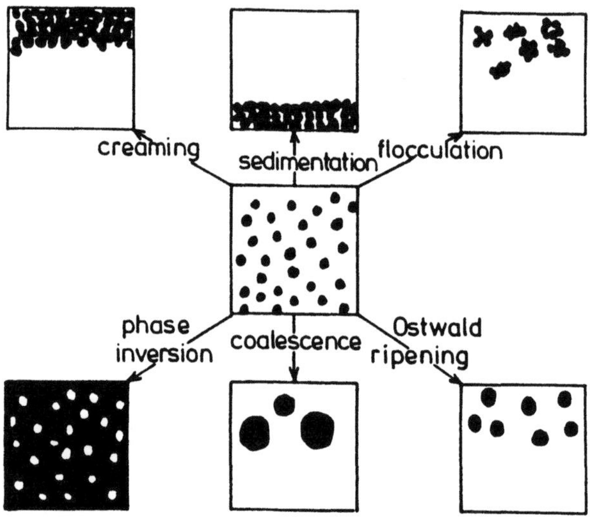

Figure 6.1 O/W emulsion stability problems.

suspoemulsion (SE) formulations and also as a means of producing stable emulsion concentrates (EW) which are water based and therefore reduce or eliminate organic solvents.

O/W emulsions can have significant advantages over emulsifiable concentrates (EC) in terms of cost and safety in manufacture and use. However, they require careful selection of surfactant emulsifiers to prevent flocculation, creaming and coalescence problems. These effects are shown diagrammatically in Figure 6.1. Nonionic surfactants and polymeric surfactants are now being used to produce stable emulsions. In the case of nonionic surfactants it is sometimes useful to combine a low and a high HLB surfactant to give an average HLB of 11–16 for optimum emulsion stability.

Droplet size is also a good indicator of stability and should be below 2 μm. The emulsions are usually thickened with polysaccharides such as xanthan gum and, therefore, a preservative is added to the formulation. Sometimes polymers such as polyvinyl alcohol are used as both emulsifier and thickener.

6.2.2.4 Microemulsions (ME). Microemulsions are thermodynamically stable transparent emulsions and are stable over a wide temperature range. They have a very fine droplet size of less than 0.1 μm and consist of three components, namely:

- oily liquid or solid dissolved in solvent
- water
- surfactant/cosurfactant.

These components form a single phase containing relatively large 'swollen micelles' in which the non-aqueous phase of the active ingredient and solvent are dissolved or solubilised.

In the preparation of microemulsions two different types of surfactant are needed; one water soluble and one oil soluble. The water-soluble surfactant is usually anionic or non-ionic with a very high HLB value, and the hydrophobic part of the molecule should match the oil. The cosurfactant should be oil soluble and should have a very low HLB value, such as hexanol. The total concentration of surfactants for a microemulsion can be between 10 and 30%, compared with about 5% for an O/W emulsion (EW).

Microemulsions have relatively low active ingredient concentrations, but may have enhanced biological activity.

6.2.2.5 Controlled release formulations. The application of controlled release technology has been slow to reach commercialisation despite interesting research and development work by the major agrochemical companies over the last 10–20 years. Controlled release formulations can have a number of advantages over conventional formulations, as shown below: they

- reduce mammalian toxicity
- have longer residual biological activity
- control/reduce evaporation of pesticide
- reduce phytotoxicity to crop
- improve compatibility in the spray tank
- reduce fish toxicity
- reduce ground water leaching
- reduce solvent usage in formulation
- reduce pesticide application rate.

Controlled release pesticide formulations can be divided into four main types:

- coated pesticide granules
- matrix systems containing physically trapped pesticides
- polymer systems containing covalently bound pesticides
- polymer membrane–pesticide reservoir systems, e.g. microencapsulation.

Of the four main types of controlled release formulations, only the polymer membrane, or microencapsulation, formulations use significant amounts of surfactants. A well-known method of microencapsulation is by interfacial polymerisation. In this process the active ingredient, usually a liquid or low melting waxy solid, is dissolved into an aromatic solvent, such as C-9 and C-10 solvents used for emulsifiable concentrates. An oil-soluble monomer such as toluene diisocyanate (TDI) is dissolved in the solvent mixture. A fine emulsion of the oil phase in water is made by high shear mixing with an aqueous solution of an emulsifier and a reactive amine, such as ethylene diamine. An emulsion with droplets of 10–30 μm is formed, and polymerisation between the isocyanate and the amine occurs at the oil/water interface giving a polyurea membrane around each droplet. Alternatively the interfacial polymerisation process may be carried out by allowing the isocyanate to react with water at the interface to form an amine *in situ*, which then reacts with more isocyanate to form a polyurea membrane.[15]

The rate of release of the active ingredient can be controlled by adjusting the droplet size, the thickness of the polymer membrane and the degree of cross linking or porosity of the polymer. This is, therefore, an example of a diffusion controlled process.

A typical microencapsulated suspension (CS) formulation is shown:

	% By weight
Active ingredient	10–30
Emulsifier	1–5
Polymer	10–15
Solvent	5–15
Anti-settling agents	1–3
Water	to 100

Microcapsule suspensions need to be stabilised with surfactants and thickeners in the same way as suspension concentrates and emulsions, and similar additives are used, including preservatives.

A number of microencapsulated products are now on the market, including selective herbicides to reduce volatility, insecticides to reduce toxicity and to increase residual activity, and pheromones to maintain the required vapour concentration over a period of 10–14 days. The benefits of microencapsulated products over conventional formulations in terms of bioavailability may be demonstrated graphically as shown in Figure 6.2, where the optimum level of pesticide availability can be maintained over a much longer period than with conventional formulations.

6.2.2.6 Built-in-wetter formulations. There is increasing pressure from regulatory authorities and for marketing reasons to include surfactant adjuvants in the formulation in order to optimise biological activity and to reduce the rate of active ingredient usage. In some cases, the regulatory

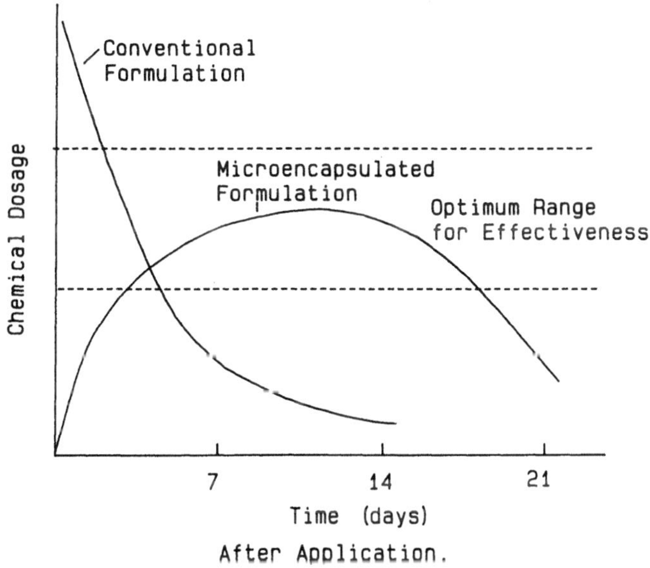

Figure 6.2 Bioavailability of microencapsulated formulation compared with conventional formulations.

authorities require specific data on the formulation which includes the biological enhancing wetter.

The potential effects of built-in-wetters to formulations are:

- better foliar wetting and spreading
- better adhesion of the droplets
- reduced droplet size of the spray
- increased drying time and water retention
- increased uptake and translocation in plant.

Non-ionic surfactants are often used as built-in-wetters to give the above benefits. They can increase the solubility of the active ingredient in the droplet by micellisation, thus making it easier for the active ingredient to enter the target. Built-in-wetters are useful for hydrophilic active ingredients, such as paraquat and glyphosate, to enhance their uptake into the leaf surface. They may also improve the physical compatibility of different pesticide formulations in the spray tank mixture.

No universal surfactant wetter exists for all pesticides and it is necessary to carry out stability tests and biological activity tests with a range of different surfactant wetters to find the optimum system. However, surfactants such as nonylphenol ethoxylates, linear fatty alcohol ethoxylates and fatty amine ethoxylates are often used. The mechanism of action of surfactant adjuvants in contact with the target organism is not fully understood, but it seems that lowering the interfacial tension, reducing the contact angle and increasing the movement of pesticide through the leaf surface are all important processes.[16]

6.3 Surfactants for agrochemicals

Surfactants are essential components for the formulation of most agrochemical products. They have several functions the most important of which are:

- wetting
- dispersing
- emulsifying
- solubilising
- bioenhancement.

Surfactants are able to wet powders into water by lowering the surface and interfacial tensions so that concentrated premixes can be made. They also help in the particle dispersion process by adsorbing onto the freshly formed surface and preventing reaggregation. Surfactants can emulsify oils into water and in some cases can increase the concentration of active ingredients by solubilisation of the material in the surfactant micelles. Surfactants play a major role in the stabilisation of pesticide formulations to impart good shelf life stability. During the spray application process they enable solid products to wet out and disperse into the spray tank dilution, and liquid products to emulsify and disperse. Surfactants are also used by themselves or as components of adjuvants for tank mixing with pesticide products. A knowledge of the physicochemical properties of surfactants is essential for the successful design of agrochemical formulations and adjuvants.[17] Agrochemical formulations usually contain 1–10% of surfactant or a mixture of surfactants. For spray applications surfactants are sometimes added to the spray tank at levels of 0.01% to 0.1% to improve droplet wetting and adhesion on the foliage. In recent years higher concentrations of surfactants up to 1–2% are being used to enhance the biological performance of the pesticide by increasing uptake into the plant and translocation within the plant. It has been estimated that the world consumption of surfactants for agrochemical use was about 230 000 t in 1993, representing about 3.3% of the total consumption of surfactants for all end uses.[18]

The simplest surfactant molecule comprises a lipophilic part which prefers an oil phase, attached to a hydrophilic head group which prefers water. A common and simple example is sodium dodecyl sulphate; $C_{12}H_{25}SO_4^- Na^+$.

Surfactants are classified into the following types:

- anionic: negatively charged hydrophilic headgroup,
- cationic: positively charged hydrophilic headgroup,
- nonionic: uncharged hydrophilic headgroup,
- amphoteric: negatively and positively charged hydrophilic headgroup.

Some common examples of the different types of surfactants used in agrochemical formulations are shown schematically in Figure 6.3. A wide range of surfactants is available to enable the formulator to make the best choice for a particular formulation. Surfactants are used primarily as wetting agents, emulsifiers and dispersing agents but also have uses as antifoaming agents and anticaking agents and an increasingly important use as agents to enhance the biological activity of pesticides and herbicides by improving capture by and penetration of the biological target. Anionic and nonionic surfactants are much more commonly used with agrochemical formulations than cationic and amphoteric surfactants in order to prevent flocculation problems with other anionic formulation additives. However, where they are used, cationic surfactants may also exhibit bactericidal properties. Amphoteric surfactants are rarely used in agrochemical formulations, but in some cases they can have interesting effects at different pH values. For agrochemical formulations anionic surfactants comprise about 50% of the total surfactant usage, whereas for spray application adjuvants nonionic surfactants comprise about 75% of the total surfactant usage.

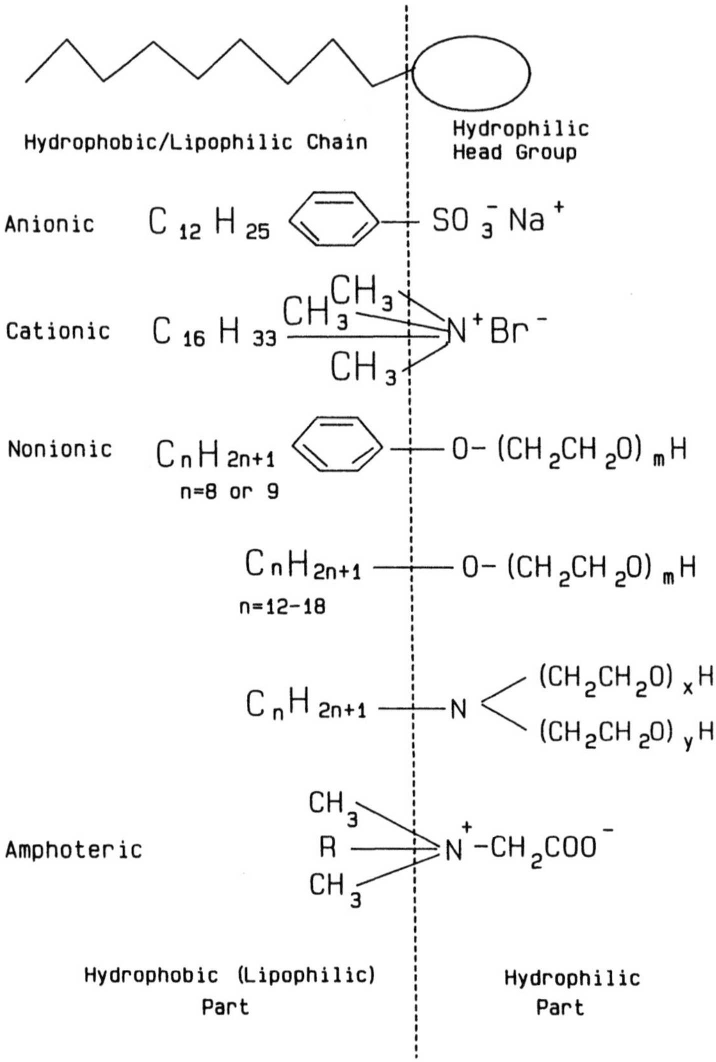

Figure 6.3 Surfactant classifications and examples.

6.3.1 Conventional surfactants

Wettable powder formulations usually contain a wetting agent such as sodium lauryl sulphate or a sodium sulphosuccinate derivative:

$$C_{12}H_{25}SO_4^- \ Na^+$$

$$\begin{array}{l} \quad\; CH_2COOC_nH_{2n+1} \\ \quad\;\; | \\ ^-SO_3{-}CHCOOC_nH_{2n+1} \quad Na^+ \qquad n = 6\text{–}8 \end{array}$$

The most commonly used dispersing agent for wettable powders is sodium lignosulphonate. Another popular dispersing agent is naphthalene sulphonic acid formaldehyde condensate sodium salt. The structures of these two complex polyelectrolyte anionic dispersing agents are shown in Figures 6.4 and 6.5.

Both of these polyelectrolyte anionic dispersing agents are also useful for the preparation of suspension concentrates. They are sometimes combined with nonionic surfactants such as alkylphenol ethoxylates or long chain alcohol ethoxylates with the typical structures shown:

C_nH_{2n+1} 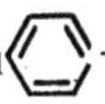 $-O(CH_2CH_2O)_m\ H$ $n = 8–9$ $m = 6–20$

$C_nH_{2n+1}O(CH_2CH_2O)_m\ H$ $n = 12–17$ $m = 6–20$

Increasing the number of ethylene oxide units (EO) in the molecule increases the hydrophilicity of the surfactant and reduces its lipophilic tendencies, e.g. solubility in oils.

By changing the mass of the lipophile or hydrophile, what is known as the hydrophilic/lipophilic balance (HLB) can be changed, thereby modifying the surface activity and solubility of the molecule. For example, increasing the ethylene oxide chain length increases the cloud point of the surfactant and can prevent flocculation problems during storage at high temperatures.

Another common group of surfactants used in suspension concentrate formulations is based on polypropylene oxide as the hydrophobe and polyethylene oxide as the hydrophile. These are formed as ABA blocks where A is the polyethylene oxide unit and B is the polypropylene oxide unit. A large number of surfactants having a wide range of properties can be obtained by changing the A to B ratio and the molecular weights of A and B. The number of ethylene oxide units can range from two to a few hundred. Polypropylene oxide chains below about 12 units are not really hydrophobic and can range from this minimum to a few hundred units.

One of the advantages of nonionic surfactants is the way that their properties can be modified by changing the level of ethoxylation, i.e. the hydrophile/lipophile balance (HLB). Products in the HLB range 1–4 are likely to be immiscible in water at room temperatures, HLB range 4–7 form unstable dispersions, HLB range 7–9 give opaque stable dispersions, HLB range 10–13 give hazy solutions and HLB 13–20 clear solutions. Nonionic surfactants in the HLB range 2–7 are likely to form water-in-oil (W/O) emulsions and in the range 7–18 oil-in-water (O/W) emulsions. Wetting, foaming and defoaming properties are also HLB dependent. A wide range of degrees of ethoxylation is available, normally in the range 4 EO to 50 EO units. Some properties are summarised in Table 6.2 for nonylphenol (NP) ethylene oxide condensates with 8–20 moles of ethylene oxide.

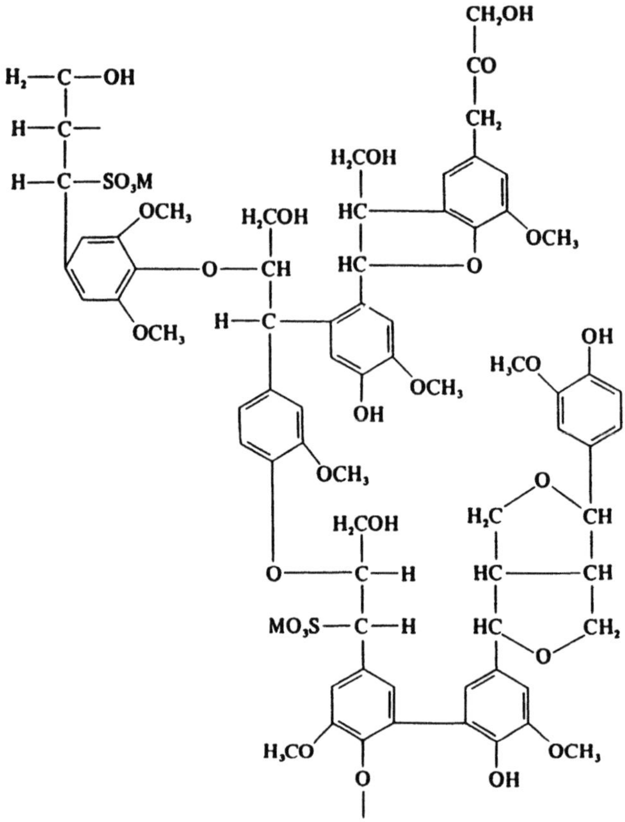

Figure 6.4 Structure of a typical section of polymeric lignosulphonate salt. Lignosulphonates are anionic polyelectrolytes whose molecular weight varies between 1000 and 20 000. Their organic structure has not been completely determined, but it is known that the basic lignin monomer unit is a substituted phenylpropane.

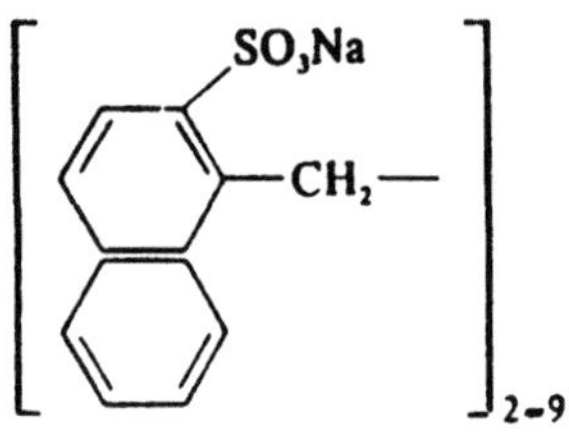

Figure 6.5 Structure of naphthalene sulphonic acid formaldehyde condensate sodium salt. Naphthalene sulphonate formaldehyde condensates are a mixture of low polyelectrolytes in the approximate molecular weight range 500–2200 (corresponding to a naphthalene nucleus content of 2–9 per molecule). Major components of the mixture are believed to have the structure shown.

Table 6.2 Properties of nonylphenol ethoxylates

Product	HLB	Cloud point (°C)		Pour point (°C)	Surface tension (0.1%) (mN m^{-1})
		Water	10% NaCl		
NP8	12.3	31	–	<0	29
NP12	13.9	82	54	14	35
NP15	15.0	97	67	21	33
NP20	16.0	–	73	30	42

Table 6.3 Properties of linear alcohol ethoxylates

Degree of ethoxylation (moles EO)	7	11	20
Physical form	viscous liquid	soft paste	hard wax
Pour point (°C)	21	27	37
Cloud point (°C)	45–49	84–89	100
HLB	12.2	13.9	16.2
Surface tension (0.1%) (mN m^{-1})	28.6	32.8	41.3

The properties of long chain alcohol ethoxylates vary in a similar way to the alkylphenol derivatives as determined by the degree of ethoxylation. Data on physical properties are shown in Table 6.3 for some linear alcohol ethoxylates based on C-13–C15 synthetic fatty alcohols.

A very important usage of surfactants is for the emulsifier blends which are added to emulsifiable concentrate formulations to produce stable emulsions in the spray tank. Careful selection of a 'balanced pair' of emulsifiers is necessary to ensure that spontaneous emulsification takes place and stable emulsions are formed in water with varying degrees of hardness and temperature. Popular emulsifier pairs are mixtures of nonionic and anionic surfactants such as nonylphenol ethoxylates and the calcium salt of dodecyl benzene sulphonic acid, shown:

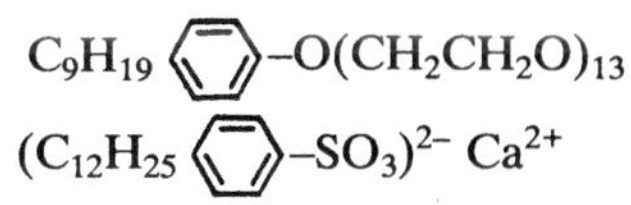

The total concentration of the emulsifier blend is usually 5–10% of the formulation. The ratio of non-ionic to anionic surfactant is determined by using the HLB system as a guide and carrying out dilution and emulsion stability tests. The preferred range of HLB for good emulsion stability is 8–18 to avoid the problems of creaming and coalescence.

6.3.2 Recent surfactant developments

In recent years there has been considerable interest in the preparation of 'tailor made' surfactants with improved properties to suit specific functions in formulations and spray adjuvants. For instance, the trend towards aqueous suspension and emulsion formulations has provided a need for surfactants with better adsorption characteristics and long term stabilisation properties. The conventional surfactants have molecular weights of about 1000–2000 and generally give incomplete coverage of particle surface area. They may also desorb from the surface, leading to flocculation and problems of increased viscosity on storage. The need for improved long term stability has led to the development of polymeric surfactants with molecular weights of 20 000 to 30 000 for the hydrophobic 'backbone', which has multiple anchoring points for adsorption onto surfaces. These polymeric surfactants can have more than ten times the adsorption affinity of conventional surfactants, and are much less likely to desorb from the surface.

Polymeric surfactants, therefore, impart better stability and allow higher volume concentrations of particles to be suspended in water without increasing viscosity.[19] A good example of a polymeric surfactant of this type is a graft 'comb' copolymer which has a hydrophobic backbone of polymethyl methacrylate/methacrylic acid onto which are grafted 'teeth' of polyethylene oxide.

Examples of other surfactant developments are the silicone-based and the fluorinated surfactants giving very low surface tensions, which can be as low as 15–20 mN m^{-1} compared with 30–40 mN m^{-1} for conventional surfactants.

There is also considerable interest in using surfactants which are as environmentally friendly as possible. The trend in the future will be towards surfactants which are fully biodegradable and have low toxicity to mammals and fish. Surfactants based on sugars or polysaccharides are being developed as alternatives to the conventional alkylphenol ethoxylates and other synthetic surfactants. Sugar ethers or alkyl polysaccharides are nonionic surfactants and may be suitable alternatives in that they give a similar effect to petroleum-based surfactants, but are more acceptable from a toxicology and environmental standpoint. Preservatives may have an increasing role to play in these developments.

6.4 Preservatives for agrochemicals

Microorganisms which cause spoilage of formulated products can enter formulations in a number of ways, such as by using unclean plant and equipment, contaminated raw materials and water, or by exposure to the atmosphere. In order to grow and develop, microorganisms need a nutrient source. Most agrochemical pesticide types do not provide a nutrient source for microorganisms and are not susceptible to microbial degradation. Indeed, some fungicides act as preservatives themselves to prevent the growth of fungicidal spores in formulated products. However, because of the need to formulate agrochemicals to give products which have good biological activity and long term storage stability, it is necessary to add preservatives to protect the formulation additives from microbial attack and prevent spoilage of the final product. Surfactants and carbohydrate-based thickeners may sometimes act as nutrient sources for microorganisms and generally require preservatives to protect them.

Dry products such as dusts, wettable powders and granules, and solvent-based formulations such as emulsifiable concentrates do not contain enough water to support the growth of microorganisms. These formulations, therefore, are not susceptible to biodegradation and do not require the use of preservatives.

Aqueous-based formulations such as solution concentrates, suspension concentrates and O/W emulsions are particularly vulnerable to attack by microorganisms and this is the area where the bulk of preservatives are used. It is likely that the use of preservatives will increase in the future because of the trend towards safer aqueous-based formulations and also the use of formulation additives which are more biodegradable. The problems that can occur if preservatives are not added to aqueous-based formulations are summarised:

- production of gas
- production of bad odours
- discoloration
- pH change
- viscosity change
- phase separation
- sedimentation.

All the above problems need to be avoided to give products which perform satisfactorily and have long shelf life.

6.4.1 Spoilage microorganisms

Once present in a formulation, a viable population of microorganisms may grow very rapidly. A few organisms will normally become dominant and this is determined by the nature of the formulation components, and other parameters such as pH, temperature and oxygen and water concentration. Microorganisms tend to grow best at around neutral pH and from 15 to 40°C, although significant growth can occur between pH 4 to 9 and temperatures of 7 to 60°C. Examples of the wide range of microorganisms typically associated with spoilage of aqueous-based formulations are shown in Table 6.4.

The presence of a suitable carbon source is necessary to support the growth of microorganisms. Many microorganisms secrete enzymes which

Table 6.4 Microorganisms in formulations

Bacteria	*Pseudomonas aeruginosa* *Pseudomonas putida* *Escherichia coli* *Enterobacter cloacae* *Proteus vulgaris* *Serratia marcescens* *Rhodopseudomonas capsulata* *Staphylococcus aureus* *Streptococcus lactis* *Streptococcus faecalis* *Bacillus subtilis*
Fungi	*Aspergillus niger* *Penicillium notatum* *Aureobasidium pullulans*
Yeast	*Saccharomyces cerevisiae* *Rhodotorula rubra* *Endomycopsis albicans*

break down polymeric molecules into smaller fragments to act as a source of carbon. These enzymes may persist after the organism has been killed, making it important that microorganisms are controlled throughout the manufacturing process. Good plant hygiene is essential to minimise the likelihood of microbial contamination of agrochemical products. However, in order to avoid problems occurring during long term storage, it is necessary to add a suitable preservative.

6.4.2 Types of preservatives

Preservatives are usually required for aqueous-based formulations and baits or pellets used as molluscicides and rodenticides. The general requirement for preservatives is that they should have the widest possible spectrum of antimicrobial activity and should be effective at very low concentrations.

For many years formaldehyde, used as Formalin (a 40% solution in water and ethanol) has been added to formulations as a general bactericide. However, in recent years its use has been very much reduced due to the potential mammalian carcinogenicity of formaldehyde. Cationic surfactants, especially quaternary salts such as cetyl trimethyl ammonium bromide or cetyl pyridinium chloride, are also well known as antimicrobial agents. However, many agrochemical formulations contain anionic surfactants or anionic polysaccharide thickeners which would be destabilised by cationic surfactants.

Other preservatives which are sometimes used for aqueous-based formulations are:

- propionic acid and its sodium salt
- sorbic acid and its sodium or potassium salts
- benzoic acid and its sodium salt
- *p*-hydroxy benzoic acid sodium salt
- methyl *p*-hydroxy benzoate.

These preservatives are regarded as quite safe to use and some are used as food product preservatives.[20] However, problems have sometimes occurred due to deactivation of the bactericide during storage of the formulated product. This effect is thought to be due to the presence of nonionic surfactants which can solubilise the preservative inside the surfactant micelle, thus reducing the availability of the bactericide.[21]

In recent years some new preservatives have been developed which are active against a wide range of microorganisms, effective over a wide range of pH values and not deactivated by surfactants in the formulations. One of the most popular of these newer bactericides is 1,2-benzisothiazalin-3-one which is sold under the trade name 'Proxel' by Zeneca Biocides (formerly ICI Biocides). 1,2-Benzisothiazolin-3-one is compatible with most anionic and nonionic surfactants and also with polymeric and clay-type antisettling thickeners used in aqueous suspension and emulsion formulations. It is available as a suspension in water or as a solution in propylene glycol, and is normally added to formulations at about 0.03 to 0.06%. 1,2-Benzisothiazolin-3-one has very good long term stability.

Another isothiazolin-3-one in common use is Kathon, a mixture of 5-chloro-2-methylisothiazolin-3-one and 2-methylisothiazolin-3-one. This is available from Rohm and Haas.

Other newly developed bactericides which are sometimes recommended are:

- 6-Acetoxy-2,4-dimethyl-*m*-dioxane,
  sold as 'Giv-Gard DXN' by Givaudan,
  use rate about 0.03 to 0.1%.
  This is an active aldehyde release agent. A number of formaldehyde release agents are also sometimes used (local regulations permitting). They are:

- Hexahydro 1,3,5,-tris(2-hydroxyethyl)-*sym*-triazine,
  sold as 'Glokill 77' by Rhône-Poulenc,
  use rate about 0.04 to 0.2%.
- Sodium hydroxymethyl glycinate,
  sold as 'Integra 44' by ISP,
  use rate about 0.05 to 0.2%.
- *N*-(Hydroxymethyl)-*N*-(1,3-dihydroxymethyl)-2,5-dioxo-4-imidazolidinyl)-*N*-(hydroxymethyl) urea,
  sold as 'Integra 22' by ISP,
  use rate about 0.05 to 0.3%.

In the formulation of aqueous suspension concentrates and emulsions it is sometimes necessary to make a stock solution of the polysaccharide thickener, and this is often done in the case of xanthum gum. These thickeners are stable in the dry state but when made up as 2% stock solutions the use of a preservative is necessary to prevent the growth of microorganisms which eventually lead to spoilage and a reduction in viscosity. This applies to stock solutions which may be held for longer than 24 h before use.

It is difficult to make general recommendations on the use of preservatives since many factors affect their efficiency. The choice depends on the composition of the product being made, on the type of microorganisms likely to be present and on the required period of stability. Each candidate preservative must be tested under conditions of manufacture and storage in order to determine the most suitable type. In some cases the use of more than one type may be necessary.

Solid agrochemical formulations such as molluscicide baits and rodenticide pellets are a special case, where the product may be attacked by soil-borne or air-borne microorganisms during prolonged exposure after application. In these cases preservatives are sometimes used to inhibit mould growth. However, the presence of a preservative can sometimes lead to 'bait shyness' which prevents the target pest from eating the bait.

Some of the active ingredients used in biocides are also used in agrochemical products such as rodenticides, avicides, insecticides and acaricides and repellents.

6.4.3 Testing and regulatory protocols

Agrochemical formulations are normally subjected to a series of storage tests at different temperatures to check chemical and physical stability and predict shelf life stability under normal conditions of use. A shelf life of at least two years at ambient temperature in a sales pack is usually required. Accelerated storage tests are also carried out at 54°C for two weeks, and sometimes at intermediate temperatures such as 35 or 40°C. Chemical and physical stability data are usually required in order to register the product for sale in many countries.

If spoilage of a formulation by microorganisms is thought to be a problem, it may be possible to identify the type of bacteria or fungi which is causing the problem by incubating a sample of the diluted formulation on agar plates. Various preservatives can then be tested to determine the minimum concentration of biocide which will prevent growth after incubation for 48 h for bacteria and 120 h for fungi.

At present the regulation of agrochemicals for sale in various countries is controlled by the relevant government authority in each country, who may request companies to provide sufficient data on the active ingredients and the formulated products to show that they are safe to use, and unlikely to have a deleterious effect on the environment. In some countries, data on product shelf life and biological efficacy may also be requested.

In the USA pesticides are regulated by The Environmental Protection Agency (EPA) under two statutes; Federal Insecticide, Fungicide and Rodenticide Act (FIFRA) and Federal Food, Drug and Cosmetics Act (FFDCA).

Recognising that, while pesticidally inert, some formulation additives may present a toxicological or environmental risk, the EPA is categorising inerts into four lists:

List 1. Inerts of known toxicological concern.
List 2. Inerts of potential toxicity requiring further testing.
List 3. Inerts with insufficient data.
List 4. Inerts known to be of minimal concern and generally recognised as safe (GRAS).

Wherever possible only those inerts which are included in List 4 should be used, and inerts included in List 1 should be avoided.

Until recently, individual countries in the European Union (EU) were responsible for registering products in their own countries using their own procedures and test data. However, the member states of the EU have now agreed on a system to harmonise registration procedures and testing throughout the EU following the issue of Directive 91/414/EEC in 1991, which regulates the placing on the market of pesticide products. Under Directive 91/414/EEC, the EU will develop a positive list (Annex I) of active ingredients, for a ten-year period, which may be used in any member state, and the individual countries will be responsible for approving the sale of formulated products of the active ingredients. Data generated for registration in one member state will be allowed for registration of the same formulation in other member states. Other Annexes will define the testing protocols for toxicology, biological efficacy, product chemistry and physical properties (Annex II for active ingredients and Annex III for formulations.)

It is proposed that the scope of Directive 91/414/EEC should be extended to include inert materials added to the formulation and adjuvants added to the spray tank. It seems likely that inert formulation additives such as surfactants and preservatives will continue to be considered only as part of the pesticide-formulated product for the purposes of product safety and biological efficacy. The EU has recently published a proposal concerning the placing of biocidal products on the market (Official Journal of the European Communities, Volume 36, No. C239/3, 1993).[22] However, the proposal states 'Plant protection products and other biocides that are already covered by Community requirements are excluded from scope.'

References

1. Sugavanam, B. (1990) UNIDO's activities on pesticides. In *Recent Developments in the Field of Pesticides and their Application to Pest Control*, Holly, K., Copping, L.S. and Brooks, G.T. (eds), UNIDO, Vienna, pp. 262–71.
2. Green, M.B., Hartley, G.S. and West, T.F. (1987) *Chemicals for Crop Improvement and Pest Management*, Pergamon Press, Oxford.
3. Karsa, D.R. (ed.) (1987) *Industrial Application of Surfactants I*, Royal Society of Chemistry, Cambridge.
4. Karsa, D.R. (ed.) (1990) *Industrial Application of Surfactants II*, Royal Society of Chemistry, Cambridge.
5. Karsa, D.R. (ed.) (1992) *Industrial Application of Surfactants III*, Royal Society of Chemistry, Cambridge.
6. Valkenburg, W. van (ed.) (1973) *Pesticide Formulations*, Marcel Dekker, New York.
7. Chow, P.N.P., Hinshalwood, A.M. and Simundsson, E. (eds) (1989) *Adjuvants and Agrochemicals*, Vols. 1 and 2, CRC Press, Boca Raton, Florida.
8. Foy, C.L. (ed.) (1992) *Adjuvants for Agrichemicals*, CRC Press, Boca Raton, Florida.
9. Tadros, T.F. (ed.) (1987) *Solid/Liquid Dispersions*, Academic Press, London.
10. Heath, D., Knott, R.D., Knowles, D.A. and Tadros, T.F (1984) Stabilisation of aqueous pesticidal suspensions by graft copolymers. In *Advances in Pesticide Formulation Technology*, ACS Series 254, Scher, H.B. (ed.), American Chemical Society, Washington D.C., pp. 11–28.
11. Holden, W.T.C. (1992) Future formulation trends – the likely impact of regulatory and legislative pressures, *Brighton Crop Protection Conference*, Vol. 1, BCPC, Brighton, pp. 313–20.
12. Seaman, D. (1990) Trends in the formulation of pesticides – an overview. *J. Pesticide Sci.*, **29**, 437–49.
13. Capes, C.E. (1980) *Particle Size Enlargement*, Vol. 1, Elsevier, Amsterdam.
14. Tadros, T.F. (1988). In *Proceedings of 2nd World Surfactants Congress*, Section D, ASPA, Paris, pp. 271–83.
15. Scher, H.B. (1983) Human welfare and the environment. In *IUPAC Pesticide Chemistry*, Miyamote, J. and Kearney, P.C. (eds), Pergamon Press, Oxford, pp. 295–300.
16. Holloway, P.J. and Stock, D. (1990). Factors affecting the activation of foliar uptake of agrochemicals by surfactants. In *Industrial Applications of Surfactants II*, Karsa, D.R. (ed.), Royal Society of Chemistry, Cambridge, pp. 303–37.
17. Porter, M.R. (1991) *Handbook of Surfactants*, Blackie, Glasgow.
18. Reports on *Agricultural Surfactants and Related Materials*, Vols. I and II, (1994), Hewin International, Amsterdam.
19. Knowles, D.A. (1994) Trends in the use of surfactants in agricultural formulations. In *Proceedings of 8th IUPAC International Congress of Pesticide Chemistry*, American Chemical Society, Washington D.C. (to be published).
20. Smith, J. (ed.) (1991) *Food Preservatives*, Blackie, Glasgow.
21. Lehmann, R.H. (1988) Synergisms in disinfectant formulations. In *Industrial Biocides*, Payne, K.R. (ed.), published for SCI by John Wiley, London, pp. 82–87.
22. EU (1993). *Official J. European Communities*, **36**, No. C239/3.

7 Preservation of personal care products

D.K. BRANNAN

7.1 Introduction

7.1.1 History of personal care products and preservation concerns

The Sumerians, Babylonians, Hebrews and Egyptians used cosmetics for ceremonial, medicinal and ornamental purposes as early as 4000 BC. Their cosmetics were mainly face and body paints, skin oils, ointments packaged in pots or jars, and pigments contained in sticks (crayons) or pencils. Their focus was on the eyes. Not only could cosmetics highlight and emphasize these expressions of the soul, but they had a practical side as well by protecting the eyes from flies and the Sun's glare. Lashes, lids and eyebrows were painted black with kohl. The lower lid was edged with a green paste made from malachite. Malachite green is an excellent antibacterial dye and so this may have been the first instance of a preserved cosmetic.[1]

But cosmetic history goes back even farther to a more primitive time. Men and women seem to have a basic primal need to change their appearance using cosmetics. Archaeologists claim humans have used cosmetics since the Paleolithic period nearly 50 000 years ago exploiting pigments blended with fats to decorate their bodies. The most commonly used pigments were made from white lead, chalk or gypsum for white coloration. For black coloration, they used charcoal or manganese ores. Or they used kohl, a paste made from soot and antimony, or galena, a lead ore. For red, orange and yellow they used ocher from crushed and powdered iron ores such as the hydrous iron oxide limonite for yellow or the red iron oxide crystal hematite for red. Heating the ocher powders produced the reddish brown pigment known as burnt ocher.[1]

Life was tougher in the Paleolithic period; nature was mysterious and frightening. Magic, wizardry and the supernatural pervaded the society. The hallmark of being human seems to be the ability to anticipate how one's actions today will affect our and others' futures. Ritual helped reduce the anxiety of anticipating the future. It gave them the ability to believe they had control over their fate. During these rituals, primitive man apparently used the pigments for body painting. Perhaps the painting transformed the participants into entities no longer associated with self and gave them a sense of defying fate. The Judeo–Christian creation story of Adam and Eve mimics this idea of hiding one's true self. Once they defied the rule God gave them (eating forbidden fruit from the tree of knowledge of good and evil), they covered themselves to hide who they really were. Hiding who we really are has ever since been ingrained in our psyche. Cosmetics are the tools we use to accomplish this. Cosmetics are the cures for the disease of being someone whom we do not want to be.

Pigments remained in popular use by Neanderthals, a variety of *Homo sapiens*. They painted their dead with them before burial to return the pink blush of life into the body. When Cro-Magnons appeared, they continued to use the mixtures of pigments and fat. But when Egyptian culture flourished, the simple pigments were replaced with elegant cosmetics and hair dyes. They used henna to dye everything from hair and fingernails to palms and soles. Scents and essences from almond, thyme, oregano, frankincense and myrrh, saffron and rosewater were compounded into perfumes, creams and lotions. The upper classes even used rouges, whitening powders, bath oils and lipsticks. The Greeks used imported perfumes, depilatories, rouges and kohl eye paint. Greek women even dyed their hair or lightened it with colored pigments. The Romans used perfumes, henna, rouge, face paint, hair dye and bleaches as status and wealth symbols to differentiate between upper and lower classes.[1,2]

The refined cosmetics were now luxuries. The lower classes could not afford them. No longer was the use of cosmetics needed for rituals to ease the anxiety of the future. Now, the upper class defied their common fate with the commoners (death) by claiming they were special and separate from them. How better to announce this separation from the lower classes than by a simple changing of one's appearance with cosmetics? Modifying our appearance is integral to our human nature and cosmetics have been with us throughout the development of culture.

In medieval Europe, face paint was fashionable but restricted to noble males, high born ladies and courtesans. The medieval ideal of feminine beauty was skin as white as the lily and cheeks as red as the rose. While the rich had special cosmetics to achieve this, the commoners used wheat-flour powder and beet-juice rouge. In 18th century Europe, the beauty mark (made from a velvet patch) was invented to cover smallpox blemishes. What started out as simple dots, the patches became large and shaped as birds, stars, flowers or as symbols of the wearer's occupation or avocation.[1]

Microbial contamination of cosmetics, however, did not become much of an issue until about 50 years ago.[2] The first microbial contamination was mold spoilage which affected cosmetics once large scale manufacture began. Parabens were used to provide adequate protection. As World War II ended, we entered an antibiotic euphoria and bacterial phobia. Everywhere there were new antibacterial skin creams, toothpastes, mouthwashes, deodorants, deodorant soaps and shampoos. Microbiologists frenetically screened the latest biocide. There were as many different preservative challenge test protocols as there were different companies. Since most of these tests were for the use of the companies rather than contributions to science, little of the work was published. This meant that none of the test protocols ever underwent the scientific process of peer review.

As we entered the space age and became moonstruck, a more scientific paradigm began emerging. Both industry and medicine became aware of bacterial resistance particularly among the Gram-negative organisms. Contamination by *Escherichia*, *Klebsiella*, *Enterobacter*, *Serratia* and *Pseudomonas* spp. became a major issue in non-sterile drugs, cosmetics, disinfectants and hand cream dispensers.[3–6] During 1966, 1967 and early 1968 the US Food and Drug Administration (FDA) recalled 25 microbially contaminated cosmetics.[7] In two US surveys conducted in 1969, anywhere from 20–24% of the cosmetics tested were contaminated with microorganisms; 3.6–6.4% of the cosmetics were contaminated with *Pseudomonas* spp.[8,9] The Cosmetic, Toiletry and Fragrance Association (CTFA) quickly established a form of peer review known as the Microbiology and Quality Assurance committees to develop technical guidelines for the industry.

More effective and more responsible preservation practices began as a result of this collaboration between microbiologists from the various companies. The 1970s saw a flurry of literature contributions. Not all the contributions were in prestigious or even recognized scientific journals but at least cosmetic microbiology was out of the closet. The CTFA issued technical guidelines covering all aspects of good manufacturing and microbiological practices. This time (1972–1975) when surveys of cosmetics were conducted by FDA and again by Wolven, only 3.6–5% of marketplace cosmetics showed contamination.[10,11]

But we still were not out of the woods. By mid-1970s, eye cosmetics came under special scrutiny.[12,13] There were several cases of blindness occurring due to *Pseudomonas*-contaminated mascaras.[14] The finding was that most products reached the consumer in good microbiological condition, but they could not withstand the insult of organisms added to the product during use.[15–20] This was the main issue to be addressed over the next few years. As a result, the FDA gave a contract to Ahearn and Wilson at Georgia State University in 1975 to develop eye area challenge tests.[21] Coming out of this contract was the 'direct contact membrane' method for preservative efficacy testing of mascaras[22] which was further tested in another FDA contract given to the University of California at San Francisco in 1977. The CTFA became involved quickly by setting up an Eye Area Task Force. This group arranged a collaborative study comparing the direct contact membrane method for evaluating preservative efficacy with the CTFA direct inoculation procedure. Both tests were proved to be equally satisfactory for measuring preservative activity.[23]

The FDA has continued its campaign to develop preservative challenge methods that predict the risk of consumer contamination. Since cosmetics are viewed by FDA as providing a zero health benefit, they must present zero risk. In 1985, the FDA gave a contract to Schering-Plough and the University of North Carolina to develop preservative challenge testing methods for creams and lotions that were predictive of consumer contamination. The FDA has still not published the data from this study. In 1990, the CTFA published the results of a survey to determine if companies tried to correlate their challenge test data with consumer use data.[24] Nearly all companies claimed they had correlation programs in place.[25] None presented their claim support data, however. There are only two challenge test methods that have published validation data proving their ability to predict the in-use potential for consumer contamination.[26–29] Both tests are modifications of the CTFA test.

In 1990, the FDA established a joint program with CTFA and the Association of Official Analytical Chemists (AOAC) to develop standard preservative challenge tests. This represents the first major step in developing a statistically based protocol for preservative efficacy testing. Rather than rely on a compendial method, the CTFA/AOAC/FDA collaborative intends to be a study that infuses reproducibility and reliability into the method. The collaborative test will be conducted using the same inoculum (pools of similar organisms), the same four types of products and with the same materials at 17 different laboratories. The four products to be evaluated are preserved and unpreserved versions of a shampoo, a conditioner, a water-in-oil emulsion and an oil-in-water emulsion. There will be three replicates of each product to be subjected to the challenge organisms. The testing phase should have been done by the end of summer 1994 and a first draft available for review in mid-1995. The major negative of the test method developed by this collaborative is that the goal of having a challenge test correlate with the risk of consumer contamination will continue to be absent. Despite the tests' stated purpose 'To demonstrate the ability of products to withstand microbial insult which may occur during intended use', there will be no consumer-use studies conducted prospectively to validate this modified CTFA method.

The United States Pharmacopeia (USP) is also looking at ways to refine their preservative effectiveness test. The final revision should be available by 1995. Some of the more significant changes under consideration by the panel will be to eliminate the phenol coefficient test (replacing it with a five subcultures away from the original source rule) and eliminating the requirement for a 56 day evaluation (demoting that requirement to a suggestion by moving it to an information chapter). Finally, the 150% variability rule between sampling times will be replaced with a 0.5 log varibility.

Perhaps the final goal of all these revisions should one day be to combine all the preservative efficacy tests from all the various nations and organizations into one coherent method that is scientifically and statistically sound and that is truly a validated predictor of consumer contamination potential.

7.1.2 A look at the future

The world is continuing to become a global economy. The breaking down of the Berlin Wall, the establishment of the European Commerce Commission, the North American Free Trade Agreement and moves trying to re-establish most favored nation trade agreements with China will continue to help us along the way to a global economy. In light of this movement, microbiologists must be concerned wtih developing tests, manufacturing practices, labeling requirements and other details that are uniform throughout all the nations and compendial sources rather than maintain a parochial attitude. This blending of the methods will be the biggest demand we face in the future.

The best way to accomplish globally valid methods is to stop developing compendial methods based on a panel of industry experts who eventually come to consensus on a protocol. Instead, we should rely on scientifically sound statistically valid procedures that go through rigorous peer review and testing by a variety of laboratories. The industry is now alert to the fact that it is responsible for its products during manufacturing and consumer use. Good industrial microbiologists anticipate microbial resistance and adaptability. A high awareness of asepsis, hygiene and sanitation now mitigate our once overdependence on antimicrobials. To continue on this road of continuous improvement, we must develop new globally valid approaches and methods if we are to persist in winning microbiological battles.

7.2 Importance of preservation

7.2.1 Types of personal care products needing preservation

Cosmetics include skin care and hair care products. Skin care products can be as ordinary as deodorants, antiperspirants (which are also drugs) and hand and body lotions. Hair care products can be as ordinary as shampoos and conditioners. But cosmetics also include a plethora of other products: make-up foundations, eye shadows, eye liners, mascaras, lipsticks, rouges, powders, perfumes, mousses, styling gels, finishing sprays, night creams, eye creams, sunscreens (which are also drugs), nail treatment oils, nail polish, cuticle removers, nail polish binder and nail fortifiers, nail polish remover, depilatories, hair spray, permed hair treatments and hair coloring agents. Of these products, the ones that most need preservatives are those with the most water; shampoos, conditioners, emulsions including make-up creams and water-based mascaras and gels. Products with low water activity ($A_w \leq 0.90$) usually need little more than antifungals such as methyl or ethyl parabens to provide adequate preservation. Table 7.1 provides the water activity limits for microorganisms.[30,31]

Most of the products needing biocide preservatives are emulsions. The common characteristic is the presence of some type of surfactant and a water activity above 0.95. Whenever water is present, so is life. Regardless of temperature, pH or any other environmental factor, if water is in a

Table 7.1 Water activity and the potential for adaptation[31]

Water activity	pH	Microbial adaptation potential	Problem organisms capable of growth	Examples of cosmetic products
0.98–1.00	Any pH	High	Most Gram-negatives	Shampoos and emulsion products
0.95–0.97	Above 5.5	Moderate	Most Gram-negatives	Liquid make-ups and eye area products
	Below 5.5	Low	Some Gram-negatives	Some hair conditioners
0.92–0.95	Above 5.5	Low	Few Gram-negatives	Some pressed powders
	Below 5.5	Very low	Spoilage bacteria	
0.90–0.92	Any pH	None	Gram-positive Lactobacilli	Some rouges (non-water based)
0.80–0.90	Any pH	Potential for *Staphylococcus*, mold and yeast adaptation	*Staphylococcus* Molds Yeasts	Lipsticks (non-water based)
0.70–0.80	Any pH	Some potential for yeast and mold adaptation	Molds Yeasts	Some talcs
0.65–0.70	Any pH	Very little potential for microbial adaptation	Osmotolerant yeasts	
0.60–0.70	Any pH		Osmotolerant molds	Some anti-perspirants
Below 0.60	Any pH	None (all microbes)	None	

liquid form, a microorganism will live there. The goal then is to add biocides in the hope of limiting the organisms that adapt to the aqueous environment while the customer uses the product. To think that one can develop the fail-safe biocide or process that will prevent contamination by microorganisms is the ultimate delusion. In the words of Winogradsky, 'Microbes are everywhere, the environment only selects.' They will eventually grow in even the most hostile of environments provided water is available in liquid form.

7.2.2 *Surfactants used*

Some of the most common surfactants used in the manufacture of cosmetics include ammonium lauryl and lauryl ether sulfates, sodium lauryl and lauryl ether sulfates, triethanolamine lauryl sulfates, ammonium xylene sulfonate, mono- and di-alkyl quaternary compounds and diethanolamides. The general categories of surfactants used are α-olefin sulfonates, betaines, lauryl sulfates, silicone-based surfactants, amide-based surfactants, quaternary compounds, ethoxylated surfactants, ethers and alcohols. Nearly all surfactants require preservatives with a few exceptions where the pH extremes are below 3.5 or above 10. The most common biocide preservative for surfactants is formaldehyde at about 100 to 200 ppm. This is added as a processing aid and therefore does not have to be included on the cosmetic label. Other preservatives used are dimethyloldimethyl (DMDM) hydantoin and the isothiazolinones. The formulator should always be very careful not to use the same preservative in the surfactant as he or she does in finished product. Adaptation and contamination occur much easier in the raw surfactant. A carryover of those adapted organisms into a product using the same preservative results in contamination and thus recalls due to the contaminated finished product.

7.2.3 *Consequences of not adding a preservative*

The person who believes that a thorough microbiological testing or monitoring program will ensure product quality simily does not understand that biological systems are constantly evolving. Bacterial evolution is so rapid that the agents which kill today can become bacterial food within less than a year after exposure to the microbial world. Bacteria adapt rapidly. Preservative failure should be expected and anticipated. As a result, at least two suitable preservative systems should be identified for each new product early in the development process. One way to slow the adaptation process is to protect cosmetics from contaminants by strict sanitation rules. Items to enforce rigidly are cleanliness of process water, raw materials, manufacturing and personnel hygiene. Nevertheless, one still needs to be ready for the surprises that the microbial world has in store.

Microbial contamination presents a substantial risk to product quality, regulatory compliance and consumer health. In terms of affecting product quality, contamination will produce adverse changes in product odor, color and viscosity.[32] The microorganisms metabolize and digest the product ingredients to produce their waste products in the process. They do this by producing a wide variety of hydrolytic enzymes. These changes happen rapidly if the invading microbes have an environment conducive for high reproduction. They happen slowly if the microbes reproduce more slowly. In fact, microbial contamination is usually not all that obvious. At first, there is no immediate evidence of growth in the product. Months can go by before the insidious creatures affect the product.

By far, the most critical issue associated with microbial contamination of cosmetics is the risk to consumer health. Fortunately, health-related incidences involving contaminated cosmetics are very rare. These incidences include infection from a hand lotion,[33] eye infections from use of eye area cosmetics[14,34] and the death of one immunocompromised individual.[35]

Microbiological safety of a cosmetic is also a regulatory responsibility administered by the US Food and Drug Administration. The Food, Drug and Cosmetic Act,[36] gives FDA the authority to protect the public against adulterated cosmetics. Adulterated products contain injurious or pathogenic microorganisms or their metabolites.[37–40] Adulteration can be real or even just perceived. By the wording of the law, even if conditions are such that contamination 'may have occurred' this perception gives the FDA the right to claim adulteration. Determination of adulteration will result in product recalls. It may even lead to seizure of the cosmetic resulting in significant economic consequences for the manufacturer.

Cosmetics do not need to be sterile, but they must be adequately preserved to withstand consumer use and manufacturing processes. Manufacturing contamination can be reasonably controlled, but consumer use and abuse cannot be controlled. Consumers repeatedly challenge the cosmetic with microorganisms in saliva, dirty hands and in tap water. The bathroom, where most cosmetics and toiletry articles are kept, is a perfect incubator providing heat and humidity ideal for stimulating microbial growth.[41,42] With all these potential insults and our litigious society, it only makes sense to be sure that cosmetic products can withstand consumer use and resist becoming contaminated.

7.2.4 Microorganisms encountered

Cosmetics can be contaminated with spoilage microorganisms and a variety of organisms found in the household environment.[43] Table 7.2 lists the types of organisms that contaminate shampoos and skin lotions.[44] The contaminants include bacteria, yeasts and molds. Most are spoilage

Table 7.2 Types and percentages of microorganisms contaminating cosmetics after use.[44]

Organisms	Isolated from shampoo	Isolated from skin lotion
Citrobacter freundii	18	0
Enterobacter spp.[a]	37	9
Klebsiella spp.[b]	9	9
Pseudomonas spp.[c]	9	21
Serratia spp.[d]	18	4
GNR[e] (non-fermentative)	0	4
GNR (fermentative)	9	0
CDC serotype IVC2	0	4
Bacillus spp.	0	4
Staphylococcus epidermidis	0	4
Propionibacterium sp.	0	4
Sarcina sp.	0	4
Diphtheroid	0	4
Yeasts and molds	0	29

[a]*E. aerogenes*, *E. agglomerans* and *E. cloacae*. [b]*K. pneumoniae* and *K. oxytoca*. [c]*P. putida*, *P. fluorescens*, *P. paucimobilis*, *P. aeruginosa* and *P. maltophilia*. [d]*S. liquefaciens*, *S. odorifera* and *S. rubidaea*. [e]GNR, Gram-negative rod.

organisms, however, even non-pathogenic spoilage microbes in a cosmetic may cause disease under appropriate conditions. A product contaminated with spoilage organisms may invade and create disease if it is applied to a break in the skin.[45] Use of a shampoo that irritates the eye will compromise it and allow microbial contaminants to infect the eye. With more and more immunocompromised individuals in the population from the pandemic of AIDS, the problem becomes critical from a product liability standpoint.

The biggest contamination concern, however, are pathogens that present a frank health risk. An historic example of the organisms involved here are the pseudomonads.[46] Cosmetics intended for eye area use are particularly susceptible to providing health risks. The cornea, when compromised, is highly vulnerable to infection. There have been several instances of mascara contamination from *Pseudomonas* spp.[14,31] These cause infection as well as product discoloration and odor formation. Other typical contaminants of cosmetics include *Enterobacter* spp., *Klebsiella* spp., *Serratia* spp. and a variety of *Pseudomonas* spp. (see Table 7.2).

7.2.5 *Considerations for preservation*

In order to understand fully how to preserve a product, the microbiologist must consider all the variables that contribute to a product being well preserved against microbiological contamination. Adding a biocide to a product is not all there is to preservation. Even disinfectants with some of the most powerful biocides have been contaminated[47–49] and recalled. Minimally, one should use the biocide that gives the product the best ability to withstand the microbial challenges to which it will be subjected during manufacturing and consumer use. However, a good microbiologist goes beyond this minimal approach. Microbiologists must be holistic practitioners of preservation if they expect to formulate a cosmetic product with appropriate antimicrobial hostility. He or she will also ensure that the biocide is safe, stable, compatible with the product and its container, inexpensive, readily available, approved by appropriate regulatory agencies, have a positive consumer perception and is environmentally friendly.

In addition, preservation includes having good manufacturing practices that insist on only the cleanest of manufacturing conditions. Those cleanliness standards should even include the surfactant supplier being under good manufacturing practices (GMP) and strict sanitation practices. Raw material quality, container and cap design, expected shelf life and exposure conditions, and even how the consumer will use and misuse the product should be considered. Consideration of all these elements is critical to maintain the microbiological quality of a product.

The entire product matrix should be considered as the preservative. A biocide may be adversely affected by the other ingredients in the product. For example, some of the suspended solids in a formulation such as carbonates and silicates, talc and metal oxides, cellulose and starch react by absorbing out preservatives. Minor pH changes can inactivate some preservatives especially weak acids (e.g. sorbic acid) which rely on hydrogen ions that disassociate to disrupt chemiosmotic membrane balance.[50–52] Minor shifts in ionic strength or changing the buffering system in a product can also alter a bacterium's susceptibility to a biocide or affect how a preservative partitions between the water matrix and the microbial cell.[53,54] Preservatives are usually water soluble and so exist in the water phase of emulsions. Some parabens, however, present unique formulation challenges because they have an affinity for the oil phase. This presents a particular problem for water-in-oil emulsions since the microbes live in the water phase only.[55] Even the surfactant system used can affect biocide performance.[52,56–58] In fact, nonionic surfactants are used to neutralize some preservatives (e.g. parabens)[59–61] however they tend to enhance quats.[62] Finally, high levels of protein (often used in conditioners and lotions) will also reduce the antimicrobial activity of many preservatives[63–65] and the presence of hydrophilic polymers will affect others.[66]

In considering choice of container, the microbiologist should also consider compatibility with the preservative.[67,68] Most preservative manufacturers do not include a statement of container compatibilities. Many of the negative container effects are due to long term storage. The preservative may either be absorbed into the container material in the case of lipid-soluble preservatives or inactivated because of complexation of the preservative with the dyes used in the plastic or lost because of the volatility of the preservative (e.g. phenoxyethanol). In one instance, 50% of the parabens used to preserve a mascara were absorbed into the polyethylene container after storage and could only be extracted from the container in 50% methanol (personal communication, S. Richards). The microbiologist should test long term stability of the preserved product in the container expected for marketing. Sometimes, these tests are done under accelerated conditions (high temperatures) making use of the assumption that the inactivation will follow Arrhenius kinetics. Chemical analyses for the preservative should be included as well as conducting preservative efficacy tests. In this way a failure in the preservative efficacy test can be traced to a breakdown or loss of the preservative. If the preservative is still present, then some other factor is acting to make it unavailable to kill the organisms such as partitioning out of the water phase.

When considering containers, one should not overlook the impact that dispensing closures have in preventing microbial contamination especially during consumer use. Flip caps and pump tops provide more protection for shampoos and skin lotions than screw caps.[44] Some closures can even inactivate the preservative.[69] Preservation is a function of formula and even container, not just the biocide. The wise microbiologist inserts herself or himself into the product development team early on to help decide upon the formula, the package and even specific instructions for use on the label.

7.3 Preservative efficacy testing

7.3.1 General considerations of preservative efficacy testing (PET)

Test protocols for determining preservation efficacy in cosmetics vary.[32,67,70,71] The protocol used in Europe is one found in the European Pharmacopeia. Although Britain has her own pharmacopeia, it is likely she will adopt the European Pharmacopeia. The three protocols that are fairly widely used in the United States include the protocols from the CTFA, the American Society for Testing and Materials (ASTM) and the USP. The AOAC is also attempting to define a protocol in collaboration with CTFA. The reason for such a plethora of methods is the types of products for which they are designed. Each trade organization must address a variety of different products. The CTFA focuses on cosmetics, the ASTM on commercial products such as cutting oils and paints and USP focuses on an assortment of drugs.

Many of the specific differences among the procedures are a result of the unique concerns of the organization developing the test. Some of the differences are due to lack of agreement on what the purpose of the preservative efficacy test is. Sitting on both the CTFA and the USP committees responsible for developing such tests has reminded me of the maxim 'one does not want to know how laws and sausage are made.' At other times, the cooperation and concern shown by the participants to develop an equitable method was impressive. The logic and arguments that go into establishing these protocols are primarily based on consensus of what is being done by the various participants on the panel developing the protocol. Certainly, these efforts represent 'state-of-the-art', but they do not necessarily represent rigidly controlled protocols subjected to multiple laboratory-controlled experiments and statistical analysis. To develop these kinds of tests, a company must employ a microbiologist who provides the scientific expertise needed to develop a validated 'in-house protocol' that is specific for the company.

The primary purpose of a PET might be to have a test that predicts consumer contamination. Two other purposes of a PET might be that it demonstrate the presence of a biocide or that it predict the change of manufacturing contamination. Regarding these latter two purposes, the first can be met by chemical analysis, the second can be met by controlling manufacturing insults to the product through good manufacturing practices and sanitary processing. However, no controls exist to provide protection from the consumer except what can be designed into the product itself and into its container. Several publications have already shown that a modified version of the CTFA preservative efficacy test is a valid predictive model of the risk of consumer contamination.[26–29]

The FDA has tried several times to develop a PET that predicts consumer contamination as described in detail above. If the FDA had such a standard test, then products failing it would be subject to recall. The FDA is now collaborating with CTFA and AOAC to develop a standardized laboratory preservative efficacy test complete with multi-lab comparisons and statistical analysis. However, the purpose of having the test correlate with the risk of consumer contamination will continue to be absent. Despite the AOAC/CTFA/FDA collaborative's protocol stating that its purpose is 'to demonstrate the ability of products to withstand microbial insult which may occur during intended use,' it does not have this ability to predict consumer contamination since no correlative consumer testing is planned using the same products. The omission of this goal was not entirely unplanned. If the test is developed with this purpose in mind, then it might be used to enforce a recall on those products that fail it. Regardless of which philosophy one adopts to define the purpose of a PET, at a minimum the goal should be to develop a database to rank the antimicrobial hostility of the company's products.

Although preservative efficacy (or challenge) tests are deceptively simple in concept, they are complex to perform. All the protocols involve introducing microorganisms into the cosmetic, sometimes referred to as 'challenging the product.' All the protocols have several other similar steps: product preparation, inoculum selection, inoculation, incubation, estimating the surviving microorganisms and interpreting the data. The ideal PET does not exist. Nevertheless, the cosmetic industry has been fairly successful without a standard PET. The rarity of contamination incidents speaks well for an industry that properly uses the right methods, the right manufacturing and the right packaging controls to prevent microbial contamination.

7.3.2 *Common issues shared by the PET methods*

7.3.2.1 Product sample. The product sample used in a PET should replicate formulation, packaging, manufacturing conditions and other factors significant to preservation. All raw materials should be of the same quality and from the same source when testing preservation options. As the product becomes refined, minor changes will be made to adjust viscosity, color, perfume or pH. These minor changes can alter the antimicrobial capacity of the product. They require the product to be reassessed for preservative efficacy since even small composition changes in individual ingredients can adversely affect the hostility of a product. Even the quality of the process water affects a product's antimicrobial activity. Changing raw material suppliers whose standards and processes are different should prompt a re-evaluation of the product's hostility since even the presence of trace chemicals in the new supplier's materials can be the cause of preservative inactivation.

Even going from lab to pilot plant to plant (scale-up), may cause preservative inactivation. As a result, going from small batch to continuous large volume manufacturing requires that the product be tested for its preservative efficacy. Scale-up provides the opportunity for inadequate mixing times, inappropriate temperature and pH changes in localized areas of the mixers and slow cooling times that may inactivate the preservative. These variables need to be identified and controlled. The final nationally delivered product produced and packaged for retail sale should, once again, be evaluated for its preservative efficacy.

7.3.2.2 Microbial inoculum. An appropriate microbiological challenge of the product is the most critical factor in determining the validity of a preservative efficacy test. All the tests currently specify a set of inoculum microorganisms.[32,67,70,71] Some of the methods list specific strains from American Type Culture Collection (ATCC). Both the CTFA and USP tests also allow the manufacturer to include any other organisms that can be likely contaminants. Most cosmetic manufacturers include the recommended organisms from the compendial tests. However, they also use their own selection of resistant strains recovered from consumer-used product samples, raw materials or manufacturing sites. These microorganisms are sometimes considered more resistant to preservatives and other hostile conditions. This kind of parochial thinking, however, should be re-evaluated if the company has not maintained a rigorous program of preserving the originally isolated culture.

Most companies use Gram-negative bacteria such as *Enterobacter*, *Serratia*, *Pseudomonas*, *Klebsiella*, Gram-positive bacteria such as *Staphylococcus aureus* and fungi such as *Candida*, *Aspergillus*, *Fusarium*, *Penicillium* and *Saccharomyces*. The isolates are chosen based on their resistance to preservatives or their adaptation potential and their potential as opportunistic pathogens. Resistance to antimicrobials includes resistance to even what might be referred to as whole cell poisons such as chlorine.[72,73] However, rather than a genetic mode of resistance development (e.g. plasmid acquisition or mutation) the development of chlorine resistance is more likely to be a result of population shifts in the amount of production of capsule or even physical community developments within biofilms where certain organisms exist as protector guilds for other organisms. *S. aureus* (ATCC 6538) is used to challenge frequently used cosmetics since it is an opportunistic pathogen that inhabits the skin.[71] *P. aeruginosa* is a non-fermentative Gram-negative rod used in all three methods. The ASTM and USP use ATCC strains 9027 while CTFA uses strains 15442 and 13388. *Pseudomonas* spp. are ubiquitous and show high resistance to many preservatives as well as being potentially pathogenic. Both CTFA and USP use *Escherichia coli* (ATCC 8739), a fermentative Gram-negative rod used as an indicator of fecal contamination in water. Within the same family of organisms (the enterobacteriaceae) are *Enterobacter*, *Klebsiella* and *Proteus*. These may be adequate replacements of *E. coli*. For example, ASTM uses *Enterobacter aerogenes* (ATCC 13048) as the inoculum. ASTM, CTFA and USP all use *Candida albicans* for the yeast inoculum. It is an opportunistic pathogen. ASTM and USP use ATCC strain 10231; CTFA makes no specific recommendation on strain. Filamentous fungi may also contaminate cosmetics. All three methods use *Aspergillus niger* (ASTM and USP use ATCC strains 16404; CTFA uses ATCC 9642). CTFA also suggests *Penicillium luteum* ATCC 9644 be used to represent fungal potential for contamination.

Finally, CTFA recommends using microorganisms that are indigenous to the eye, clinically significant isolates and product isolates for challenging eye cosmetics. They also include Gram-positive spore formers such as *Bacillus subtilis*. However, when using *B. subtilis* for a challenge inoculum, most of the cells should be in the vegetative form rather than the spore form when the challenge is done. The practice of using spore-forming bacteria as a challenge is questionable at best. If the challenge inoculum consists of spores and a product is developed to be sporicidal, the product will probably be unsafe for use.

Most of the tests make recommendations regarding the organisms to use. However, they disagree on which organisms to use, whether the challenge should be pure or mixed, the inoculum concentration and how to maintain the organisms. These and other issues related to inoculum will be discussed next.

7.3.2.2.1 Use of ATCC cultures as the inoculum. The ATCC cultures used may or may not have the specific biocide resistance that the manufacturer needs for its products. The ATCC is a repository for researchers to deposit cultures. As a result, it depends on the depositor to correctly characterize the originally deposited strain and its origin. The ATCC cannot be expected to verify the claims of all depositors. It can, however, be expected to maintain the resistance of the culture by ensuring that subcultures sold are never more than two to three transfers removed from the original strain deposited.

Using as an example some of the preservative strains deposited in 1940, the ATCC does a remarkable job. The strains were deposited as lyophilized vials that were kept at −70°C. In the 1960s, when ATCC began using liquid nitrogen, the original strain was cracked open, grown up and a population was lyophilized and frozen in liquid nitrogen as the main stock. The lyophilized vials serve as the working vials from which cultures are distributed to investigators today. The cultures received today, therefore, are never more than three transfers away from the original culture deposited. The ATCC also uses a population biology approach for preserving cultures. After verifying that the culture is pure, they put up an entire population derived from the original culture rather than a single cloned representative from the population. This population approach, which is not the typical clinical microbiology approach, enhances the likelihood of receiving the originally deposited strain. Thus, the manufacturer with its own preservative-resistant strains should rely on submitting cultures to ATCC for maintenance.

An alternative approach in lesser culture houses would be the following nightmare. If more culture requests are made than the original set of vials preserved, then the last remaining vial is subcultured. From this vial an isolated colony (obtained by streaking for isolation) is selected to grow up and harvest. This is preserved in another set of lyophilized vials. This process represents a departure from the originally deposited culture because the progeny of only a single individual was selected to represent the original population. If the requests are very high, several subcultures using this process may occur. If enough subcultures occur, the risk of losing the original resistance is high due to loss of resistance factors from the artificial selection process known as 'cloning from an isolation streak.'

7.3.2.2.2 Maintaining preservative resistance of the inoculum. Maintaining culture strains in their original biocide-resistant state represents quite a challenge for the cosmetic microbiologist. Routine subculturing on non-selective growth media will eventually cause the loss of this resistance since the selective pressure to keep the resistance is no longer present. Since most resistance genes are plasmid located, the organism may lose the plasmid unless selective pressure is kept on the population to maintain the resistance. The routine practice of pure culture techniques in microbiology is to isolate a clone for subsequent transfer to a new tube of medium. This approach is invalid when trying to maintain a population of resistant organisms. The likelihood of selecting out the organisms without the resistant factor is high when using traditional 'streak for isolation' pure culture concepts. Instead, one should rely on assessing the purity of the population by its homogenous appearance on a lawned agar plate. Preferably this should be done on a medium that has the preservative in an active state (not neutralized) incorporated into the agar. Subcultures should be done from a heavy inoculation off this lawn. The concept here is that a population of organisms is being employed most of which will carry the resistance genes. A sampling of the entire population is needed to ensure that the organisms within the population that carry the resistance genes are capable of producing plenty of progeny that will maintain the resistance within the population. In addition, the organism should be lyophilized or kept in liquid nitrogen as a back-up.

The working cultures should never be subcultured more than a set number of passages away from the original strain. The USP is planning to recommend a five passage limit. This recommendation is not based on any particularly scientific rationale, it just seems reasonable and is pragmatic. In fact, the five passage limit was an arbitrary number that originated in the Bureau of Biologics (FDA) for virus passages for vaccines.

In order to maintain a culture collection of resistant organisms a 'house organism' is isolated from a contaminated product. The organism is grown and checked to see if it looks pure (the colonies are all identical). Then the population is grown up as a lawn spread on an agar plate incorporating the preservative to which the organism is resistant. The entire lawn is harvested for freezing or lyophilization. The culture should be frozen in at least twenty vials in liquid nitrogen and another twenty vials put up for lyophilization. When the organism is scheduled for use in inoculation, one of the lyophilized vials is used by subculturing the entire vial and using plates or slants as the working stock (this is the first subculture from the original). If a PET is conducted each week, then four more weekly transfers take place before a new lyophilized vial must be cracked open. With this schedule, a culture collection can last two years on the lyophilized vials. At that time, a new liquid nitrogen vial is cracked open and another twenty lyophilized vials set up. The liquid nitrogen vials will be used up over the next forty years. By that time, the company will probably be using new kinds of preservatives anyway and the resistant strains to today's preservatives will be obsolete! Alternatively, another culture collection can be set up for another forty years and still only be one subculture away from the original strain.

Contaminated product may also be used directly as the inoculum. The advantage of this approach is that the organism can be maintained in a biocide-resistant state because the population contaminating it is continuously selected by subculturing it into fresh product. However, direct use of the contaminated product can create the problem of carrying over the growth product into the test product. If the two products are not compatible (e.g. a shampoo with anionic surfactants and a conditioner with quaternary surfactants), then confounding results may be obtained.

7.3.2.2.3 Growth and harvesting of the inoculum. The method of harvesting the culture and preparing it may also influence the organism's viability. Challenge inocula may be grown in broth or on solid media.[74,75] When grown on broth, the challenge is often adjusted to a standard concentration (measured with optical density) with sterile saline or water added directly to the broth. When grown on solid media, the culture is washed off the slant or plate with sterile saline or water and adjusted to the standard optical density. The ASTM and USP methods mention harvesting techniques generally. Both of these methods use solid agar medium to grow the inoculum. The ASTM uses sterile water while the USP uses sterile saline when washing the organisms from the transferred stock culture. The organisms are then diluted to 1×10^8 colony forming units (cfu) per ml. The CTFA does not specify any recommendations for harvesting. None of the methods discusses properly buffering the harvesting solution to the correct pH or osmolarity. The ideal approach would be to resuspend the cells in a solution of the same pH and osmolarity as the culture medium.

Greater resistance to preserved product has been described for broth-grown cultures compared to cultures grown on solid media.[74,75] However, this result may have been due to the carry-over of broth into the product acting as a neutralizing agent of the preservative in the product rather than any intrinsic resistance gained by the bacteria by growing in broth.

Another area needing further research is investigating the importance of the growth phase of the challenge inoculum. The growth phase affects the physiological state of the organisms used as the inoculum. For example, Holm-Hansen found that ATP per cell is decreased as cells reach stationary.[76] This physiological change and other potential changes may affect an organism's resistance to preservatives.

Finally, considerable guidance is needed in the preparation of the fungal inoculum for any PET. One of the most difficult steps in the challenge methods is to obtain standardized fungal spores. In the USP test, the fungi are harvested in sterile saline supplemented with 0.05% polysorbate 80 to help disperse the spores. No one has assessed the effect this non-ionic detergent has on inactivating preservatives when spores harvested by this method are used as inoculum. The spore count is adjusted to 1×10^8 cfu ml^{-1} by standard methods. The suspensions should be used immediately. The biggest problem is the ability to harvest a high enough concentration of spores. ASTM recommends harvesting fungal spores by putting distilled water onto the surface of the plate and rubbing the lawn of growth on the plate with a sterile inoculating loop or sterile glass hockey stick to dislodge the spores. The liquid is poured off and filtered with sterile non-absorbent cotton. This removes the hyphae to give pure spores. One may also refer to several US patents on how to harvest fungal spores (3 294 647; 3 300 390; 3 357 895; 3 616 246; 4 046 593; 4 280 000). The other area of concern is how to keep the spores from germinating. Within 2 h, some of the fungal spores will germinate if at room temperature. Refrigeration of the fungal preparation will slow this germination down. Light also has an effect on germination.

7.3.2.2.4 Concentration of inoculum and rechallenge. The bacterial concentration of the inoculum should be about 1×10^8 cfu ml^{-1}. If 20 g of product are inoculated with 0.2 ml, then the final recommended concentration of 1×10^6 colony forming units per gram (cfu g^{-1}) product is obtained. All of the tests recommend this level of organisms in the product. However, USP will allow the inoculum to drop to 1×10^5 cfu g^{-1} product. Any 1:100 ratio of inoculum to product should be adequate; the ASTM recommends using 100 g of product rather than 20 g. The goal is to keep the dilution of product by the inoculum to a minimum; a good rule is to not dilute the product over 1% with the inoculum. In like manner, fungi and yeast are introduced into the product. However, their concentration ranges from 1×10^4 cfu g^{-1} product in the CTFA method to 1×10^5 or 1×10^6 cfu g^{-1} product in the USP and ASTM methods. Assuming a 1:100 ratio of inoculum to product, the concentration of the inoculum must be 1×10^6 cfu ml^{-1} (CTFA) or 1×10^7 to 1×10^8 cfu ml^{-1} (USP and ASTM). The assumption that these counts represent fungal spores may not be valid since hyphae can also give rise to fungal colonies.

How to standardize the concentration of the inoculum is left up to the microbiologist in the USP and CTFA methods. Only the ASTM method provides specific recommendations regarding spectrophotometer, spectrophotometer tubes and wavelength of light used. The transmittance is measured at 425 nm to yield 1.0×10^8 cfu ml^{-1} (generally about 30–40% transmittance). ASTM suggests use of a hemocytometer count to adjust the spore level to 1.0×10^7 cfu ml^{-1}. However, any reference to standard microbiological methods will provide the specific detail for determining microbial concentrations.[77] A qualified microbiologist should have no trouble developing a valid method.

Rechallenge is the addition of fresh inoculum to a product that has already killed off the first challenge after an appropriate time. CTFA provides for a rechallenge if desired but does not require it; ASTM recommends a rechallenge if the product is expected to be used repeatedly by the consumer. Some studies suggest this practice does not provide any more information than single challenges.[78] The manufacturer, however, may be able to make a case for multiple challenges. For example, mascaras are commonly subjected to repeat insults by the consumer. If this is the case, the microbiologist should select a challenge level that is reasonable and likely from consumer use (1×10^2 cfu gm^{-1}) rather than the high levels recommended in the compendial methods. These levels could be determined by allowing people to use unpreserved products and then immediately analyze the level of organisms introduced into the product after that use.

7.3.2.2.5 Pure versus mixed challenge inoculum. There is a considerable debate regarding whether pure or mixed challenges should be used. Part of the debate is confusion over semantics. The term mixed challenge does not refer to use of impure cultures. Instead, it is the use of several pure cultures which have been mixed together after they have been grown up and harvested. Use of this mixed inoculum may be more representative of actual conditions of contamination; microorganisms do not exist as pure cultures in nature but as interacting populations within communities of microorganisms. Some feel that the mixed cultures provide greater stress to the preservative system.[79] This view is intrinsically logical if one assumes that cometabolism or synergism occurs within a community biology dynamic. In fact, cometabolism, vitamin and cofactor synthesis help stimulate mixed interactions within communities of microorganisms.[80,81] However, no published information exists to support the hypothesis that these interactions take place during a preservative efficacy test of a cosmetic. The idea is supported by observations by Henrlette *et al.*[82] who described a mixed bacterial community that developed in disinfectants and antibiotics. None of the individual species was resistant to the antimicrobials. Only the community showed resistance.

It is the theory that mixed populations may be more robust, however, that forms one of the objections to mixed challenges. The opponents of mixed challenges claim that it introduces the variable of microbial population dynamics into the challenge test. Therefore a failure may occur since the mixed challenge was more capable of contaminating the product. Alternatively, there is the theory that the mixed community may be less stringent than pure challenges because one organism may produce metabolic factors which are antagonistic to other microorganisms in the challenge.[83] Another idea is that the organisms will compete with each other for limiting substrates and growth factors such as iron.[84] Resolution of the issue will take more research, possibly more than is practical in an industrial microbiology laboratory.

Some methods permit the use of mixed challenges composed of similar groups of microorganisms. For example, a group of Gram-positive bacteria and yeasts or a group of Gram-negative bacteria is permitted for the challenge in the CTFA and ASTM methods. This permission is not based on scientific concerns but is purely pragmatic. It is far less work to challenge one container with multiple microorganisms than it is to challenge multiple containers with each of the microorganisms.

7.3.2.3 Product concentrations. Most published preservative methods test full strength (100%) product.[24] However, Brannan *et al.*[26,27] and Cooke *et al.*[85] added diluted product to the challenge test as well. The dilutions stress the product and also provide a means of ranking the product. For example, products that can be diluted and continue killing the inoculum may be classified as well preserved whereas products that kill the inoculum only when the product is at full strength are classified as marginally preserved. In addition, this approach is another way of mimicking expected use patterns. For example, products that are diluted during normal use such as shampoos should be able to continue killing microbial challenges. One may even make a case for the product being diluted due to manufacturing error. The product should, it is argued, remain hostile with water added since microorganisms can adapt easier to low levels of biocide in the diluted product. Once acclimated, they are better able to adapt and grow in full strength product.

Microbial adaptation to biocides is certainly a possibility.[86] However, several papers claiming to have demonstrated this phenomenon are really either a case of neutralization of the biocide (by carry-over of the growth medium) or a case of saturating the biocide with more organisms than available biocide to the point of inactivating it.[87,88] A mistaken idea is to think of chemical biocides as enzymes. In reality, one mole of biocide reacts with ‘one mole’ of microorganisms to give one mole of spent biocide and ‘one mole’ of dead organism. Certainly, biochemical mechanisms, true adaptation, for resistance to biocides exist in the case of formaldehyde and parabens.[89,90] However, resistance to all biocides by similar biochemical mechanisms must not be assumed for all microorganisms. In some cases, the resistance claimed may be a result of physical factors such as capsule formation or existing within a clump of organisms or within a biofilm. The most naive idea is that the resistance mechanisms against biocides are similar to those mechanisms found in antibiotic resistance. Whereas antibiotic resistance can be described based on molecular activity at a single site, the resistance to biocides cannot. Biocides attack at multiple sites.

7.3.2.4 Plating methods and neutralization. One of the easiest ways of conducting a challenge test is to place the product in large 50 ml plastic throwaway pill vials with snap on lids. Most of these are made with the same polypropylene or polyethylene plastics from which the majority of cosmetic packages are made. One can then easily use sterile tongue depressors to stir the product before sampling. In the ASTM method, the product samples are put in lidded glass jars. The other two methods specify that the test be conducted in the product container or at least a container made of similar material. The product (20 g for USP and CTFA; 100 g for ASTM) is put into the containers. After inoculating, the inoculum should be thoroughly mixed into the product. The inoculated product is then incubated for at least 28 days at 20–25°C. These storage conditions mimic shelf conditions but they also eliminate the potential that some preservatives are actually more biocidal at elevated temperatures.[91]

Aliquots of inoculated product are taken (after thorough mixing) at specific times. Sampling of viscous materials should be done with a syringe. Usually a 1 ml sample is pulled and diluted into 9 ml of the proper neutralization broth in order to inactivate the preservative. Both physical dilution and chemical inactivation operate to neutralize the preservative. The number of surviving microorganisms is then determined by plating the diluted product onto suitable media. This estimate of the numbers of survivors is typically done by plating 1 ml of diluted sample into melted agar for pour plates or by spread plating 0.1 ml of the diluted sample onto prepoured plates of agar (a special gel-like microbial growth medium). After the organisms grow for a few days, they appear as visible colonies. The colonies are then counted as colony forming units (cfu) and multiplied by any dilution factors to estimate the numbers of organisms in the original sample.

The problems with this traditional approach are the assumptions that we microbiologists never question. Does one organism give rise to one colony? Are the organisms evenly distributed as single cells or do they exist as clumps? Are they all equally exposed to the product? Do they all grow well in the medium provided?

It may be that these assumptions are false. The paradigm of organisms existing as non-uniformly distributed clumps that later break up into individual cells helps explain the anomalous results which are occasionally obtained in preservative efficacy testing. The following scenario is sometimes seen. An initial kill occurs at 7 days (seen as a decrease in cfu) but is followed by an increase in cfu at 14 days, followed by another decrease at 21 days.

Using the paradigm of microbes existing in clumps, this occurrence is now explainable (Figure 7.1). The initial kill at 7 days may have been due to killing of the cells in smaller clumps where the entire clump of cells is killed but the larger clumps have a few cells within them that remain alive because they were protected. Here, our model is that cfu are really derived from clumps rather than individual cells. The surviving cells in clumps then disperse to result in single cells that give a higher cfu at 14 days but now are more susceptible to the biocide allowing a reduction to follow at 21 days.

The paradigm of organisms existing as attached cells to each other and to any other surface will also help explain 'grow-back' without resorting to claims of microbial adaptation, injured cells or the fanciful term of 'the phoenix phenomenon'.[92] Without this model and continuing with the false

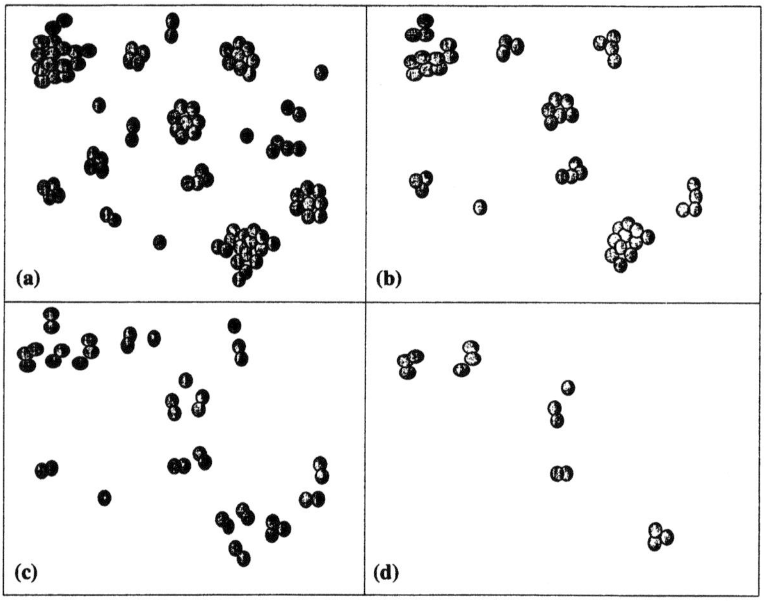

Figure 7.1 The paradigm of bacteria existing as clumps helps explain anomalous results in PET testing. Microorganisms exist in clumps in Poisson distribution. Each clump gives rise to a colony rather than each individual doing so. After 7 days, the CFU/ml is reduced due to the death of organisms existing outside of the protection from living within a clump. A few of the organisms at the periphery of the clump are killed but, since it is clumps that produce a single colony, the clump shows as an individual colony and the fact that some of the organisms in the clump died is not detected upon plating. After 14 days, the clumps break up to provide more CFU. Now the individuals are no longer within the protection of the clump and are more susceptible to exposure to the biocide. Therefore, at day 21, the total CFU is decreased. (a) At time $T = 0$ days, population = 18 cfu ml^{-1}; (b) $T = 7$ days, population = 10 cfu ml^{-1}; (c) $T = 14$ days, population = 21 cfu ml^{-1}; (d) $T = 21$ days, population = 6 cfu ml^{-1}.

assumption of organisms existing as individuals rather than attached to each other, we will continue to have what appear to be anomalous results. Certainly there are cases where adaptation occurs, antibiotics for example and some food preservatives. However, where adaptation is claimed for preservatives that have multiple modes of action, the clump paradigm is more likely.

Traditionally, only those dilutions giving countable plate growth of 30–300 cfu/plate are included. This dogma can be accepted or rejected depending on one's expertise. A well-trained technician can count plates accurately with considerably more than 300 cfu/plate. Additionally, if dilutions are made meticulously, it is possible to rely on numbers less than the 30 cfu/plate value. Use of good judgement is a better guideline than the antiquated, clinically based, dogma of 30–300 cfu/plate.

Either pour plates or spread plates can be used to determine how many cfu gm^{-1} survive. In pour plates, the diluted product (about 1 ml) is vortexed into a test tube of about 15 ml of melted agar at 46 ± 2°C. The agar is then poured into a petri dish. With spread plates, the diluted product (about 0.1 ml) is spread onto prepoured solidified agar plates using a bent glass rod. Spread plating allows easy processing of samples. The main advantage is that it avoids exposing microorganisms to heated media. However, pour plating allows for more exposure to neutralizing agents in the agar. Some published information finds that the two methods give similar results.[93,94]

Appropriate use of neutralizers is often overlooked when conducting preservative efficacy tests. Dilution of the product allows for biocide neutralization. Chemical neutralization can also be used. Filtration is another approach but is limited to those products that can be filtered. Singer paraphrases work done by the CTFA Microbiology Committee where they describe some of the chemical neutralizers useful for preservatives.[95] Sutton's work also describes a number of neutralization methods for preservatives.[96–98] Russell describes several neutralizers for disinfectants.[99] Both CTFA and ASTM suggest using neutralizers but neither provides extensive guidance regarding which ones are effective. ASTM does, however, offer a method to determine if a neutralizer is non-toxic and effective. First the organism's maximum tolerated concentration (MTC) of neutralizer is determined. This is essentially a comparison of growth medium to the medium containing various increasing levels of the neutralizer. The MTC is the largest amount of neutralizer put into the medium that did not decrease the microbial count as compared to the normal growth medium which serves as the control. Once the MTC is determined, then the maximum amount of biocide that the neutralizer can inactivate is determined. This is done by adding various concentrations of biocide to the neutralizer at its MTC. With this approach, one is using microorganisms as biological indicators of a neutralizer–biocide reaction that results in inactivated biocide. The colony counts of the neutralizer–biocide combinations (square-root transforms) should not differ significantly ($p = 0.05$) from the counts in the growth medium/neutralizer controls using a *t*-test.

In addition, ASTM offers a retroactive check for neutralization of the preservative. If a plate shows no growth, it is streaked using a nutrient broth culture at about 10^2 cfu ml^{-1}. Lack of growth after incubation suggests the biocide was not neutralized. A positive growth result, however, does not prove that neutralization actually occurred. Retroactive procedures like these tell us only that neutralization had finally occurred by the time the plate was streaked with the test organisms. We have no idea if the product was neutralized at the time it was originally plated. Therefore, positive growth using a retroactive check is not proof that biocide neutralization took place at the time of sampling. This kind of retroactive check on neutralization is invalid.

Neutralizers can be used in the diluent or the plating medium, or both. The goal of a neutralizer is to inactivate the biocide before the biocide inactivates the microorganism and thus allow uninhibited microbial growth. Failure immediately to inactivate the biocide upon sampling causes overestimation of the biocide's killing potential. This failure is actually a measure of the kill that continues even in the recovery of plating medium because active biocide is carried over into the medium.[100] Of course the neutralizer itself must not be toxic to the organism. For example, sodium bisulfite is a good neutralizer for formaldehyde and glutaraldehyde but it inhibits the growth of some bacteria and the germination of spores. In order to determine neutralizer toxicity the number of survivors after exposure to the neutralizer are compared to those surviving after exposure to the growth medium. Lecithin, polysorbate 80 and sodium thiosulfate are neutralizers for quaternary ammonium compounds, phenolics and halogens, respectively. Studies by Sutton and Brannan suggest that the type of neutralizer used for each biocide and microorganism combination is unique.[101,102] However, a fairly effective all purpose (universal) neutralizing medium is the Dey–Engley. It contains sodium thioglycollate, sodium thiosulfate, sodium bisulfite, lecithin and polysorbate 80.[100,103] It is available in broth or agar preparation. Some preservatives only require dilution to be inactivated (e.g. some alcohols, organic acids and aldehydes).

7.3.2.5 Recovery of inoculum after apparent kill. All of the recognized tests require long incubation periods (28–56 days). These long periods are supposed to account for the phenomenon of adaptation. After a lag phase the microbes 'grow back' to high enough levels to be detected again. The mechanism(s) for this regrowth are not well understood. Perhaps it is due to survival and adaptation. Perhaps it is due to *in situ* recovery of injured organisms.[104] It may be due to container-associated organisms that slough off into the product.[105,106] It may even be due to inadequate mixing and inconsistent plating methods, since bacteria display Poisson distribution in the sample. Bacteria rarely, if ever, are evenly distributed in a sample despite the vigorous shaking to which it is subjected. Thus the adapting microorganisms may be located in a portion of the product where they are never sampled. Unless the entire test volume is cultured at each sampling interval complete kill of the inoculum cannot be certain. Late sampling intervals must be completed even when total and rapid kill of the challenge inoculum seems to have occurred.

Finally, the question of bacteria living not as isolated cells but in clumps (perhaps in communities) should not be ignored as previously addressed. We need to throw off the dogma that bacteria exist as single evenly dispersed entities that can be isolated from one another. We even need to look at microbial contamination from a community of several species viewpoint rather than a solitary species population standpoint. Even though historically, most contamination incidences have appeared to be due to individual species, it may be more likely that what is observed is just a transient successional event set up by previous contaminants. The cause and effect relationships of pure cultures that cause disease (Koch's postulates) apply when studying isolated populations that are selected by a host's defenses. However, the pure culture, clinical microbiology paradigms do not apply to environmental microbiology problems including cosmetic contamination. We should be studying multi-species communities as contaminants rather than isolated species in pure cultures.

7.3.2.6 Enrichment. Enrichments of the product can be performed to detect low levels of potentially recoverable organisms.[107] The sensitivity or detection limit of typical dilution and plate count methods is usually from 10 to 20 cfu g^{-1} of product. In enrichment, at least 10 g of product is put into 1 l of broth and incubated. Any turbidity (or color change if one uses either a pH or redox indicator) indicates at least one organism was present in the 10 g sample. This approach makes the detection limit 1 cfu per 10 g of product (theoretically 0.1 cfu gm^{-1}). Thus the sensitivity is increased by 100 times compared to traditional plate count methods. It is most useful in determining if, after the 28 day period, low levels of inoculum still exist that may still be capable of adapting and growing later on.

7.3.2.7 Interpretation of data. All the compendial tests prescribe how to interpret the data. This interpretation is not related to any field data on how the product should behave in the hands of the consumer. Instead, it is based on anecdotal evidence and opinion regarding how long a product should take to reduce the numbers of the challenge inocula.

The best way of interpreting the data is to compare how the test product performs compared to known well-preserved and poorly preserved control products. The well-preserved products should be defined as those that do not become contaminated during consumer use and the poorly preserved products should be those that do become contaminated (on a statistical basis using a chi-square analysis with $p = 0.05$). This allows a test of the inoculum to be sure it is capable of contaminating a product that is poorly preserved but not so robust as to grow in well-preserved product.

Some microbiologists use the PET to predict the likelihood of manufacturing contamination. They develop their preservatives to withstand manufacturing abuse from a dirty environment or contaminated raw materials. This is the wrong approach for designing a PET. The microbiologist should spend more time focused on bringing the plant into sanitary control than developing preservatives to hide sloppy manufacture.

The performance criteria that characterize an acceptably preserved product vary according to the compendial method used. The data are in the form of cfu g^{-1} product. For most PETs, the data points are the plate counts from samples taken at 0, 7, 14, 21 and 28 days after inoculation. USP does not include the zero day count. CTFA also includes a day one count and, for eye cosmetics, a three day count. In the CTFA and ASTM methods, the vegetative bacterial counts should be reduced to 0.1% of their original numbers within seven days of inoculation. This is a 99.9% (or 3 log) reduction (from 10^6 to 10^3 cfu g^{-1}). The counts should remain at 10^3 or continue to decrease for the rest of the test. The USP (for category I-A products) allows 14 days for a 3 log reduction of bacteria to occur and then the final 28 day count should be no more than 150% of the 14 day count. The yeast and fungal counts must be reduced by 90.0% (a 1 log reduction) after seven days for the CTFA and after 28 days for the ASTM. The USP (for category I-A products) expects that by 14 and 28 days the level of yeasts and fungi will be 'not more than 150% of the initial inoculum.' Other USP product categories (IB-D and II) require less stringent microbial reductions. The 150% variability rule is being reconsidered for a change to 'not more than a 0.5 log increase over the initial inoculum.' The USP does not give details how to treat the seven day data at this time.

7.3.3 Other PET methods

7.3.3.1 D-value methods. Long term tests are intended to mimic incubation times that reasonably parallel the life of the product while in use by the consumer. Within individual companies, however, rapid tests are sometimes used for quick impressions of which preservatives to use in a product. One such method is the *D*-value method. Aside from one author, no one else claims *D*-value methods are valid for final testing of nationally distributed product.[108] In fact, there is evidence that *D*-value methods are inappropriate for at least some consumer products[109] since *D*-values rely on first order rate kinetics and biocide kills are not described by first order rate kinetics.

The *D*-value method uses short sample times and estimates the final response at 28 days by linear regression. The claim is that each organism has a characteristic rate of death. By multiplying the death rate by the log of the inoculum challenge, the method is supposed to predict the time it takes to inactivate the entire challenge. This method is an adaptation from food microbiology's heat or radiation destruction *D*-values. These two methods of microbial kill do indeed follow first order rate kinetics and therefore the *D*-values determined for them are quite valid. The major weakness with applying the *D*-value technique to preservatives and other chemical means of disinfection is that these follow second order rate kinetics where the microorganism must first react with the disinfectant to result in the product: dead microbe and spent disinfectant. When disinfecting with heat or radiation, the reaction does not depend on the microbe and the agent interacting first before the 'reaction' product of dead microbe can take place. The only case where a second order reaction can approach pseudo-first order rate kinetics is when the second reactant is present in such large excess that it is virtually in constant concentration. It is rarely the case however that preserved products have an excess of preservative in such great abundance that the biocide remains in constant concentration. The second criticism of the *D*-value technique is that it extrapolates beyond the measured data by falsely assuming a linear relationship between biocide exposure time and the number of surviving microorganisms.

In defense of the rapid *D*-value methods, it allows at most a preliminary screening of preservatives despite its invalidity. This approach assumes that appropriate reproducible controls are in place such that the various *D*-values can be ranked amongst a wide variety of products and then the data can be correlated to full scale PET results on the same products. However, the microbiologist should never rely on *D*-values as the sole method of testing for preservative efficacy.

7.3.3.2 Capacity tests. A capacity test determines how many bacterial challenges are needed before the product begins growing microorganisms.[110] After each challenge the products are sampled and challenged again until the product either receives 15 challenges without showing growth (a well preserved product) or until three consecutive positive results occur (a lesser preserved product). The goal is for the product to reduce the number of viable organisms in the inoculated formulation by 3 logs (99.9%) in 48 h. With each subsequent challenge, this ability diminishes as a result of dilution, neutralization and reaction with the inoculum. The value of such a test may be for multi-use products.

However, to obtain a more predictive test of consumer contamination potential, the challenge inocula should probably be lowered to levels likely to be encountered during use. Once done, the capacity test may be a more quantitative and valid way of understanding a product's ability to handle contamination by use.

7.4 Preservatives available for use

7.4.1 Mode of action of preservatives

The modes of action of antibiotics are known at the molecular level because of their highly specific biochemical reactions. With preservatives and biocides, the modes of action are far more generalized with points of attack at numerous sites. With few exceptions, nearly all biocides work by denaturing proteins throughout the cell or by affecting membrane permeability so that either transport or energy generation is blocked. Biocides such as formaldehyde, formaldehyde releasers, isothiazolinones, bromine and chlorine compounds act by protein denaturation. The parabens and weak acids like sorbic, benzoic and dehydroacetic acid act by disrupting control of membrane electrical potential and its permeability to block energy generation and loss of transport, respectively. Of course, any compound that denatures proteins will also denature membrane proteins to inhibit transport of essential materials via permeases.

With some biocides, it is almost as simplistic to ask what is the mode of action of heat in killing microorganisms. The effect of chlorine, for example, is to oxidize every reduced site that exists in a cell and, as a result, denaturation of proteins takes place. Chlorine is a virtual poison. Thus, very few cases of true chlorine resistance occur. Instead, the resistance seen is really a population or community effect of cells existing within the protection of a biofilm or surrounded by an incredible capsule that excludes the chlorine.

One compound that is not technically a biocide but rather an adjuvant to most biocides is ethylenediaminetetraacetic acid (EDTA). This chelating agent and others remove magnesium divalent cations from the cell wall and cell membranes which are needed for stability. Once destabilized, they permit easier access of biocides into the cell. One illustrative event personally encountered was a formaldehyde-resistant *Enterobacter* spp. in ammonium lauryl sulfate. A simple capsule stain indicated that this organism had an incredible capsule surrounding the cell that was nearly five times as thick as the organism's width. The capsule was scavenging out the formaldehyde to allow the organism to grow. A simple addition of 0.1% EDTA (disodium salt) permitted the formaldehyde once again to be effective in this surfactant apparently because the EDTA removed the capsule. Had the organism been a *Pseudomonas* spp. with a formaldehyde dehydrogenase we would not have been so fortunate with simple EDTA addition.

When bacterial cells are exposed to parabens, nutrient transport is inhibited.[111] Thus the parabens apparently inhibit nutrient uptake by shutting down permeases, disrupting porin channels, or by disrupting the membrane pH gradient or electrical charge potential across the membrane to prevent substrate transport and ATP generation. This inhibition is apparently reversible and is consistent with other observations that the mode of action of parabens is by disruption of the membrane electrical potential.[112]

Organic acids probably work in the same fashion[112,113] however, they may even be enzyme inhibitors as well.[114,115] Typically, they are only biocidal at pH values below their pK_a. In this protonated form, they pass through the membrane and the hydrogen ion dissociates from the weak acid to decrease the cytoplasmic pH. As a result, both substrate transport and oxidative phosphorylation are uncoupled from the electron transport system. This effectively starves the cell of needed substrate and energy derived from ATP kinase driven by hydrogen ions.

Phenolics also disrupt the proton motive force of the cell membrane[116,117] but they have the ability non-specifically to denature cytoplasm, cell walls and cell membranes.[118] The more lipophilic phenolics have the greater antibacterial capacity, perhaps because of a greater ability to partition out of the water phase and into the lipid membrane.[119,120] Alcohols likewise disrupt the membrane causing permeability loss,[121] but they also appear to inhibit enzymes.[122]

Perhaps one of the most widely used of the newer preservatives are the isothiazolinones. These are usually compounded into a single product composed of chloromethylisothiazolinone and methylisothiazolinone, but they can also include benzyl-type compounds as well.[123,124] The isothiazolinones inhibit glucose oxidation and active transport without significantly affecting membrane integrity.[125] In fact, these compounds denature thiol-containing proteins and enzymes (e.g. ATPase and glyceraldehyde-3-phosphate dehydrogenase). The isothiazolinones inactivate proteins and enzymes that contain thiol groups (e.g. asparaginase).[125] Thus, the isothiazolinones attack proteins with thiol-containing amino acids. Initially, the isothiazolinone forms a disulfide link with the thiol group on the amino acid. Occasionally, the chloromethylisothiazolinone may facilitate linkage with another thiol group to establish a new disulfide linkage and release the biocide as a mercaptoacrylamide. This mercaptoacrylamide can tautomerize to a thioacyl chloride which may react again by denaturing nucleic acids.[126]

Formaldehyde also denatures protein but by alkylating amino and sulfhydril groups; it can also alkylate the nitrogens of purine rings to denature DNA.[127] Most of the formaldehyde donors (e.g. Glydant®, Germall®, Dowicil®, Nuosept®, etc.) act in this basic manner since these compounds release formaldehyde into the microbial cell. Differences seen between the formaldehyde donors may exist as a result of when or what triggers the compound to release formaldehyde. For example, a compound with a long hydrophilic chain connected to the formaldehyde donating region (e.g. Nuosept®) may release formaldehyde only when the long chain enters into the LPS section of the membrane.

Brominated compounds such as bromonitropropanediol and bromonitrodioxane act by oxidation of thiol groups[128–131] or by causing thiols to convert to disulfides[132,133] where the thiol group first becomes brominated and then reacts with another thiol group to yield a disulfide and free bromide. As a result, enzymes involved in respiratory activity (e.g. dehydrogenases) and nucleic acid synthesis are inhibited, cell membrane integrity is compromised, and the cell wall may even be affected.[133,134]

7.4.2 Selection of preservative

The microbiologist uses a variety of chemical preservatives to prevent contamination by pathogens or spoilage microorganisms. The ideal preservative would be broad spectrum, safe and completely free of any sensitization issues, completely water soluble, completely stable to all extremes of pH and temperature, completely compatible with all ingredients and packages, impart no color or odor to the product and be inexpensive. This ideal does not exist. Instead, a preservative must be selected based on empirical testing. The only approach bordering on a theoretical basis for choosing a preservative is a qualified microbiologist's intuition finely honed by experience. There are published lists of available preservatives[135] and certainly no dearth of advertisements in trade magazines.[136] These are good places for getting ideas of what might work in a formulation, but only the novice relies entirely on these lists or places much faith in the advertising claims when choosing a preservative. Every formulation must be considered unique. Factors such as the physical and chemical nature of the product, how it is to be used, the container type and the shelf life must be considered when choosing the preservative. Often the selection of a preservative must be a compromise between efficacy, stability and safety.

7.4.3 Safety considerations of preservatives

One must always balance the risk of microbial contamination with the risk that a biocide may give to a product. For example, many eye area products were permitted to contain phenyl mercuric acetate because the risk of infection to the eye was greater than the risk of exposure to the compound.

The key consideration is to judge whether the product will be safe for the consumer under normal use and foreseeable misuse conditions. This section will not make a microbiologist into a good toxicologist any more than reading this chapter would make a toxicologist a good microbiologist. Instead, it will allow both to have a basic understanding of the other's responsibility to allow a positive and productive interaction.

One of the first considerations of a preservative is its acute toxicity. In cosmetics, the route of exposure is dermal but often oral routes of acute toxicity are also needed in cases of accidental ingestion. Occular irritation, subchronic and chronic toxicity tests are performed via the expected consumer exposure route to determine at what level the preservative can exhibit any irritant, toxic or carcinogenic properties. These levels will then be compared with typical use exposure levels to determine safety ratios.

Perhaps more important than these tests are the skin responses to biocides. Basic irritant responses can be a result of corrosion, acute irritation, cumulative irritation or photoirritation. In cosmetic use of a preservative, most testing involves placing the biocide onto the skin using an occlusive patch dosed with the compound. After a time, the patch is removed and the skin graded as to how irritated it became.

Skin sensitization is another key concern when using biocides. Nearly all biocides used today will elicit sensitization in some people. Sensitization testing is performed in much the same way as irritant patch testing except much lower concentrations are used. Even in-use sensitization studies can be used as well.

Another concern for biocides is mutagenicity testing to determine if the biocide has the potential to mutate somatic or germ cells. In addition to this testing, embryological (or developmental) toxicity testing is done to determine if the biocide may be a teratogen capable of causing birth defects.

In all these tests, the results must be compared to the ordinary use and foreseeable misuse exposure levels to give a reasonable risk assessment. Often the test data are expressed as no observed effect level (NOEL – the highest dose without causing a toxic effect). Consumer exposure levels of one-tenth the NOEL (a 10× safety factor) based on human testing may be considered safe to use under some circumstances. This safety factor may increase up to 1000 times depending on whether the data used are animal data or where insufficient chronic exposure data are lacking. The definition of reasonable risk must include considerations based on the benefits of using the biocide, the ability to use less risky biocides for the same use, the economic benefits of using the biocide (can the biocide help prevent costly recalls due to contamination?), even how the biocide may affect quality of life, environment and public opinion about the company.

7.5 Conclusion

The personal care product (cosmetic) microbiologist must balance a variety of factors to provide safe, unspoiled quality products.[137] In addition to knowing preservatives, he or she must understand microbial physiology, pathogenic microbiology and microbial ecology. In addition to microbiology, he or she must understand organic and physical chemistry, toxicology, engineering, manufacturing and processing, sanitation, and regulatory/environmental law. The cosmetic microbiologist must use all this education and knowledge within the context of the business needs of the company and be able to balance risk/benefit to the consumer using the product. The microbiologist that best fits this description is the one who has a good solid liberal arts undergraduate education followed by post graduate training in microbiology (preferably in microbial physiology and ecology). He or she must be a critical thinker, a philosopher dedicated to lifelong learning, and a true renaissance man or woman interested in a broad range of subjects pertaining to the field. Finally and most importantly, this person must have the highest of ethical standards. They must consider themselves as part of the cadre of health care providers in the world dedicated to serving humankind via the mission of providing them with microbially safe and efficacious products.

Acknowledgements

I would like to thank the many colleagues who have helped me prepare this chapter and, in particular, those who allowed me to base parts of this chapter on drafts they have prepared for other publications. Their insight has been very valuable. I thank Drs. Phil Geis, Bill Apel, Scott Sutton and Tom Sox especially. I also thank Ms. Jan Curry, Mr. Stephen Richards and Mr. Rich Hennessy.

References

1. Gunn, F. (1974) The artificial face: A history of cosmetics. In *Multimedia Encyclopedia*, ver. 5.01, Grolier and The Software Toolworks, Novato, CA.
2. Curry, J.C. (1983) Cosmetic microbiology 50 years of change – An historical perspective of microbiology and cosmetics. *HAPPI*, **8**, 44–52.
3. Kallings, L.O., Ringertz, O., Silverstone, L. and Ernerfeldt, F. (1966) Microbial contamination of medical preparations. *Acta Pharm. Suec.*, **3**, 219–228.
4. Morse, L.J., Williams, H.L., Grenn, F.P. Jr., Eldridge, E.E. and Rotta, J.R. (1967) Septicemia due to *Klebsiella pneumoniae* originating from a hand cream dispenser. *New Engl. J. Med.*, **277**, 472–473.
5. Morse, L.J. and Schornbeck, L.F. (1968) Hand lotions – a potential nosocomial hazard. *New Engl. J. Med.*, **278**, 376–378.

6. Noble, W.C. and Savin, J.A. (1966) Steroid cream contaminated with *Pseudomonas aeruginosa*. *Lancet*, **1**, 347–349.
7. Dunnigan, A.P. (1968) Microbial control of cosmetics. *Drug Cosmet. Ind.*, **102**, 43–45; 152–158.
8. Dunnigan, A.P. (1970) Report of a special survey: Microbiological contamination of topical drugs and cosmetics. *TGA Cosmet. J.*, **2**, 39–41.
9. Wolven, A. and Levenstein, I. (1969) Cosmetics – contaminated or not. *TGA Cosmet. J.*, **1**, 34–37.
10. Wolven, A. and Levenstein, I. (1972) Microbial examination of cosmetics. *Am. Cosmet. Perfum.*, **87**, 63–65.
11. Unpublished report on an FDA survey from 1975 where 2541 units of 501 domestic and imported cosmetics were tested.
12. Wilson, L.A., Kuehne, W., Hall, S.W. and Ahearn, D.G. (1971) Microbial contamination in ocular cosmetics. *Am. J. Ophthalmol.*, **71**, 1298–1302.
13. Ahearn, D.G. and Wilson, L.A. (1976) Microflora of the outer eye and eye area cosmetics. *Develop. Indust. Microbiol.*, **17**, 23–25.
14. Wilson, L.A. and Ahearn, D.G. (1977) *Pseudomonas*-induced corneal ulcers associated with contaminated eye mascaras. *Am. J. Ophthalmol.*, **84**, 112–119.
15. Ahearn, D.G., Wilson, L.A., Jilian, A.J., Reinhardt, D.J. and Ajello, G. (1974) Microbial growth in eye cosmetics: Contamination during use. *Develop. Indust. Microbiol.*, **15**, 221–216.
16. Wilson, L.A., Jilian, A.J. and Ahearn, D.G. (1975) The survival and growth of microorganisms in mascara during use. *Am. J. Ophthalmol.*, **79**, 596–601.
17. Anderson, D.W. and Ayers, M. (1972) Microbiological profile of selected cosmetic products with and without perservatives after use. *J. Soc. Cosmet. Chem.*, **23**, 863–73.
18. Anderson, D.W., McConville, J.F. and Anger, C.B. (1973) Microbiological profile of used eye cosmetics by examination of product only. *Cosmet. Perfum.*, **88**, 25–27.
19. Anderson, D.W., McConville, J.F. and Anger, C.B. (1973) Some comments on the microbiological profile of used automatic eye cosmetics by examination of both applicator and product. *Cosmet. Perfum.*, **88**, 29–30.
20. Dawson, N.L. and Reinhardt, D.J. (1981) Microbial flora of in-use, display eye shadow testers and bacterial challenges of unused eye shadows. *Appl. Environ. Microbiol.*, **42**, 297.
21. Food and Drug Administration, *Microbial Contamination in Ocular Cosmetics*. Contract Nos. 71–74 *et seq.*, Ahearn, D. and Wilson, L., principal investigators.
22. Tran, A.T., Hitchins, A.D. and Collier, S.W. (1990) Direct contact membrane method for evaluating preservative efficacy in solid cosmetics. *Internat. J. Cosmet. Sci.*, **12**, 175–183.
23. CTFA Preservation Subcommittee (1983) Preservative testing of aqueous liquid and semi-liquid eye cosmetics. In *CTFA Technical Guidelines*, Haynes, C.R. and Estrin, N.F. (eds), Cosmetics, Fragrance and Toiletries Association, Washington, D.C.
24. CTFA Microbiology Committee (1990) CTFA survey: preservative test methods companies use. *Cosmet. Toiletries*, **105**, 79–82.
25. Spielmaker, R.J. (1985) Report of a cosmetic industry survey concerning correlation of preservative challenge and consumer use test results. *J. Soc. Cosmet. Chem.*, **36**, 5.
26. Brannan, D.K., Dille, J.C. and Kaufman, D.J. (1986) Correlation of *in vitro* challenge testing with consumer-use testing for cosmetic products. *Abstr. Am. Soc. Microbiol.*, Session Q41, 290.
27. Brannan, D.K., Dille, J.C. and Kaufman, D.J. (1987) Correlation of *in vitro* challenge testing with consumer use testing for cosmetic products. *Appl. Environ. Microbiol.*, **53**, 1827–32.
28. Lindstrom, S.M. (1986) Consumer use testing: assurance of microbiological product safety. *Cosmet. Toiletries*, **101**, 71–73.
29. Lindstrom, S.M. and Hawthorne, J.D. (1986) Validating the microbiological integrity of cosmetic products through consumer-use testing. *J. Soc. Cosmet. Chem.*, **37**, 481–88.
30. Beucht, L.R. (1983) Influence of water activity on growth, metabolic activity and survival of yeasts and molds. *J. Food Protection*, **46**, 135–141.
31. Curry, J. (1985) Water activity and preservation. *Cosmet. Toiletries*, **100**, 53–55.

32. Geis, P. (1988) Preservation of cosmetics and consumer products: rationale and application. *Develop. Indust. Microbiol.*, **29**, 305–315.
33. Morse, L.J., Williams, H.L., Grenn, F.P. Jr., Eldridge, E.E. and Rotta, J.R. (1967). Septicemia due to *Klebsiella pneumoniae* originating from a hand-cream dispenser. *New Engl. J. Med.*, **277**, 472–473.
34. Reis, F.R. and Wood, T.O. (1979) *Pseudomonas* corneal ulcer: the causative role of eye cosmetics. *Arch. Ophthalmol.*, **97**, 1640.
35. Fainstein, V., Andres, N., Umphrey, J. and Hopfer, R. (1988) Hair clipping: another hazard for granulocytopenic patients. *J. Infect. Disease*, **158**, 655.
36. Federal Food, Drug, and Cosmetic Act of 1938, as later amended, Secs. 201–702, 21 U.S.C. 321–392.
37. Eierman, H.J. (1984) Cosmetic product preservation: safety and regulatory issues. In *Cosmetic and Drug Preservation; Principles and Practices*, Kabara, J.J. (ed.), Marcel Dekker, New York, pp. 559–569.
38. Madden, J.M. and Jackson, G.J. (1981) Cosmetic preservation and microbes: viewpoint of the Food and Drug Administration. *Cosmet. Toiletries*, **96**, 75–77.
39. Federal Food, Drug, and Cosmetic Act of 1938, Sec. 301, 21 U.S.C. 331.
40. Federal Food, Drug, and Cosmetic Act of 1938, Sec. 601, 21 U.S.C. 361.
41. Finch, J.E. and Hanksworth, M. (1978) A bacteriological survey of the domestic environment. *J. Appl. Bacteriol.*, **45**, 357–364.
42. Scott, E., Bloomfield, S.F. and Barlow, C.G. (1982) An investigation of microbial contamination in the home. *J. Hygiene*, **89**, 279–293.
43. Smart, R. and Spooner, D.F. (1972) Microbiological spoilage in pharmaceuticals and cosmetics. *J. Soc. Cosmet. Chem.*, **23**, 721–737.
44. Brannan, D.K. and Dille, J.C. (1990) Type of closure prevents microbial contamination of cosmetics during consumer use. *Appl. Environ. Microbiol.*, **56**, 1476–1479.
45. Bassett, D.J.C., Stokes, K.J. and Thomas, W.R.G. (1970) Wound infection with *Pseudomonas multivorans* a water-borne contaminant of disinfectant solutions. *Lancet*, **i**, 1188–1191.
46. Tenenbaum, S. (1967) *Pseudomonas* in cosmetics. *J. Soc. Cosmet. Chem.*, **18**, 797.
47. Anderson, R.L., Berkelman, R.L., Mackel, D.C., Davis, B.J., Holland, B.W. and Martone, W.J. (1984) Investigations into the survival of *Pseudomonas aeruginosa* in polaxmer-iodine. *Appl. Environ. Microbiol.*, **47**, 757.
48. Ansell, J.M. (1984) Contamination of germicidal and nongermicidal agents. *Am. Clin. Pathol. Rev.*, **1984**, 32.
49. Craven, D.E., Moody, B., Connolly, M.G., Kollisch, N.R., Stottmeier, K.D. and McCabe, W.R. (1981) Pseudobacteremia caused by povidone-iodine solution contaminated with *Pseudomonas cepacia. New Engl. J. Med.*, **305**, 621–623.
50. Bandelin, F.J. (1959) The effect of pH on the efficacy of various mold-inhibiting compounds. *J. Am. Pharm. Assoc. Sci. Ed.*, **46**, 691–694.
51. Entrekin, D.N. (1961) Relation of pH to preservative effectiveness. *Internat. J. Pharmaceut. Sci.*, **50**, 743–746.
52. Jacobs, G., Henry, S.M. and Cotty, V.R. (1975) The influence of pH, emulsifier, and accelerated aging upon preservative requirements of O/W emulsions. *J. Soc. Cosmet. Chem.*, **26**, 105–117.
53. Bean, H.S., Konnig, G.H. and Thomas, J. (1970) Significance of the partition coefficient of a preservative in cosmetic emulsions. *Am. Perfum. Cosmet.*, **85**, 61–65.
54. Croshaw, B. (1978) Preservation of emollient and moisturizing preparations: effect of partition coefficients. *Cosmet. Toiletries*, **93**, 42–49.
55. Bean, H.S., and Herman-Ackah, S.M. (1964) Influence of oil:water ratio on the activity of some bactericides against *Escherichia coli* in liquid paraffin and water dispersions. *J. Pharm. Pharmacol.*, **16** (suppl.), 58T–67T.
56. Bean, H.S., Konnig, G.H. and Malcom, S.A. (1969) A model for the influence of emulsion formulations on the activity of phenolic preservatives. *J. Pharm. Pharmacol.*, **21** (Suppl.), 173S–181S.
57. Coates, D. (1973) Interaction between preservatives and surfactants. *Mfg. Chem. Aerosol News*, **44**, 41–42.
58. Horn, N.R., McCarthy, T.J.K. and Price, C.H. (1971) Interaction between preservatives and suspension systems. *Am. Perfum. Cosmet.*, **86**, 37–40.

59. Beckett, A.H. and Robinson, A.E. (1958) The inactivation of preservatives by nonionic surface active agents. *Soap Perfum. Cosmet.*, **31**, 454–459.
60. Blaug, S.M. and Ahsan, S.S. (1961) Interaction of sorbic acid with nonionic macromolecules. *J. Pharm. Sci.*, **50**, 138–141.
61. Evans, W.P. (1964) The solubilization and inactivation of preservatives by non-ionic detergents. *J. Pharm. Pharmacol.*, **16**, 323–331.
62. Schmolka, I.R. (1973) Synergistic effects of nonionic surfactants upon cationic germicidal agents. *J. Soc. Cosmet. Chem.*, **24**, 577–592.
63. Bean, H.S. (1967) Preservative for pharmaceuticals. *J. Soc. Cosmet. Chem.*, **23**, 703–720.
64. Russell, A.D. (1992) Factors influencing the efficacy of antimicrobial agents. In *Principles and Practice of Disinfection, Preservation and Sterilization*, 2nd edn, Russell, A.D., Hugo, W.G. and Ayliffe, G.A.J. (eds), Blackwell Scientific, London, pp. 89–113.
65. McCarthy, T.J. (1984) Formulated factors affecting activity of preservatives. In *Cosmetic and Drug Preservation; Principles and Practice*, Kabara, J.J. (ed.), Marcel Dekker, New York, pp. 359–388.
66. Miyawaki, G.M., Patel, N.K. and Kostenbauder, H.B. (1959) Interaction of preservative with macromolecules, III. Parahydroxybenzoic acid esters in the presence of hydrophilic polymers. *J. Am. Pharm. Assoc. Sci. Ed.*, **48**, 315.
67. Parker, M.S. (1992) Preservation of pharmaceutical and cosmetic products. In *Principles and Practice of Disinfection, Preservation and Sterilization*, 2nd edn, Russell, A.D., Hugo, W.B. and Ayliffe, G.A.J. (eds), Blackwell Scientific, London, pp. 335–350.
68. McCarthy, T.J. (1970) Interaction between aqueous preservative solutions and their plastic containers. *Pharm. Weekbl.*, **105**, 557–563; 1139–1146.
69. Lachman, L., Urbanyl, T. and Weinstein, S. (1963) Stability of antibacterial preservatives in parenteral solutions. IV. Contribution of rubber closure composition on preservative loss. *J. Pharm. Sci.*, **52**, 244–249.
70. Parker, M.S. (1984) Design and assessment of preservative systems for cosmetics. In *Cosmetic and Drug Preservation; Principles and Practice*, Karbara, J.J. (ed.), Marcel Dekker, New York, pp. 389–402.
71. Cowen, R.A. and Steiger, B. (1986) Antimicrobial activity: A critical review of test methods of preservative efficacy. *J. Soc. Cosmet. Chem.*, **27**, 467–481.
72. Stickler, D.J. and King, J.B. (1992) Bacterial sensitivity and resistance: A. Intrinsic resistance. In *Principles and Practice of Disinfection, Preservation and Sterilization*, 2nd edn, Russell, A.D., Hugo, W.G. and Ayliffe, G.A.J. (eds), Blackwell Scientific Publications, London, pp. 211–224.
73. Russell, A.D. (1992) Plasmids and bacterial resistance. In *Principles and Practice of Disinfection, Preservation and Sterilization*, 2nd edn, Russell, A.D., Hugo, W.G. and Ayliffe, G.A.J. (eds), Blackwell Scientific, London, pp. 215–229.
74. Al-Hiti, M.M. and Gilbert, P. (1983) A note on inoculum reproducibility: A comparison between solid and liquid culture. *J. Appl. Microbiol.*, **55**, 173.
75. Orth, D.S., Lutes, C.M. and Smith, D.K. (1989) Effect of culture conditions and method of inoculum preparation on the kinetics of bacterial death during preservative efficacy testing. *J. Soc. Cosmet. Chem.*, **40**, 193–204.
76. Holm-Hansen, O. (1973) Determination of total microbial biomass by measurement of adenosine triphosphate. In *Estuarine Microbial Ecology*, Stevenson, L.H. and Colwell, R.R. (eds), University of South Carolina Press, Columbia, pp. 73–89.
77. Koch, A.L. (1981) Growth measurement. In *Manual of Methods for General Bacteriology*, Gerhardt, P. (ed.), American Society for Microbiology, Washington, D.C., pp. 179–207.
78. Preservation Subcommittee, Cosmetic Toiletry and Fragrance Association Microbiology Committee. (1981). A study of the use of rechallenge in preservation testing of cosmetics. *CTFA Cosmet. J.*, **13**, 19.
79. Parsons, T. (1990) A microbiological primer for the microbiology manager. *Cosmet. Toiletries*, **105**, 73–77.
80. Beam, H.W. and Perry, J.J. (1974) Microbial degradation of cycloparaffinic hydrocarbons via cometabolism and commensalism. *J. Gen. Microbiol.*, **82**, 163–169.

81. Horvath, R.S. (1972) Microbial co-metabolism and the degradation of organic compounds in nature. *Bacteriol Rev.*, **36**, 146–155.
82. Henriette, C., Petitdmange, E., Ravel, G. and Gray, R. (1991). Mercuric reductase activity in the adaptation to cationic mercury, phenyl mercuric acetate and mixed antibiotics of a Gram-negative population isolated from an aerobic fixed-bed reactor. *J. Appl. Microbiol.*, **71**, 439.
83. Fredrickson, A.G. and Stephanopoulos, G. (1981) Microbial competition. *Science*, **213**, 972–979.
84. Neilands, J.B. (1981) Microbial iron compounds. *Ann. Rev. Biochem.*, **50**, 715.
85. Cooke, P.K., Gandhi, U.R., Lashen, E.S. and Leasure, E.L. (1991) Preservative evaluation: designing an improved test system. *J. Coat. Technol.*, **63**, 33–39.
86. Levy, E. (1987) Insights into microbial adaptation to cosmetic and pharmaceutical products. *Cosmet. Toiletries*, **102**, 69–74.
87. Favet, J., Fehr, A., Griffiths, W., Amacker, P.A. and Schorer, E. (1987) Adaptation of *Escherichia coli*, *Pseudomonas aeruginosa*, and *Staphylococcus aureus* to Kathan CG and Germal II in an O/W cream. *Cosmet. Toiletries*, **102**, 75–85.
88. Orth, D.S. and Lutes, C.M. (1985) Adaptation of bacteria to cosmetic preservatives. *Cosmet. Toiletries*, **100**, 57–59.
89. Johnson, P.A. and Quayle, J.R. (1964) Microbial growth on C_1 compounds. 6. Oxidation of methanol, formaldehyde, and formate by methanol-grown *Pseudomonas* AM1. *Biochem. J.*, **92**, 281.
90. Nash, T. and Hirch, A. (1954) The revival of formaldehyde-treated bacteria. *J. Appl. Chem.*, **4**, 458–463.
91. Gililand, D., Li Wan Po, A. and Scott, E. (1991) The bactericidal activity of a methyl and propyl paraben combinations: isothermal and non-isothermal studies. *J. Appl. Bacteriol.*, **72**, 252.
92. Orth, D.S. (1993) Microbial injury and the phoenix phenomenon. In *Handbook of Cosmetic Microbiology*, Marcel Dekker, New York, pp. 119–46.
93. Beerens, H., Romond, C. and Lemarie, D. (1976) Techniques for the enumeration of microorganisms in cosmetic products, *Soaps Perfum. Cosmet.*, **49**, 262–263.
94. El-Zanfaly, H.T. and Sabbour, M.M. (1984). Evaluation of pour and surface plate methods for *Staphylococcus aureus* enumeration. *Arch. Lebensm. Hygiene*, **34**, 133.
95. Singer, S. (1987) The use of preservative neutralizers in diluents and plating media. *Cosmet. Toiletries*, **102**, 55–60.
96. Sutton, S.V.W., Wrzosek, T. and Proud, D.W. (1991) Neutralization efficacy of Dey-Engley medium in testing of contact lens disinfecting solutions. *J. Appl. Bacteriol.*, **70**, 351–354.
97. Proud, D.W. and Sutton, S.V.W. (1992) Development of a universal diluting fluid for membrane filtration sterility testing. *Appl. Environ. Microbiol.*, **58**, 1035–1038.
98. Sutton, S.V.W. (1992) The evaluation of biocide neutralization in the determination of microbiocidal efficacy of contact lens solutions. *Internat. Contact Lens Clinics*, **19**, 167–173.
99. Russell, A.D. (1981) Neutralization procedures in the evaluation of bactericidal activity. In *Disinfectants: Their Use and Evaluation of Effectiveness*, Collins, C.H., Allwood, M.C., Bloomfield, S.F. and Fox, A. (eds), Academic Press, London, pp. 45–59.
100. Dey, B.P. and Engley, E.B. Jr. (1983) Methodology for recovery of chemically treated *Staphylococcus aureus* with neutralizing medium. *Appl. Environ. Microbiol.*, **45**, 1533–1537.
101. Rachui, S.R., Sutton, S.V.W. and Brannan, D.K. (1992) Evaluation of disinfectant neutralizers show that not all are effective. *Abstr. Amer. Soc. Microbiol. Session* Q 351, p. 394.
102. Brannan, D.K. and Rachui, S.R. (1995) Specific neutralizers required for each microorganism-biocide permutation. In preparation.
103. Engley, F.B. and Dey, B.P. (1970) A universal neutralizing medium for antimicrobial chemicals. In *Proceedings of the 56th Mid-Year Meeting of the Chemical Specialties Manufacturers Association*, Chemical Manufacturers Association, New York, pp. 100–106.
104. Russell, A.D. (1991) Injured bacteria: occurrence and possible significance. *Lett. Appl. Microbiol.*, **12**, 1.

105. Van Loosdrecht, M.C.M., Lyklema, J., Norde, W. and Zehnder, A.J.B. (1990) Influence of interfaces on microbial activity. *Microbiol. Rev.*, **54**, 75–87.
106. Delaquis, P.J., Caldwell, D.E., Lawrence, J.R. and McCurdy, A.R. (1989) Detachment of *Pseudomonas fluorescens* from biofilms on glass surfaces in response to nutrient stress. *Micro. Ecol.*, **18**, 1999.
107. Curry, J. (1987) Thoughts on preservation testing of water-based products. *Cosmet. Toiletries*, **102**, 93–95.
108. Orth, D.S. (1979) Linear regression method for rapid determination of cosmetic preservative efficacy. *J. Soc. Cosmet. Chem.*, **30**, 321–332.
109. Sutton, S.V.W., Franco, R.J., Mowrey-McKee, M.F., Busschaert, S.C., Hamberger, J. and Proud, D.W. (1991) *D*-value determinations are an inappropriate measure of disinfecting activity of common contact lens disinfecting solutions. *Appl. Environ. Microbiol.*, **57**, 2021–26.
110. Barnes, M. and Denton, G.W. (1969) Capacity tests for the evaluation of preservatives in formulations. *Soap, Perfum. and Cosmet.*, **42**, 729–733.
111. Eklund, T. (1980) Inhibition of growth and uptake processes in bacteria by some chemical food preservatives. *J. Appl. Bacteriol.*, **48**, 423–432.
112. Eklund, T. (1985). The effect of sorbic acid and esters of *p*-hydroxybenzoic acid on the protonmotive force in *Escherichia coli* membrane vesicles. *J. Gen. Microbiol.*, **131**, 73–76.
113. Freese, E., Sheu, C.W. and Galliers, E. (1973) Function of lipophilic acids as antimicrobial food additives. *Nature*, **24**, 321–325.
114. Martoadiprawito, W., and Whitaker, J.R. (1963) Potassium sorbate inhibition of yeast alcohol dehydrogenase. *Biochim. Biophys. Acta*, **77**, 536–544.
115. Troller, J.A. (1965) Catalase inhibition as a possible mechanism of the fungistatic action of sorbic acid. *Can. J. Microbiol.*, **11**, 611–617.
116. Prindle, P.F. (1983). Phenolic compounds. In *Disinfection, Sterilization, and Preservation*, 3rd edn, Block, S.S. (ed.), Lea and Febiger, Philadelphia, pp. 197–224.
117. Hugo, W.B. (1978) Membrane-active antimicrobial compounds – a reappraisal of their mode of action in light of the chemi-osmotic theory. *Internat. J. Pharm.*, **1**, 127–131.
118. Hugo, W.B. (1983) Mode of action of non-antibiotic antibacterial agents. In *Pharmaceutical Microbiology*, 3rd edn, Hugo, W.B. and Russell, A.D. (eds). Blackwell Scientific, London, pp. 258–64.
119. Fogg, A.H. and Lodge, R.M. (1945) The mode of antibacterial action of phenols in relation to drug-fastness. *Trans. Faraday Soc.*, **41**, 359–65.
120. Richardson, E.M. and Reid, E.E. (1940) Alpha, gamma-di-*p*-hydroxyphenyl-alkanes. *J. Am. Chem. Soc.*, **62**, 413–415.
121. Silver, S. and Wendt, L. (1976) Mechanism of action of phenethyl alcohol: breakdown of the cellular permeability barrier. *J. Bacteriol.*, **93**, 560.
122. Gilbert, P., Beveridge, E.G. and Crone, B.P. (1977) The lethal action of 2-phenoxyethanol and its analogues upon *Escherichia coli* NCTC 5933. *Microbios*, **19**, 125–141.
123. Law, A.B., Moss, J.N. and Lashen, E.S. (1984) Kathon CG®, a new single-component, broad spectrum preservative system for cosmetics and toiletries. In *Cosmetic and Drug Preservation; Principles and Practice*, Kabara, J.J. (ed.), Marcel Dekker, New York, pp. 129–141.
124. Collier, P.J., Ramsey, A., Austin, P. and Gilbert, P. (1990) Growth inhibitory and biocidal activity of some isothiazolone biocides. *J. Appl. Bacteriol.*, **69**, 569–577.
125. Fuller, S.J. (1985) The mode of action of 1,2-benzisothiazolin 3 one on *Staphylococcus aureus*. *Lett. Appl. Microbiol.*, **1**, 13.
126. Collier, P.J., Ramsey, A., Waigh, R.D., Douglas, K.T., Austin, P. and Gilbert P. (1990). Chemical reactivity of some isothiazolone biocides. *J. Appl. Bacteriol.*, **69**, 578.
127. Grossman, L., Levine, S.S. and Allison, W.S. (1961) The reaction of formaldehyde with nucleotides and T2 bacteriophage DNA. *J. Molec. Biol.*, **3**, 41–60.
128. Shepherd, J.A., Buerby, M.R., Pemberton, D., Gilbert, P. and Waigh, R.D. (1987) The reactions of Bronopol® (2-bromo-2-nitropropan-1,3-diol) with thiols. *J. Pharm. Pharmacol.*, **39**, 116P Supplement.
129. Zsolnai, T. (1961) Versuche zur entdeckung neuer fungistatika II. Nitroverbindungen. *Biochem. Pharmacol.*, **5**, 287–304.

130. Croshaw, B., Groves, M.J. and Lessel. (1964) Some properties of Bronopol®, a new antimicrobial agent active against *Pseudomonas aeruginosa*. *J. Pharm. Pharmacol. Suppl.*, **16**, 127T–130T.
131. Croshaw, B. and Holland, V.R. (1984) Chemical preservatives – use of Bronopol® as a cosmetic preservative. In *Cosmetic and Drug Preservation; Principles and Practice*, Kabara, J.J. (ed.), Marcel Dekker, New York, pp. 31–62.
132. Ghannoum, M., Thompson, M., Bowman, W. and Al-Khalil, S. (1986) Mode of action of the antimicrobial compound 5-bromo-5-nitro-1,3-dioxane (Bronidox®), *Folia Microbiol.*, **31**, 19.
133. Stretton, R.J. and Manson, T.W. (1973) Some aspects of the mode of action of the antibacterial compound Bronopol® (2-bromo-2-nitropropan-1,3-diol). *J. Appl. Bacteriol.*, **36**, 61–76.
134. Naito, R., Itoh, T., Hasegawa, H., Arimura, H., Fujita, Y., Hasegawa, K., Inaba, T., Kagitani, Y., Komeda, S., Matsumoto, T., Okamoto, H., Okano, K., Oguro, Y. and Ogushi, T. (1974) Bronopol® as a substitute for thimersal. *Develop. Biol. Stand.*, **24**, 39–48.
135. Wallhäusser, K.H. (1984) Antimicrobial preservatives used by the cosmetic industry. In *Cosmetic and Drug Preservation; Principles and Practice*, Kabara, J.J. (ed.), Marcel Dekker, New York, pp. 605–745.
136. One trade magazine that routinely publishes what they call a 'Cosmetic Preservative Encyclopedia' is *Cosmetics and Toiletries* published by Allured Publishing Corp. IL. The last of these was published in their October 1993 issue.
137. Brannan, D.K. (1993) Cosmetic Microbiology. In *Encyclopedia of Microbiology*, Lederberg, J. (ed.), Academic Press, San Diego, pp. 593–603.

8 Preservation of paint formulations

R.D. VORE and J.A. PIERCE

8.1 Introduction

Cost effective use of modern biocides in paint formulations requires an understanding of the susceptibility of the formulation, consequences of degradation, laboratory optimization and proper plant hygiene. Poor hygiene, the use of spoiled raw materials and ineffective preservative selection can lead to intermittent spoilage and produce long term contamination problems. Understanding the relationship between these factors is essential. This chapter is intended to provide the formulation chemist or microbiologist with the background necessary to produce high quality, cost effective product in light of the possibility of microbial contamination.

8.2 The chemistry and manufacture of paint

The terminology used to describe paint formulations is quite varied depending on the company, nationality and industry involved. Powder finishes are dried mixtures of pigment and binder that are electrodeposited on the substrate. Bake-on, UV curables, epoxy systems (both one and two component) and industrial finishes are normally solvent-borne systems. With reductions in solvents being mandated under environmental pressures most major manufacturers are developing water-borne systems. Solvent-borne alkyd paints are rapidly being replaced by water-borne latex systems. Durability and application characteristics of the water-borne systems are beginning to match critical characteristics of solvent-borne alkyds in many areas. The latex area itself is divided between exterior and interior use.

Dispersants, wetting agents and coalescing agents are functional surfactants in paints. Paint, reduced to its simplest form, consists of pigment, a binder and a solvent. Pigments are any material that imparts a final color. Pigments may be either mineral or organic. Mineral pigments include various iron oxides, calcium carbonate and titanium dioxide. The binders or film formers cross link after application to form a solid surface trapping the pigments in place. In latex paint these are usually acrylic and vinyl acrylics though polyvinyl acetates are also used. Solvents are designed to aid in application and delivery of the system to the intended target. Prior to the introduction of water-borne systems hydrocarbon solvents were used in paint manufacturing. These solvents also acted as preservatives. In-can spoilage of solvent-based paints is quite rare. This chapter will be limited to spoilage of water-borne latex paints. Water-borne latices require the use of a co-solvent and contain up to 38% ethylene and propylene glycol. The exception to this is low volatile organic carbon (VOC) paints which contain minimal volatile organic solvents. The binder and solvent together constitute the vehicle for the pigment. Numerous modifications to the basic formulations are possible. White washes are formulations consisting of pigment, calcium carbonate and water. Varnishes are binders and solvent without pigment. Solid materials added to the formulation that do not impart color are called extenders or thickeners.[1] Cellulosic extenders are inexpensive materials and are used in place of expensive pigments. They also provide high viscosity to hold ingredients in a homogeneous blend and provide rheological properties important during application. Clays and associative thickeners are also used.[2]

Paint manufacturing consists of two steps: dispersion or grind and let down. During the manufacturing of mineral pigments, large blocks of material are reduced in physical size to produce uniform small particles. In shipping and storage the primary particles absorb moisture and air on to their surfaces and then clump together to form agglomerates or secondary particles and aggregates or tertiary particles. Agglomerates are primary particles in edge-to-edge contact. Aggregates are primary particles in face-to-face contact. Agglomerates are stabilized by weak electrostatic charges whereas aggregates are stabilized by strong charges. In the dispersion step of paint manufacture the agglomerates are disrupted to produce a homogeneous slurry of primary pigment particles. Aggregates are difficult to disperse without imparting large kinetic energies that may destroy the crystal structure of the primary particles. The dispersion process relies both on chemical dispersants and high shear forces. Dispersants are added to electrically charge the pigment particles and thus allow uniform distribution.[3] Tamol 731 is commonly used. Tamol 731 is a sodium salt of a polymeric carboxylic acid. Wetting agents are used to allow uniform contact of the aqueous phase over the surface of the solid particles. Commonly used wetting agents are Igepals from Rhone Poulenc and Tritons from Rohm and Haas.[4] Igepals are nonionic surfactants of nonylphenoxypoly(ethyloxy)ethanol. Tritons are polyoxyethylene ethers. Dispersants and wetting agents reduce the kinetic energy input required to produce a uniform product. Once the pigment and mineral fillers have been incorporated the binders and extenders can be added. These are added at low speed and low shear. Additional dispersants are added at this time.

Paints must be shelf stable emulsions and dispersions. Excess settling will alter the application rheology and coverage of the product. Viscosity stability during storage is thus of vital importance. Rheology is maintained by incorporating thixotropic agents and the correct selection of dispersants. Paints rely on more than one pigment component. Organic pigments have very large hydrophobic surface areas compared to the smaller hydrophilic surfaces of mineral pigments. The stability of the emulsion requires that all of the particles of varying surface tensions and charges be maintained in dispersion at relatively constant concentrations, otherwise selected particles may either flocculate or settle and the performance of the paint will be diminished. Combinations of dispersants and wetting agents are required to achieve adequate stability.

Properties of the paint formulation during application are critical to user acceptance. The binder resins must blend smoothly with one another during drying; this is referred to as coalescing. Coalescing agents are added during the let down. During the initial drying process water evaporation leaves large voids in the film. The coalescing agent allows the film to maintain fluidity and thus form a uniform film of the binder and pigment phases. Coalescing agents are high boiling point solvents that are miscible with the liquid phase. The coalescing agents thus act as surfactants by coupling the binder and liquid phase while wet and the binder and pigment during the drying period. Texanol and butyl carbitol are commonly used. Butyl carbitol is diethylene glycol monobutyl ether.

Reduced solvent latex paints are being introduced by a number of major companies in the USA and Europe. These include both architectural paints and water-borne latices to replace solvent-borne marine maintenance coatings. In the United States VOCs are measured by the imprecise US Environmental Protection Agency Method 24 where solvent contents are inferred by drying loss rather than quantitatively measured. This rather poor method has caused the definition of Zero VOC to be questioned scientifically and hotly debated among raw material suppliers and paint producers. Simultaneously, both the total volatile content and the types of volatiles released are being scrutinized for additional legislation.[5] Present labeling requirements in the United States reveal little useful information on the actual solvent content of paints.[6] It is generally agreed that solvent levels will be reduced in future latex paints. How this trend will affect their microbial susceptibility remains to be seen.

8.3 Consequences of microbial spoilage

The preservation and spoilage of paint concerns two quite different physical states of the material. The wet state is an aqueous state prior to use and desiccation associated with final application. The spoilage, or more properly the defacement, of the dried paint is the result of surface growth on the applied and cured film itself. The organisms involved and the substrates utilized in these two different types of contamination are considerably different. In the wet state water is the dominant solvent with lesser amounts of surfactants and binders also being present. In the wet state spoilage is normally a result of bacterial growth with yeast as an occasional factor. The microorganisms utilize either the surfactants or the cellulosic thickeners during spoilage of the wet state. The defacement of dried paint films is apparently the result of a complex community of microorganisms including bacteria, fungi and algae. This succession of organisms has only recently begun to be investigated.[7]

Exterior paints are subjected to continuous UV damage, changes in temperature and humidity and nearly every microorganism that can possibly survive in the soil. Interior paints are applied, used and survive in the much more hospitable environment of the indoor human habitat. As such, the extremes of environmental conditions are lacking. Formulations of components utilized to survive in these two environments are quite different. These changes are significant in that they affect not only the performance of the dried film but also the resistance of wet paint to microbial degradation. The defacement of exterior paints is the result of either fungal or algal growth. Therefore, exterior latex paints incorporate a fungicide. The fungicide is designed to leach slowly from the film and inhibit fungal growth. Commonly used fungicides are presented in Table 8.1.

Algal growth is a recognized problem in masonry paints. Both algae and fungal growth may appear as black stains. The actual amount of algal defacement on house paint in unknown but it is now under extensive investigation.[8] The defacement of dried films does not involve the degradation of surfactants since these have been volatilized during the process. Defacement processes are beyond the scope of this work and are not addressed further. The investigations and studies of fungal defacement

Table 8.1 Commercial paint film fungicides used in the USA

Chemical name of active agent	CAS number of active agent	Trade name	Producer
Diiodomethyl-*p*-tolylsulfone	20018–09–1	Amical®	Angus
3-iodo-2-propynyl butyl carbamate	55406–53–6	Troysan® Polyphase	Troy Chemical, Inc.
Tetrachloroisophthalonitrile	1897–45–6	Nuocide® 960	Hüls
2-*n*-octyl-4-isothiazoline-3-one	26530–20–1	Skane®	Rohm and Haas

and control are numerous. Several articles are cited for further reference.[9–13]

To prevent UV damage to latex film many exterior formulations use zinc oxide. The presence of a fungicide with or without zinc oxide in exterior formulations makes these paints naturally resistant to most bacterial attack in the wet state. Neither of these are true antibacterial agents in the wet state but they do raise the bioresistance level of the paint. Since it is bacteria that are responsible for wet state spoilage the exterior paints of most US companies are able to survive with approximately half of the preservative required in interior paints. This rule holds true independent of the preservative type utilized. Spoilage patterns of interior paints are also subject to formulation vagaries. Gloss, semi-gloss and egg shell formulations contain more binder than pigments. Formulations where the binder content exceeds that of the pigment binder are called under critical. Flat formulations contain more pigment than binder. Where the pigment exceeds the binder content the paints are referred to as overcritical paints. Quality factors, component types and variation of the ratio of components allow the production of first, second and third lines of paints. This characterization refers not only to cost but also to the tangible factors such as scrubability. The demands on the surfactant in these two conditions are different. Flat formulations are cheaper to produce since they are based on more inexpensive components. Not surprisingly the microbial susceptibility of gloss and egg shell paints is quite different from that of flat paints. This appears to be the rule whether acrylic or PVA-based paints are examined.

Spoilage of paint in the wet state is noticeable to the untrained observer as an off odor, off color, bulged container, broken emulsion and loss of viscosity. The symptoms observed depend on the type of species of microorganism involved. Bacteria are responsible for the vast majority of in-can spoilage. In measuring contamination levels the standard unit of measure is colony forming units per gram (cfu g^{-1}). In most cases a single colony of bacteria arises from one bacterial cell. Some bacteria clump or form chains, such as *Staphylococcus* and *Bacillus*, respectively. In these cases bacterial levels may not be equivalent to colony forming units. For other than philosophical reasons the distinction between the two terms is meaningless. Paint is normally sterile. Vegetative growth in the can is indicative of some spoilage. Bacterial levels below 10^5 cfu ml^{-1} will have little long term impact on the paint. As levels rise above 10^5 the rate of deterioration becomes a relationship between level and time. At populations of 10^6 it may take weeks or months before any detectable change is noticed. Off odors are noticed as the spoilage proceeds and bacterial levels rise past 10^6. With populations of 10^8 the complete and irreversible destruction of the product may occur in several days. Growth rates and associated spoilage rates are also temperature dependent. In pragmatic terms this results in three or four times faster spoilage rates during hot summer months than the cold winter months. Once the emulsion has been broken the paint is no longer salvageable. Interventions with appropriate antimicrobials prior to breaking the emulsion are possible and used commonly by manufacturers. The production of organic acids may occur in both bacterial and yeast spoilage. If sufficient acid production occurs the pH of the paint will be compromised. If the formulation has a pH-sensitive rheology it may be necessary to make corrections after biocidal treatment. A comprehensive review of biodeterioration of paint is mentioned for further reference.[14]

Can bulging and copious out-gassing are normally associated with yeast spoilage due to the genus *Candida. Candida* spp. are able to degrade a number of hydrocarbon compounds under the anaerobic conditions that exist inside of a sealed paint can. The out-gassing observed with *Candida* spoilage is a result of the liberation of carbon dioxide during the fermentation process. The gas pressure developed in the sealed cans may be several atmospheres and the resulting explosion at can failure is dramatic. The controlling factor in pressure buildup is the integrity of the container, not biological since few metabolic systems are inhibited by even hyperbaric pressures. *Candida* induced spoilage is associated with a sweet, apple-like aroma of rising bread.

The loss of viscosity in-can is a substantially different spoilage problem. Viscosity loss is almost always the result of degradation of the cellulosic extenders. Cellulose is a biologically intractable polymer of glucose linked through *B*1,4 bonds. Few organisms are able to degrade this particular linkage, but among those that can are the soil fungi *Trichoderma* and members of bacterial genera *Bacillus*.[15] The actual degradation of cellulose is the result of secretion of the extracellular enzyme cellulase. Cellulases are stable enzymes that are able to depolymerize considerable quantities of cellulose with no biological cofactors or viable cells present. Spoilage due to cellulose degradation is rare in US paints but stories of this problem have been elevated to mythical levels. Industry lore is abundant with stories of viscosity loss during the late 1970s through the mid 1980s. Most of these stories are not substantiated by detailed microbiological reports. Lack of documentation has contributed to the lore. Furthermore, non-microbiologists are often baffled by cellulase degradation problems since viable cells need not be present.

Cellulase spoilage incidents appear to occur in similar patterns. Paints are produced and held without an adequate in-can preservative, allowing the development of spoilage microorganisms. As the spoilage is noticed the bulk storage tanks are treated with a biocide to kill all viable cells and thus recover the lot. This approach is not appropriate for cellulose degradation since enzyme degradation is the culprit and biocides are designed to kill viable cells. Biocides may interfere with specific enzymes but these are endogenous enzyme systems vital to cellular function not *exo*-enzymes liberated from the cells which are of little consequence to the continued viability of the cell. The inappropriate use of biocides to recover cellulase-damaged paints is problematic. After 'recovery' the paint technologist may believe that spoilage incident is under control since standard practice is to measure spoilage by the number of viable microorganisms present. But in this case recovered batches will still contain active cellulase but no viable bacteria. The recovered material is then either packaged or blended off with new paints. In the latter cases the dilution approach allows one lot of cellulase-containing paint to comingle with a large number of otherwise good lots. In either system the cellulase continues to function and the viscosity of the paint approaches that of the solvent, water. Cellulase-degraded paints are seldom recoverable for this reason. In typical incidents of this type, production plant personnel try repeated conventional approaches to kill the offending bacteria over a period of several months. It is the hours of fruitless efforts and usually total loss of all material produced during these incidents that allow the lore of cellulase.

Effective control of cellulase spoilage can only be regained by complete cleaning of the entire production facility or by switching to sterically hindered thickeners. The potential for viscosity loss remains so significant in manufacturers' minds than any possible indication results in a rapid and significant knee-jerk reaction. Present manufacturing facilities and formulations now in commercial production are not particularly susceptible to cellulose degradation. Concerns about cellulose degradation are misplaced since preservatives used for in-can protection also provide activity against both surfactant and cellulose-degrading species. As long as adequate preservative levels and types are maintained in cellulose slurries incidents of viscosity loss are unlikely to occur. Since this pattern of spoilage is dependent on cellulose degradation the surfactant systems are not involved. Further discussion on cellulose degradation and viscosity loss is beyond the scope of this chapter.[16]

The commercial implications of paint spoilage are of great interest to manufacturers. Both professional and do-it-yourself users demand that paint be of acceptable quality in the price range they are using. Off-specification paint is returned to the point of purchase and replaced by the sales agent. Profit margins in paint manufacture can only be maintained by steady and high plant outputs. Spoilage incidents require both administrative and production personnel to modify their work patterns until the situation is resolved. Contaminated materials must either be treated with additional antimicrobial agent or disposed of. The material must be held until then. If the material is recovered it is usually held in bulk tanks. Thus, even small spoilage incidents result in reduced plant production, reduced profit-oriented labor and increased biocide use. Under present environmental legislation the disposal of non-recoverable paint is very expensive.

In major incidents companies may institute a product recall. Recalls are the nightmare of both production and marketing departments. Personnel, shipping both the bad and replacement paint, lost production and disposal costs are enormous in recalls. The loss of prestige associated with spoilage and recalls is viewed as devastating to market managers. Loss of permanent sales to established customers due to brand switching is likely if the situation is poorly handled by the paint manufacturer. For these reasons most paint companies are risk aversive where there is a possibility of spoilage.

8.4 Requirements for microbial growth

The general requirements to support microbial growth include utilizable nutrients, favorable environmental conditions including temperature, pH and a reliable water supply. Oxygen is not required by most bacteria, yeast and some fungi. Nutritional requirements for microorganisms range from simple inorganic compounds to quite complex biopolymers. With very few exceptions, microorganisms isolated from industrial products have simple nutritional requirements and do not have obligate requirements for growth factors such as vitamins. Yet these species are able to attack many surfactants and some complex polymers such as cellulose and gums. Nutrients include both carbon in appropriate structural form and all other elements. Nitrogen in the correct form is second in importance to carbon. Members of the genus *Pseudomonas* are frequent causes of industrial spoilage. This genus is able to utilize over 2000 separate carbon compounds as primary sources of energy and carbon skeletons. Appropriate temperatures for bacterial growth range from 5°C to over 50°C. The pH range of most microorganisms is about 3.5 to over 9.5. Water is critical for growth of both bacteria and fungi. Products with very high solids and very low water activity ($a_w < 0.9$) experience few bacterial problems but remain quite susceptible to fungal attack. Spackling and pre-mixed wall board compounds are two examples. Bacteria typically reproduce by doubling once every hour in industrial products. An unpreserved storage tank or paint allowed to sit overnight could have over 10^8 bacteria per ml by the next morning using a very small inoculation from previous batches or the water. Aqueous-based products contain a variety of organic compounds, are formulated in the 5.0 to 9.0 pH range, are produced, stored and used in temperatures comfortable to life processes and by their nature supply water. With this perspective, the necessity for incorporation of preservatives into paint formulations should not be surprising.

8.5 Typical flora associated with paint spoilage

Microorganisms are ubiquitous in the environment. Several thousand bacterial genera have been identified and millions of strains potentially exist. However only a handful of types of bacteria and fungi are commonly isolated throughout the world in paint formulations. Microorganisms associated with paint spoilage in practice do not present a significant health hazard to exposed personnel. The species involved are weakly pathogenic at best and lack the general infectivity of opportunistic pathogens.

8.5.1 Bacteria

Pseudomonas spp. account for about 75% of the isolates from paint. The most common species is *Ps. aeruginosa* with *Ps. cepacia* and *Ps. fluorescens* also isolated. *Ps. putida*, *Ps. putrifaciens* and *Ps. vesiculoans* are rare. *Alcaligenes* spp. are the second most common group isolated. *E. coli* and *Proteus* spp. are also occasionally isolated. Bacteria that account for 5% or less of spoilage include *Citrobacter freundii*, *Morganella morganii*, *Bacillus* spp. and *Corynebacterium* spp. Yeasts are isolated from less than 5% of spoiled paints. *Candida albicans* is the most frequent yeast isolated. *Rhodotorula* spp. and *Torulopsis* spp. are occasionally isolated.[17]

8.5.1.1 Pseudomonads. Pseudomonads produce a wide variety of enzymes and thus have the ability to metabolize many different compounds as sources of carbon and energy. Compounds such as aromatics, amines and organics in the form of thickeners, surfactants, pH regulators, antifoams and dispersants are prime candidates in a paint formulation. This includes organic acids, sugars and polymers made from those sugars, proteins, alcohols and surfactants and dispersants. Pseudomonads are found in soil, water and every wet material including most water. Swabs of facilities after rinsing with potable water should produce pseudomonads. These are usually introduced from the potable water supply. Nearly every moist environment will harbor some members of this genus. It should not be surprising that this hardy group is responsible for most industrial spoilage.

8.5.1.2 Bacillus. Members of the genus *Bacillus* are Gram-positive spore-forming bacteria mainly associated with soil. They are occasional primary spoilers of wet paint. They are readily controlled by all in-can preservatives now in use. Misunderstandings among paint chemists exist concerning the spoilage threat from *Bacillus* spores. Spores, regardless of their origin, are dormant and lack any detectable metabolic activity. Spores do not in themselves lead to spoilage. It is only after the spores germinate and lead to vegetative cells that spoilage is possible. Once germination has begun the in-can preservative is able to kill the resulting vegetative cell. Confusion occurs during microbiological examinations of paints containing spores. As stated above if the paint is preserved adequately outgrowth is not possible. During standard testing protocols the paint is diluted by the addition of water and nutrients in the biological growth medium employed. The resulting dilution decreases the in-can preservative level to below that required to kill germinating spores. Thus inadequately trained laboratory personnel may assume that spoilage due to spores has occurred or is imminent. In reality the paint is quite stable and under no threat of spoilage. Spoilage due to spores will only occur where inadequate or unstable in-can preservatives are employed. Due to both the possibility of confusion and lost laboratory time investigating spore occurrences, spores should be eliminated as soon as possible. This is easier said than done. Spores are able to survive several minutes exposure to boiling water, repeated freeze thaw cycles and attack from most chemicals including in-can preservatives. Elimination programs must include a thorough cleaning of contaminated equipment and known sporicidal compounds such as hypochlorous acid (bleach) and glutaraldehyde.

8.5.2 Yeasts

8.5.2.1 Candida. The genus *Candida* contains both industrially significant and pathogenic species. Most human infections caused by *Candida* spp. are caused by personal microbial flora.[18] *Candida* spp. are notable for their ability to transform hydrocarbon compounds under the anaerobic conditions that exist inside of a sealed paint can.[19] Can bulging and copious outgassing are normally associated with yeast of the genus *Candida*. The outgassing is a result of the liberation of carbon dioxide during the fermentation process. This is the same metabolic pathway employed by *Saccharomyces cereviseae* in the production of bread, beer and wine. The sweet, apple-like aroma from *Candida* growth is due to the accumulation of acetaldehyde during fermentation. This is the same aroma usually associated with rising bread and green beer. This aroma can be used as a presumptive indicator of yeast spoilage since no industrial spoilage bacteria will accumulate acetaldehyde in such large amounts. Yeasts are more fragile than bacteria in that they have a narrower and lower optimal pH and temperature growth range.

8.5.3 Fungi

Fungi are not usually a primary cause of in-can spoilage of paint and are almost never isolated from spoiled paint. The pH and lack of oxygen in a sealed can are formidable barriers to fungi. This is in spite of the fact that fungi are ubiquitous soil inhabitants. Examination of mineral pigments, both dry and slurries, and cellulosic fillers may reveal the presence of fungal spores.[20] In nature fungi must compete against the rapid growth of bacteria and yeast. Their competitive advantage lies in the diversity of complex organic substrates they are able to utilize. In the ecological succession of microorganisms bacteria and yeasts are primary colonizers. They deplete the relatively simple organic substrates such as sugars first. Proteins and simple polymers are degraded next. Complex polymers with multiple branch points such as cellulose and lignin are beyond the metabolic range of most bacteria and yeasts. After the initial colonization of the niche only the complex substrates are left intact. Without sufficient substrate the primary colonizers decrease in both number and diversity. This allows secondary colonizers, including the fungi, the opportunity to degrade the remaining substrates. This explains why fungi are not isolated until after considerable degradation of the system. In reviewing the records of approximately 4500 evaluations of paint in the Zeneca Biocides Laboratories we have only isolated fungi from two paint lines, both of which were from Latin America.

8.6 The problem with solvent-free systems: spoilage

Prior to the introduction of water-borne systems hydrocarbon solvents were very extensively used in paint manufacturing. These solvents acted as delivery agents and inadvertently as preservatives. The functional membranes of living cells are bilayer micelles of C_{10} to C_{18} fatty acids bound to glycerol backbones. The solvents disrupted the cellular membranes of all the microorganisms introduced during manufacturing or application. Preservatives were only required to prevent attack on applied and dried product.

The reduction and elimination of solvents exposes the paint formulation chemist to the problem of wet state microbial spoilage. Both raw materials and water itself are sources of microorganisms. Good quality drinking water may contain up to 10^3 non-pathogenic bacteria per ml.[21] Improperly stored rinse water may contain over 10^7 cfu ml^{-1}.[22] Air blowing through a manufacturing plant may have several hundred fungal spores per cubic meter. Raw materials may have 1 to 10 million bacteria per ml. In all of these cases raw materials would be viewed as good and appropriate for use in production if microbiological monitoring did not occur. Paint chemists and production personnel with years of experience in solvent-borne production report that the transition to water-borne systems is difficult if inadequate microbiological precautions are in place prior to production.

8.7 The chemistry of industrial preservatives

8.7.1 Biocide terminology

Two different functional types of biocides are used in paints today. Fungicides are designed to inhibit the surface defacement of dried films by fungi. Fungicides are also called mildewcides. Bacteriocides are designed to preserve the wet paint until application. Bacteriocides are designed to function in the can, not in the dried film. The terms in-can preservative, preservative, biocide, bicide, microbiocide and microbicide are synonymous with bacteriocide.

8.7.2 Mercurials

In the last several decades the paint industry has had experience of a wide range of products and compounds for antimicrobial preservation. Mercury was first introduced in the USA in organometallic form as phenylmercuric acetate (PMA) in the late 1930s[23] and was used extensively as both an in-can antibacterial and film fungicide for water-based paints. The biological activity of mercury is non-specific and is highly effective against both bacteria and fungi. Many paint producers utilized the broad spectrum of mercury as both an in-can aqueous and dry film single biocide.[24]

The current regulatory status of mercury depends on the environmental control agency. Following much review and many hearings with respect to the safety of using mercurials, the United States Environmental Protection Agency (USEPA) determined that use of these compounds presented an unreasonable health risk. The USEPA cancelled all registrations for use of mercury-based compounds in interior line paints in 1990.[25] Mercurials have also been banned in Japan, the United Kingdom and Germany and use is declining in Spain, yet mercurials are still used in Asia Pacific countries, Latin America and Greece.

8.7.3 Desirable properties of biocides

Factors that are of concern in selecting a biocide are persistence, mammalian toxicology (especially if volatile), cost effectiveness, spectrum of activity, ease of use, thermal stability, and its effects on color rheology and pH. Biocides must provide persistent protection during the manufacturing, storage, shipment and end use. They must not only be able to kill spoilage organisms introduced from raw materials but be persistent enough to kill organisms introduced during several use periods of final application. In the case of consumer paints these use patterns may be separated by weeks or months. This type of use pattern requires that the biocide be stable in formulated paint for at least one year and preferably two. Mammalian toxicology is important not only to the end users but also to production personnel. All biocides will cause adverse immune reactions in humans after repeated uncontrolled exposure. The human toxicology of the biocides in use varies widely (see chapter 12).

8.7.4 Biocides presently available

8.7.4.1 Formaldehyde release preservatives. Due to the extreme safety and regulatory pressure placed on mercurials, many organic synthetic antimicrobials were researched and developed. A number of the commercially available biocides today are formaldehyde condensates which rely on the release of free formaldehyde (CH_2O) from the hydrolysis of a parental nitrogen-based structure. Again, with free formaldehyde as the active mechanism for antimicrobial activity, considerations were raised around the safety of products based on these chemistries. Formaldehyde is now listed as a known human carcinogen and the US Occupational Safety and Health Administration (OSHA) has placed a limit on the concentration of free formaldehyde which is allowable in both the work environment air and in the liquid product in which it is being used. The fate, therefore, for continued and expanded use of formaldehyde-based antimicrobials in industrial products is questionable. Table 8.2 lists commercially available preservatives based on the chemistry.[26]

The exact mechanism of action of biocides in this class is poorly documented. In all cases the compounds can be shown to decompose and release significant amounts of formaldehyde into the solution. The release of free formaldehyde can be demonstrated with both the neat product and in industrial matrices. Disputes as to the precise chemistry of these compounds are ongoing.[27–29]

Table 8.2 Putative formaldehyde release biocides used in paints in the United States

Chemical name of active agent	CAS number of active agent	Trade name	Producer
2((Hydroxymethyl) amino) ethanol	34375–28–5	Troysan® 174	Troy Chemical
2((Hydroxymethyl) amino)-2-methyl propanol	52299–20–4	Troysan® 186 Troysan® 192	Troy Chemical
5-Hydroxymethyl-1-aza-3,7-dioxybicyclo(3,3,0) octane	56709–13–8	Nuosept® 95	Huls
[[(1-Methyl-2-(5-methyl-3-oxazolidinyl)-ethoxy) methoxy]methoxy]methanol	97553–90–7	Nuosept® 145	Huls
1-(3-Chlorallyl)-3,5,7-triza-1-azoniaadamantane	4080–31–3	Dowicil® 75	Dow Chemical

8.7.4.2 Isothiazolinones. Isothiazolinone biocides have been researched and developed to target defined metabolic functions in the bacterial cell. They are effective generally at low use levels (<1000 ppm), have a broad spectrum of antibacterial and antifungal activity and generally have a more acceptable portfolio with respect to toxicology.

Table 8.3 lists isothiazolinone paint preservatives. Proxel GXL is the most commonly utilized paint biocide in the United States today. Kathon LX1.5 is used in limited quantities. Promexal is the latest isothiazolinone to be commercialized. It is presently available in Europe. Commercial availability in the United States is pending EPA approval.

The chemistry and stability of isothiazolinones are quite varied. The 5-chloro-2-methyl-4-isothiazolin-3-one component in Kathon is the most biologically active of the isothiazolinones. It is used in most industrial systems at 25 ppm or less. This is compared to a typical use level of 100 to 150 ppm of 1,2-benzisothiazolin-3-one in Proxel. The higher antimicrobial activity of the Kathon system is also reflected in its higher mammalian toxicity. The stability of the actives in Proxel and Kathon are on the opposite ends of the spectrum. Whereas Proxel is stable in an environment with high amine levels and an alkaline pH, Kathon is stable at acid pHs and in more oxidizing environments. Kathon is rapidly inactivated in the presence of primary and secondary amines. The pH of most US acrylic and vinyl acrylic paints is 8.5 or above. This coupled with amines used as corrosion inhibitors and dispersants produces a very stable environment for Proxel. Research in the Zeneca Biocides laboratory has been able to detect, both chemically and biologically, more than 90% of the Proxel added after one year storage in typical paint. The stability of Kathon in paint is limited and poorly documented in published literature. The 2-methyl-4,5-trimethylene-4-isothiazolin-3-one found in Promexal displays intermediate properties between Kathon and Proxel formulations. Promexal active use levels are near that of Kathon but display chemical stability closer to Proxel. Because of paint manufacturers' desire to

Table 8.3 Isothiazolinone paint preservatives

Chemical name of active agent	CAS number of active agent	Trade name	Producer
1,2-Benzisothiazolin-3-one	2634–33–5	Proxel® GXL Proxel® BD	Zeneca
2-Methyl-4,5-trimethylene-4-isothiazolin-3-one	082633–79–2	Promexal® X50	Zeneca
5-Chloro-2-methyl-4-isothiazolin-3-one 2-Methyl-4-isothiazolin-3-one	26172–55–4 2682–20–4	Kathon® LX1.5 Kathon® LX	Rohm and Haas

maintain long shelf lives of finished product they should conduct stability studies of all actives prior to incorporation into new product lines. This should be in addition to biological challenge tests.

8.7.5 The future of biocides

As toxicological and ecological pressures surrounding the use of biocides increase, producers will be faced with developing 'greener' antimicrobials. These are likely to be lower in volatile organics (VOCs). There will need to be a careful balancing act of locating molecules that are effective at lower concentrations and are more toxicologically sound. This may be accomplished by researching and developing active agents which target more specific biological pathways in microorganisms, thus decreasing risk factors to higher organisms, including humans.

8.8 Methods to establish the correct biocide level for preservation

8.8.1 In-can preservation

Proper preservation of paint formulations requires that the biocide candidate be evaluated for resistance to bacterial insult. This is primarily accomplished by 'optimizing' the usage level using standard microbiological protocols. Evaluations should be carried out using multiple 'challenges'. This method was developed to best simulate varying plant production conditions and insult microbes. Preservatives passing this test will be able to provide the most long term protection of the formulation during plant hygiene breakdown and in field applications. In doing this, the actual paint formulation is acquired and used to serve as a substrate for bacterial growth. It is challenged repeatedly with 'typical' industrial contaminants. A number of protocols have been suggested.[30–33] Described below is the protocol presently employed by the Zeneca Biocides western hemisphere laboratory.

8.8.1.1 Challenge testing. The paint used should be relatively free from established bacterial populations. A mixture of several predominant spoilage bacteria are employed for this type of evaluation. These organisms may be acquired from actual contaminated product or stock strains from a bacterial culture collection center such as the American Type Culture Collection (ATCC). Careful maintenance is essential not only to maintain purity but also the desired biochemical properties of the strains used. Wild isolates that have undergone more than five passes, or about 25 generations, in the laboratory are essentially converted to laboratory strains. If wild isolates are employed they must be maintained as frozen stocks. Improper care of the culture collection will result in erratic results. This potential variability will eliminate the possibility of duplicating results at remote times. For this reason ZENECA has adopted the use of ATCC strains. These strains are periodically replaced with fresh vials from the ATCC. This approach also has the added advantage of reducing the maintenance requirements.

Typical organisms used in challenge tests include:

- *Pseudomonas aeruginosa*
- *Pseudomonas stutzeri*
- *Pseudomonas putida*
- *Pseudomonas cepaci*
- *Pseudomonas aureofaciens*
- *Aeromonas hydrophilia*
- *Enterobacter cloacae*
- *Serratia marcescens*
- *Escherichia coli*
- *Acinetobacter calcoaceticus*

Prior to testing microorganisms used should be grown as stock laboratory cultures. Culturing at varying pH levels on appropriate media, such as Tryptic Soy Agar, assists the development of viable controls in the challenge test. Organisms are then chosen from a stock culture pH which most closely matches the pH of the test sample. This will reduce pH shock on the organism when introduced to the industrial matrix. Lawns of bacteria are harvested from a solid medium using a small amount of sterile dilution water. The resulting bacterial suspension is then used to inoculate separate samples of the unpreserved product. The American Society for Testing and Materials (ASTM) protocol and a number of biocide suppliers inoculate paints directly with these bacterial suspensions. Zeneca has found that this does not always reflect real world spoilage patterns. To simulate more closely the manufacturing plant it is advisable to pre-adapt the bacteria to the paint. This is essentially a pre-challenge step where individual species of bacteria are repeatedly inoculated into the paint until a stable population is established. This is specifically designed to simulate a plant condition where a transitory bacterium has established an adapted population in one section of the plant. This might be in a manifold or dead-end T-fitting. The viability of these inoculations can be monitored after 24 and 48 h by evaluating the degree of persistence of the organisms on agar plates. When the majority of the organisms are found to persist at 10^7 cfu ml^{-1}, the bacterial adaptation process is completed and the various bacteria are pooled to form the combined inoculum.

In the challenge test the paint sample is then divided into aliquots and the desired preservatives are added at various levels. A biocide-free control is included as a control. If growth is not established in the control the results of the test are not valid in themselves. All samples are subsequently challenged at least three times with the adapted bacterial inoculum. The efficacy of the added preservative in controlling contamination is monitored 24, 48 and 72 h following each inoculation. During the interpretation of challenge test results the time between inoculation and measuring viable bacterial counts is very important. Formaldehyde condensates have more rapid kill rates than isothiazolinones. Isothiazolinones are biostatic on contact but may take 72 h for a complete kill at use levels. For consistency of results all challenge test comparisons must recognize this issue. The concentration of preservative affording protection (no viable growth) during the third challenge at 72 h is considered to be the effective dosage.

Results from a typical challenge test are shown in Table 8.4. Samples on day 1 and 2 after challenge are qualitative streak plates. Growth is scored on a 0 to 5 scale. The day 3 samples are standard plate counts of viable bacteria and are expressed as cfu ml^{-1}. Quantitative assessment of the effective treatment level is defined as the lowest level that provides consistent biological control in all challenges. In this example the 750 ppm provides the desired effect.

Interpolating between data points should be avoided in testing of this type. Although the 'lowest' effect level will lie between 500 and 750 ppm its exact end point is imprecisely defined. This coupled with inconsistent dose levels of the biocide during manufacture and varying biological load from raw materials makes interpolation very risky. Efforts to reduce overall biocide use would be better spent in improving plant hygiene and reducing biocide used in recovering spoiled products.

8.9 Implementing the laboratory report in practice

The trend in biocides today is to use more environmentally friendly biocides at the lowest levels practical. Because of the time and effort management spends on biocide decisions plant personnel frequently perceive biocides as exotic and dangerous materials. Indeed, few trained scientists outside of biocide suppliers understand the biochemistry of product spoilage and biocide function. This combination of general ignorance lends itself to misunderstandings. The implementation of adequate biocide usage patterns requires extensive training of personnel including management, purchasing and production. Biocides by their nature are reactive with living systems, including human beings. Changes in biocides must be accompanied by adequate training for all potentially exposed plant personnel. This training should include types of exposure hazard (inhalation or direct contact) and training on personal protection devices and clothing. Procedures for spill clean up and proper disposal of

Table 8.4 Challenge test of laboratory strains against various biocide concentrations to determine biocide level required to prevent microbial growth

Challenge test on latex wall paint											
	1st Challenge Product (cfu g^{-1}) Days post inoculation				2nd Challenge Product (cfu g^{-1}) Days post inoculation				3rd Challenge Product (cfu g^{-1}) Days post inoculation		
Concentration of biocide (ppm)	1	2	3	Concentration of biocide (ppm)	1	2	3	Concentration of biocide (ppm)	1	2	3
0	5	5	3.4×10^8	0	5	5	3.6×10^6	0	5	5	2.2×10^8
250	5	5	3.7×10^6	250	5	5	3.4×10^6	250	5	5	1.9×10^8
500	5	3	<10	500	5	4	1.7×10^2	500	5	5	1.2×10^8
750	3	0	<10	750	1	0	<10	750	2	2	<10
1000	0	0	<10	1000	0	0	<10	1000	0	0	<10
1250	1	0	<10	1250	0	0	<10	1250	0	0	<10
1500	1	0	<10	1500	0	1	<10	1500	0	0	<10
2000	0	0	<10	2000	0	0	<10	2000	0	0	<10

Streak plates are ranked from 0–5 according to the degree of microbial growth. 0 = no growth; 1 = 1–5 colony forming unit (cfu); 2 = 6–15 cfu; 3 = 16–30 cfu; 4 = 31–45 cfu; 5 = >45 cfu. Serial dilution counts are expressed as colony forming units per gram (cfu g^{-1}).

empty containers should be in place prior to use and not asked after incidents and spills have already occurred.

The time period in switching to an unfamiliar biocide is the most likely period when spoilage may occur. It is best to start new biocides at high dose levels and lower the use level with experience. When problems are encountered immediate action should be taken. Dosages used to recover spoiled material should be at least twice that used for normal preservation. Creeping the dose up in periods of sporadic spoilage will only serve to select for resistant strains. Failure to instruct plant personnel on the effective limitation of biocides is a major source of spoilage during initial conversions. Biocides vary in their mode and rate of action. Heavy metals are persistent but kill over a period of several days. Some formaldehyde donors are volatile in open systems but kill within 24 h of contact. It would be unwise to use a formaldehyde donor to preserve an open tank of warm latex for several weeks. Hypochlorous acid (household bleach) has been used for several years by a number of latex and paint companies to recover spoilage bulk tanks. This does provide a rapid and inexpensive recovery but may create more long term problems than it solves. Hypochlorite is reactive with most non-metallic biocides. The hypochlorite used not only kills the microorganism but it also leaves the material with no long term preservation and a hostile chemical environment to other biocides. Additional biocide should be added after the use of bleach but only after allowing sufficient time for all hypochlorite to dissipate. Glutaraldehyde is similarly affected by amines in alkaline conditions.

Biocide rotation programs are not required when adequate levels of biocide are maintained in raw material and finished products. Repetitive underdosing of any biocide may lead to the development of adapted bacterial populations. Eliminating these types of populations frequently requires thorough cleaning and sanitization of the entire plant. The best analogy to this situation is prescription antibiotics for humans. If the prescription is not entirely used the infection may reoccur. Upon the reoccurrence the infection is frequently more severe and more intractable to antibiotic treatment.

8.10 Plant sanitation and hygiene audits

The most critical aspect of preventive maintenance from a microbiological aspect is routine housekeeping coupled with microbiological plant audits. Fundamental practices such as cleaning up spills, keeping lids closed on storage tanks and sanitization complement the overall effectiveness of any preservative in use. The implementation of a sanitization program throughout the plant will reduce the contamination levels and should also reduce the incidence of finished product spoilage. By reducing the bioburden in the raw materials and production equipment the biocide will be more effective in preserving the quality of the finished product. A number of articles have been written on the role of sanitation and hygiene in spoilage control.[34–36] These articles tend to be brief and assume some level of comprehensive understanding of microbial physiology and ecology. Technologists desiring to understand and establish a hygiene program would be rewarded by investigating a number of articles written for the food industry.[37] A full consideration of this topic is given in chapter 3. However, consideration of how it may be specifically applied to the paint industry is outlined below.

8.10.1 Cleaning equipment

The first step in implementing a hygiene plan must be a thorough cleaning. The objective is to remove all of the accumulated material and debris. Hot water with detergents or caustic cleaning solutions are usually effective. Operator experience in cleaning should be considered when selecting the cleaning method. All of the built-up debris must be removed and bare metal or plastic should be exposed prior to sanitization or the effort will be wasted. The organics in the debris serve to deactivate sanitizers, they also serve as breeding sites for bacteria and fungi and restrict the access of sanitizers to the resident flora thus defeating the purpose of sanitization. High pressure hot water washers are ideal for industrial environments. The hot water used in cleaning by itself will not kill bacteria or fungi but may help the cleaning agents injure the cells and allow the sanitizers used later to work more effectively. Water over 160°F (71°C) is desirable.

After all of the debris is removed the equipment is sanitized. A solution of 500 ppm free chlorine in good quality water works well. A direct contact time of 10 min using 75°F (24°C) water will kill all industrially important bacteria and fungi. The 500 ppm level of free chlorine is specified because residual organics will stoichiometrically deactivate chlorine. Glutaraldehyde is also a very effective sanitizer but the smell may lead to worker complaints. The equipment should then be rinsed with drinking quality water. This is especially true with chlorine as chloride ions will cause pitting on both mild and stainless steel. Rinse waters containing residual sanitizer should be evaluated for compatibilty before being used in the manufacture of new products. Residual sanitizers may lead to inactivation of biocides, discoloration of the product or modification of the emulsion stability.

Once the equipment has been treated it should be stored so no further contamination occurs. This is not always easy when beginning a system wide hygiene plan. All tanks should have tight fitting covers to exclude contamination from dust, dripping water, birds, mice and foreign objects like hard hats. Covers should not be made of wood or soft steel. Wood retains moisture and provides large areas not treatable during surface application of the sanitizer. Soft steel will pock mark with time. Both wood and pock marked steel provide large areas for bacterial colonization that are impossible to decontaminate effectively. Water should not be allowed to stand in the plant. This includes soak tanks and troughs. Flexible transfer hoses should be cleaned, sanitized and rinsed frequently. These hoses should never be allowed to lay horizontally full of material or water. Hoses should be hung up to drain after each use. The areas that result in the most contamination are always the filtering and pumping units where different batches and products intermingle. These areas are open to air-borne and water-borne contamination.

As stated, housekeeping and plant hygiene must be a routine and regularly scheduled practice to be effective, 'once a year or so' is not adequate. If skins are not allowed to accumulate, then the time required for cleaning is minimized. Once the equipment is rinsed it is back in service again. If the time involved in a hygiene program is compared to the expense of product rework, or worse yet a product return, then plant hygiene is seen as a very cost effective quality control method.

8.10.2 Treating wash/rinse water

The wash/rinse water that is recycled through the plumbing system is frequently a major source of contamination in industrial plants. The residual chlorine level of potable water is instantaneously depleted by contact with organic material. This leaves the wash water with no bacterial protection and very large quantities of organic material, which provides an ideal situation for bacterial growth. The most important issue to keep in mind is the residence time of the water prior to re-use. If the water stands less than 8 h no treatment may be required. If wash/rinse water is allowed to stand over the weekend a program to check for bacterial growth should be started. A key factor in selecting a biocide for wash water is the interval between addition of biocide and the use of the water. Preventing growth may prove more effective than killing established populations. Over weekends growth may be prevented by the addition of adequate levels of biocide at the end of the last Friday shift.

8.10.3 Hygiene audits

Routine inspections or 'audits' are useful to determine if these practices are in place and to check for the presence of spoilage in a variety of locations throughout the production facilities. Plant hygiene audits can be performed internally by trained personnel or by external vendors who provide this type of service. Some biocide companies even provide training for in-house personnel on effective auditing programs. It is recommended by the authors that monthly internal inspections be performed and a database be generated to compile a historical perspective for a particular plant. This program should also be coupled with an external audit every year. External auditing will keep plant personnel informed as to changes in methodology. External audits frequently are able to locate areas overlooked by plant personnel due to every day familiarity.

Listed below are five of the most critical areas to monitor with respect to the potential for plant hygiene failure. Also listed are five general recommendations for building a fundamental program of plant hygiene. Table 8.5 illustrates the locations and results from a typical industrial plant hygiene audit.

Critical plant inspection locations are:

1. Bulk input manifolds and raw material tanks
2. Water system and holding tanks
3. Wash water and holding tanks
4. Rework material
5. Product filters (basket, stacked, shakers)

General hygiene recommendations are:

1. Treat wash water daily
2. Add biocide early in manufacturing process
3. Make monthly hygiene audits
4. Seek outside audit annually
5. After database is established look for patterns in (a) transient populations and (b) resident flora.

Samples indicated by S in Table 8.5 were swabs of solid surfaces. These are qualitative samples. Growth is detected by streaking the swab onto an agar plate and reading the plate following incubation. Ratings are given for the degree of growth present according to the following scale:

0 = no growth
1 = 1–5 cfu (colony forming units); lightly contaminated
2 = 6–15 cfu; lightly contaminated
3 = 16–30 cfu; moderately contaminated
4 = 31–45 cfu; heavily contaminated
5 = >45 cfu; severely contaminated.

All other samples were liquids. Samples are checked for the presence of viable microorganisms using a serial dilution plate counting technique on standard laboratory media. Zeneca Biocides uses tryptone glucose extract agar (TGEA) because this particular medium supports a very large number of organisms and is able to provide rapid recovery from weak cells. The correct biological unit is cfu g^{-1}. Results were generated after incubation for 48 h at 30°C. Counts between 10^3 and 10^5 cfu g^{-1} are considered to be

Table 8.5. An industrial paint company plant hygiene audit evaluation results

Sample no.	Location	Growth detected
	Raw materials	
1	PVA latex – rail car	<10
2	PVA latex – storage tank	<10
3	Acrylic latex – rail car	<10
4	TiO_2 – rail car	<10
5	TiO_2 – storage tank	<10
6	Clay slurry – storage tank	<10
7	$CaCO_3$ – rail car	3.7×10^4
8	$CaCO_3$ – storage tank	7.3×10^4
9	Water supply	3.1×10^3
	Production line	
10	Tank 14	1.4×10^3
11	Let down tank no. 26	<10
12	Water entering grind tanks	2.0×10^3
13 (S)	Slurry hose to grind tanks	4
14 (S)	Let down tank no. 17	0
15 (S)	Let down tank no. 15	5
16 (S)	Let down tank no. 14	5
17 (S)	Transfer hose at finishing tank	5
18 (S)	Let down tank no. 13	5
19 (S)	Let down tank no. 12	0
20 (S)	Transfer hose to let down tanks	5
21 (S)	Let down tank no. 11	3
22 (S)	Let down tank no. 10	0
23 (S)	Grind tank no. 9A	0
24 (S)	PVA line into tank no. 9A	0
25 (S)	PVA line into tank no. 10A	0
26 (S)	Tank 27 water inlet	5
	Finished products	
27	Paint retain – enamel	1.8×10^5
28	paint retain – gloss	1.72×10^5
29	Paint retain – satin	<10
30	Paint retain – eggshell	<10
31	Paint retain – flat	<10
32	Paint retain – flat	<10

S = Swab sample

substantial. Counts of 10^6 cfu g^{-1} or greater indicate that the stability of the emulsion is in danger.

8.11 When spoilage occurs

Sanitization of all tanks, mills, plumbing and transfer vessels is required to regain control of highly contaminated facilities. After the equipment has been cleaned and sanitized with hypochlorous acid, glutaraldehyde or quaternary ammonium compounds it must be rinsed with high quality water. Rinsing with potable water after sanitization is essential. Cleaning solutions contain surfactants that may affect the viscosity of products if they are not removed from the system prior to restarting production. Sanitizers may interfere with biocide action in the finished product or may lead to color changes.

8.11.1 Handling contaminated product

The best judge in handling and treating contaminated product is past experience with similar situations. The nature and the scope of the problem should be determined. Was adequate biocide added during manufacture? Is the problem unique or a repetition of historical problems? Repeated occurrences call for plant hygiene audits designed to locate point sources of inoculation and a thorough examination of all manufacturing procedures. Single occurrences, while still serious, do not necessarily require the same level of intervention and treatment as a recurrent problem.

8.11.2 Treating contaminated product

Treatment of contaminated paint requires the addition of a supplemental dose of biocide. Depending on preservative type, this may require holding the product at least 72 h after addition to allow the biocide time to kill the bacteria and verify microbiologically sound product. Quick acting disinfectants such as bleach, glutaraldehyde and some formaldehyde donor preservatives may also be used as these will produce a rapid rate of kill, but it is advised to check for formulation compatibility especially with bleach and glutaraldehyde in polymeric formulations. These compounds may alter color, viscosity or even break the emulsion. All potential candidates for such uses should be validated in all product lines as part of the in-house microbiological procedures. These efforts will save time when it becomes necessary to recover spoiled product as timing in eliminating established bacterial populations is critical to the integrity of the paint. In any case, when a quick acting biocide such as bleach is used, final dosing with a persistent preservative is required to provide subsequent protection. Quick acting biocides are usually depleted in providing antimicrobial action.

If possible, it is recommended that contaminated material be examined by a microbiology laboratory to determine an effective concentration of biocide to eliminate spoilage in the product. This is important in determining both the correct biocide and the effective level that will eliminate and prevent subsequent spoilage. Use of an ineffective biocide or sublethal concentrations will allow a multitude of future problems to develop. This is especially true if the questionable product will be eventually released for sale. Timing in laboratory examination is critical. If the spoiled product is allowed to sit for extended periods of time the change in formulation caused by bacterial growth will become irreversible. Samples should be sent by overnight express. 'Recovery' testing is one microbiological protocol employed to determine effective biocide levels. Contaminated material is simply divided into aliquots, a spectrum of concentrations is added and the material is monitored for 24 to 72 h for growth. The level of preservative which kills all viable cells is considered to be the effective level. The working concentration determined by this method is usually 100 to 200% greater than normally required to prevent contamination. Therefore, usage recommendations provided by this method are usually adequate to prevent further spoilage of the system.

Any final product that proves untreatable should be disposed of by established procedures in accordance with local, state and federal regulations. Such product should never be reworked back into the production stream. Reworking would serve to inoculate otherwise good batches with untreatable bacteria. With adequate preservative usage and preventive programs this should never be required.

8.12 Implementing a HACCP program

Hazard analysis and critical control point (HACCP) is an entire approach at manufacturing control designed to identify all areas where contamination may occur and then develop a control system to address each area. It grew out of a concern that NASA had for the health of its astronauts in orbit. If an astronaut were to become ill from contaminated food the completion of the mission and safe return from orbit of the crew would become uncertain. NASA worked closely with the food industry to develop the overall approach to food safety. The lessons learned from HACCP are easily transferable to industrial non-food manufacturing. Plant quality audit personnel responsible for establishing a hygiene system will find large numbers of well written guides on the subject.[38] Much of the above discussion is intended to introduce HACCP into this review.[39]

8.13 Summary

Industrial formulations such as paints contain constituents which are prime candidates for degradation by a multitude of microorganisms. Bacterial species will cause unprotected formulations to suffer from pH swings, viscosity loss, emulsion destabilization, off odors and off colors and gas production. Fungal organisms will cause cracking, change adhesion properties and deface the final paint after application. These effects can wreak havoc on production and marketing departments when lost time, plant contamination and product recalls develop. Ultimate losses are measured by a company's reputation suffering. Protection of industrial days of the use of mercury and its broad spectrum, more 'forgiving' qualities have passed. The use of formalin or formaldehyde releasing qualities have passed. The use of formalin or formaldehyde released preservatives is also becoming less acceptable. Today, due to ever changing environmental and regulatory pressures, manufacturers are faced with using more selective biocides which target very defined biochemical processes in microorganisms. Because of their specificity, preservatives are becoming 'greener' with more acceptable human toxicology profiles. As regulatory pressures increase both the number of biocides and the companies capable of developing new products will decrease. Producers of aqueous products are faced with efforts to reduce usage levels of these greener products for both toxicological and financial reasons. Under these conditions plant hygiene has become a major concern. Manufacturers are now faced with implementing fundamental plant hygiene practices to reduce the bioload on the preservative system in place. Frequent sanitization of production and storage facilities is required to optimize protection. In summary, protection of industrial formulations from microbial attack will remain a necessity for years to come.

References

1. Turner, G.P.A. (1988) *Introduction to Paint Chemistry and Principles of Paint Technology*, 3rd edn., Chapman & Hall, London.
2. Blake, M.D. (1987) Thickeners for waterborne coatings. In *Handbook of Coatings Additives*, Calbo, L.J. (ed.), Marcel Dekker, New York.
3. Vash, R. (1987) Wetting and dispersing. In *Handbook of Coatings Additives*, Calbo, L.J. (ed.) Marcel Dekker, New York.
4. Flick, E.W. (1988) *Water-Based Trade Formulations*, Noyes Publications, Park Ridge, NJ.
5. Tilton, H. (1992) *Shades of Health*, Chemical Marketing Reporter, Oct 1992, 3–18.
6. *House Paints and Stains* (1993) Consumer Reports, **58**: 610–615.
7. O'Neill, T.B. (1986) Succession and interrelationships of microorganisms on painted surface. *J. Coat. Technol.*, **58**, 51–56.
8. Smith, R.E. (1993) *Zeneca Biocides*, personal communication.
9. Simpson, D.K. and Kappock, P. (1994) The role of fungicides and algaecides in architectural paint. *Paint & Coatings Ind.*, October.
10. Gillat, J. and Wood, B. (1990) Prevention of organic growth on coatings. *Polymers, Paint Colour J.*, **180**, 540–541.
11. Vannoy, W.G. (1948) Mildew inhibitors for house paints. *Fed. Paint Varnish Clubs*, **277**, 163–175.
12. Goll, M., Snyder, H.D. and Birnbaum, H.A. (1952) New developments in the diagnosis of paint mildew. *Official Digest Fed. Paint Varnish Prod. Clubs*, **326**, 149–155.
13. Harnden, R.C. (1945) Mildew control in the paint industry. *Paint Oil Chem. Rev.*, **108**, 32–34.

14. Bravery, A.F. (1988) Biodeterioration of paint; a state-of-the-art comment. *Biodeterioration 7*, Selected papers presented 7th Internat. Biodeteriation Symposium, **7**, 466–485.
15. Alexander, M. (1977) *Introduction to Soil Microbiology*, 2nd edn., John Wiley and Sons, New York.
16. Springle, W.R. (1988) Liquefaction of cellulosic paint thickeners. Part 2: Quantitative aspects of enzymatic degradation. *J. Oil Colour Chem. Assoc.*, **71**(4), 109–113.
17. Zeneca Biocides, unpublished research data.
18. Boyd, R. F. and Hoerl, B.G. (1986) *Basic Medical Microbiology*, 3rd edn., Little, Brown and Company, Boston.
19. Schwartz, R.D. and Leathen, W.W. (1976) Petroleum microbiology. In *Industrial Microbiology*, Miller, B.M. and Litsky, W. (eds), McGraw-Hill, New York, pp. 384–413.
20. Carter, G. and Huddart, G. (1974) The 'in-can' preservation of aqueous paints. *Pigment Resin Technol.*, 3–10.
21. American Water Works Association (1990) *Water Quality and Treatment – A Handbook of Community Water Supplies*, 4th edn., McGraw-Hill, New York.
22. Siegert, W. (1990) Production hygiene in the manufacture of water-based coating materials. *Polymers Paint Colour J.*, **180**, 584.
23. Block, S.S. (1991) *Disinfection, Sterilization, and Preservation*, 4th edn., Lea & Febiger, Malvern, PA.
24. Matta, B.L. and Breindel, K. (1991) *Mercurial versus Non-Mercurial Biocides*, 2° Congressa Internacional de Tintas, pp. 528–540.
25. Block, S.S. (1991) *Disinfection, Sterilization, and Preservation*, 4th edn., Lea & Febiger, Malvern PA.
26. Rossmoore, H.W. (1991) Nitrogen compounds. In *Disinfection, Sterilization, and Preservation*, 4th edn., Block, S.S. (ed.) Lea and Febiger, Philadelphia, pp. 290–321.
27. Scott, C.R. and Wolf, P.A. (1962) The antibacterial activity of a series of quaternaries prepared from hexamethylenetetramine and halohydrocarbons. *Appl. Microbiol*, **10**, 211–215.
28. Rossmoore, H.W. and Sondossi, M. (1988) Applications and mode of action of formaldehyde condensate biocides. *Adv. Appl. Microbiol.*, **33**, 223–277.
29. The Dow Chemical Company (1981) *Dowicil 75 Preservative 'Formaldehyde Release' in Aquous Systems*, Form no. 192–833–981.
30. Jakubowski, J.A., Simpson, S.L. and Gyuris, J. (1982) Microbiological spoilage of latex emulsions: causes and prevention. *J. Coat. Technol.*, **54**, 39–44.
31. Ross, R.T. and Hollis, C.G. (1976) Microbiological deterioration of pulpwood, paper and paint. In *Industrial Microbiology*, Miller, B.M. and Litsky, W. (eds), McGraw-Hill, New York, pp. 309–354.
32. Carter, G. and Huddart, G. (1974) The 'in-can' preservation of aqueous paints. *Pigment Resin Technol.*, 3–10.
33. ASTM (1992) Standard method for resistance of emulsion paints in the container to attack by microorganisms (#D2574–86). *Annual Book of ASTM Standards*, Vol 6.01.
34. Carman, J.R. and Goll, M. (1970) The role of industrial hygiene in the manufacture of latex paint. *Develop. Indust. Microbiol.*, **29**, 327–329.
35. Siegert, W. (1990) Production hygiene in the manufacture of water-based coating materials. *Polymers Paint Colour J.*, **180**, 584.
36. Jakubowski, J.A., Simpson, S.L. and Gyuris, J. (1982) Microbiological spoilage of latex emulsions: causes and prevention. *J. Coat. Technol.*, **54**, 39–44.
37. Marriott, N.G. (1989) *Principles of Food Sanitation*, 2nd edn. Van Nostrand Reinhold, New York.
38. Sperber, W.H. (1991) The modern HACCP system. *Food Technol.*, **45**, 116, 118, 120.
39. Pierson, M.D. and Corlett, D.A. Jr. (1992) *HACCP: Principles and Applications*, Van Nostrand Reinhold, New York.

9 Preservation of aqueous-based synthetic polymer emulsions and adhesive formulations

M.A. CRESSWELL and K. HOLLAND

9.1 General introduction

The need for protection from microorganisms has been recognised for many centuries, but only since the late 18th and 19th centuries have orderly programmes been mounted to control this problem. Much of the early effort was concentrated in agriculture where success was mandatory for economic, social or even physical survival. The great majority of this pioneering work was directed toward preventing fungal damage on natural substances: wood, paper, other cellulosics, seeds and crops.

The development of water-based and surfactant-based systems in the past 50 years raised the additional challenge of bacterial attack and the need for antimicrobial agents with broader spectrum effectiveness.[1–15] The concern of the user was expanded from end-use protection of the finished product where fungi were long recognised as causative organisms, to preservation during processing, storage and sampling where wider varieties of microorganisms play a major role. Biological problems, never simple, became more complex, and the search for preservatives that control the myriad of organisms that flourish in many of the new formulations became more difficult.

The last 30–35 years have been characterised by a growing awareness, triggered in no small part by regulatory pressures, that to be accepted, and thereby achieve economical success, a preservative must provide the user with security beyond the old basics: efficacy, compatibility and cost effectiveness. These, although necessary, form only a few of the broad requisites for 'modern' antimicrobial agents. The other requirement is that of minimal hazard in use as well as a low toxicological, environmental and ecological impact. Trends in development of new antimicrobial compounds, their uses, advantages and disadvantages will be considered and discussed in this chapter.

Synthetic polymer emulsions play an important role in everyday products from household domestic goods to many industrial uses. Though polymer emulsions may not be apparent as the emulsion system they are still very much in abundance in our everyday life and items we use. Some examples where polymer emulsions are to be found include: home 'do-it-yourself' products (emulsions paints, wallpaper pastes, glues, grouts and cements), domestic products (food packaging, cartons etc.), synthetic fabricated goods (non-woven textile materials, waddings, insulation), as well as vast numbers of industrial uses. Some of these uses and applications will be highlighted in more detail later in the chapter.

Synthetic polymer emulsions or latices or resins are fine dispersions or suspensions of synthetic polymer particles (0.1 μm to 5–6 μm) in an aqueous stabilisation medium. The emulsion systems considered in this chapter are all of the vinyl type whereby polymerisation reactions of alkene monomers form very large, high molecular weight polymers. The process of polymerisation can be either of two types: addition polymerisation or condensation polymerisation. The main components of the aqueous stabilisation medium, apart from water, are colloids such as polyvinyl alcohol, cellulosic materials or starch materials and/or surfactant, being anionic, cationic or even nonionic depending on the polymer type. A common feature, however, with most types of these polymer emulsions is that they are susceptible to spoilage by microorganisms.

Adhesives, like polymer emulsions, play an important part in day-to-day life, from the simple glues on the back of postage stamps which are remoistened to allow their positioning on letters, to the glues used for bonding labels on bottles and the high performance adhesives designed to withstand severe stresses in automotive manufacture. Adhesives are used both in the domestic environment as well as in industrial situations in many operations.

An adhesive can be defined by many definitions that range from a 'sticky substance' to a 'material uniting other materials either temporarily or permanently', but each retains the central theme that an adhesive is capable of binding or adhering to a surface and can act simultaneously on two surfaces to hold them together. Adhesives provide an alternative to the common mechanical methods of combining materials together, e.g. screwing, nailing, riveting and fusing as well as having many advantages such as stress distribution, versatility and simplicity.

Numerous theories abound on how adhesives work and on the mechanisms of adhesion and these are covered in some of the classical texts, though the common mechanisms are by short range intermolecular interactions, physical interlocking or chemical bonding.

Formulations of adhesives vary immensely and a variety of raw materials are employed in their manufacture. A knowledge of these raw materials and their effects is necessary to understand the details of an adhesive formulation. Surfactants are used as raw materials in their own right to provide certain properties but are also frequently present in many of the other raw materials themselves.

Given the care taken in designing an adhesive to ensure its effectiveness in an application, the significance of this chapter is to show that contamination and growth of microorganisms in the adhesive can reduce its effectiveness. Several types of microorganism can be encountered in spoilt polymer and adhesive systems and their harmful effects can be judged by the deleterious results they may have on the end-use properties critical to their performance. Often the harmful effects are similar for different microorganisms and the need to identify the specific nature of the microorganism may only be important if it falls outside the spectrum of activity of the selected preservative. This chapter will consider ways of preventing microbiological spoilage in both polymer emulsions and adhesives by the judicious choice of preservative and control of the supply chain.

The prevention of microbiological contamination by the incorporation of a preservative into the formulation requires a careful choice of biocide-active agent which is governed by the raw materials in the formulation, by the chemical environment and by legislative restrictions relevant to the end use of the manufactured product. Prevention in the supply chain requires an emphasis on the importance of manufacturing plant hygiene, especially during formulation processing combined with the use of sampling and testing protocols in order to confirm the efficacy of the chosen preservative system. This involves closely monitoring each step in the chain from receipt of the raw materials through to the final use of the product and often even beyond this stage if the dry film is also susceptible to microbial attack.

9.2 Aqueous-based synthetic polymer emulsions

9.2.1 Applications and uses of polymer emulsions

Polymer emulsions are widely used as raw materials in an extensive range of industries.[16] Some of the uses of these emulsions are listed in Table 9.1.

9.2.2 Range and chemical types of polymer emulsions

There is a wide range of different types of synthetic polymer emulsions, incorporating different monomers in a variety of combinations.[16] Examples of emulsion types are listed here:

- Polyvinyl vinyl acetate (PVA) homopolymers
- Vinyl acetate/ethylene (VA/E) copolymers
- Vinyl acetate/vinyl chloride/ethylene (VA/E/VC) terpolymers
- Acrylic-containing (Ac) polymers
- Styrene/acrylic (Sty/Ac) polymers
- Vinyl acetate/vinyl versatate (VA/VeoVA) copolymers

If gaseous monomers such as ethylene or vinyl chloride are used, the production processes involve polymerisation at high pressures and the

Table 9.1 Some uses of polymer emulsions

Industry	Examples of applications and uses
Adhesives	Packaging and converting, metal and plastic foil laminating, building, wood bonding, ceramic tiles fixing, bookbinding, cold seal, automotive assembly
Emulsion paints	Emulsion paint, from full gloss to matt for exterior or purely interior use – the emulsion acts as the binding medium, contributing durability and resistance to water, scrubbing etc.
Textured coatings	Exterior and interior paints – the emulsion has the same function as in normal emulsion paints above.
Industrial finishes	Wood and metal including anticorrosive finishes – special emulsions ensure requisite hardness, solvent resistance etc. in air drying of stoving finishes.
Printing inks	Alkali-soluble emulsion polymers provide a reliable, consistent varnish base, particularly for flexographic inks.
Building auxiliaries and cement additives	Emulsions with stability to high concentrations of ions, and with polymer components resisting the weakening action of water, oils, fats, etc. are widely used for general purpose bonding, and for upgrading the adhesion and resilience of renderings and floor toppings.
Carpets	Polymer emulsions increase the tuft anchorage in woven carpets, without distracting from easy lay; in tufted carpet emulsions act as anchor coats for backing compounds, and in reinforcing and primary binders in backing.
Non-woven fabrics	Emulsions provide the full range of cost performance properties meeting the various requirements of disposable and permanent non-woven fabrics, e.g. J-cloths, disposable nappies/diapers, feminine hygiene products.
Textile finishings, printing and flocking	The versatility of emulsions covers the wide range of fabric finishing requirements, as well as the specific binding requirements for flocking printing, pigment binding etc.
Glass fibre sizes and binders	Special emulsions bind glass strands and chopped strand mat for easy handling during lay-up.
Paper and board coating	Emulsion binders for mineral coated stock enable good colour and printing properties.
Wallpaper	Emulsions are used in all front coating operations, as well as duplexing and pre pasting adhesives.
Soil and mineral stabilisation	Surface binding of loose soil and mineral deposits by emulsions is widely undertaken to improve crop yields and reduce dust and similar environmental problems.
Biological applications	Special emulsions with entrapped fungicides/mildewicides are used to coat seeds to prevent their decay prior to germination. Also, emulsions are used as cryo-preserving aids in the storage of harvested microbial cells.

polymers formed are commonly referred to as 'pressure polymers', e.g. copolymers of VA/E or terpolymers of VA/E/VC or VA/E/2-EHA. Other non-gaseous monomers may be polymerised together to form polymers in low pressure systems and these polymers are commonly referred to as 'conventional polymers' or 'atmospheric polymers', e.g. VA homopolymers, Ac polymers of methylmethacrylate. Furthermore, the use of other functional monomeric units to give the polymer specific application properties, such as cross linking in cured textile fabric applications (e.g. *N*-methylol acrylamide, acrylamide and many others), are often employed and these polymers are commonly referred to as 'speciality polymers'. Common to all polymerisation reactions the processes are usually carried out in a batch-wise system, but continuous processes can also be employed.

The majority of polymer emulsions have a pH range between 3.5 to 7.5. In most of the emulsions mentioned in this chapter the pH is acidic, between 4 to 6.5, but some emulsions, however, have an alkaline pH, e.g. acrylic types of polymers with pH of greater than 8. It is often therefore the pH range, and the fact that polymer emulsions are aqueous-based products, that determine the degree of susceptibility of the emulsion to the various classes of microorganisms. Acidic pH systems tend to favour the growth of yeasts and moulds, whereas neutral to alkaline pH systems tend to favour the growth of bacteria. Knowing this also helps when selecting and choosing the right biocide to preserve a particular polymer emulsion type.

9.2.3 The chemical nature of polymer emulsions

Apart from water, three main components are required to manufacture a polymer emulsion[16]

- surfactant and/or colloid
- monomer
- initiator

These will now be considered separately.

9.2.3.1 Surfactants. Surfactants, by definition, are substances which lower the surface tension of liquids in which they are dissolved or the interfacial tension between two or more mutually immiscible phases. It has long been recognised that the most effective surface tension depressants contain highly water-attracting (hydrophilic) and highly water-repellent (hydrophobic) groups, joined together in the same molecule.

The most important hydrophilic (polar) groups found in surfactants are, for anionics, ionised sulphonate, sulphate, phosphate and carboxyl groups; for nonionics, polyoxyethylene chains and polyols; and for cationics, ionised tertiary or quaternary ammonium groups. The hydrophobic portion of the surfactant molecule usually consists of hydrocarbon radicals with more than eight carbon atoms, i.e. alkyl–aryl or long chain alkyl. Since the hydrophobic hydrocarbon groups have considerable affinity for oils and fats they have also been termed lipophilic or fat loving. Long polyoxypropylene chains, though less hydrophobic than hydrocarbons, nevertheless possess sufficient repellency to confer strong surface activity to compounds, wherein they are linked to hydrophilic groupings, such as polyoxyethylene chains. Thus a very versatile range of surfactants consists of block copolymers of ethylene oxide and propylene oxide in which the ratios and the molecular weights of the two different types of polyether chains are varied over a wide range. Table 9.2 gives some examples illustrating the chemical make-up found in typical commercial surfactants.

The behaviour of amphiphilic compounds in either aqueous or non-aqueous media is determined by the relative effectiveness of their hydrophilic groups and hydrophobic groups. In aqueous media the hydrophilic groups are strongly associated with the water molecules whilst the hydrophobic groups are repelled and tend to concentrate at the air–water interface. In the presence of both aqueous and oily phases the affinities of both groups can be satisfied by concentration of the surfactant molecules at the oil–water interface. The surfactant molecules will be orientated in such a way that the hydrophobic groups are associated with the non-aqueous phase while the hydrophilic groups are firmly anchored in the water layer.

If the interfacial area is small, it can only accommodate a small number of molecules. When, as usual, many more surfactant molecules than this are present, the majority cannot escape from the bulk liquid to the interface and the affinities of the hydrophilic and lipophilic groups must be satisfied by other means if thermodynamic stability is to be achieved. This again occurs by a process of orientation; in an aqueous medium the hydrophobic groups turn towards and associate with one another, forming in effect their own oil phase, surrounded by the hydrophilic groups turned outwards and anchored in the water. This type of internal association and orientation has been termed micelle formation. Micelles are usually spherical in shape. The 'escape mechanism' of micelle formation only

Table 9.2 Some typical commercial surfactants

Anionics	Nonionics
Stearate soaps	Polyethoxylated alkanol
Dodecyl sulphate	Polyethoxylated nonyl phenol
Dodecylbenzene sulphonate	Polyethoxylated polypropylene glycol (pluronic)
Dioctyl sulphosuccinate	Glyceryl monolaurate
Nonyl phenol polyether sulphate	
Dodecyl polyether phosphate	

becomes operative above a certain minimum surfactant concentration. This concentration has been termed the critical micelle concentration (CMC). CMCs vary from about 5×10^{-2} mol l^{-1} for the most hydrophilic to about 5×10^{-4} mol l^{-1} for the most hydrophobic types of surfactant. They are influenced by electrolytes, especially in the case of ionic surfactants, and also by other polar/non-polar chemical compounds such as alcohols, amides and, of course, other surfactants.

The most important consequence of micelle formation in the process of emulsion polymerisation is the solubilisation of organic compounds in aqueous media. Since the association of lipophilic groups inside a micelle leads to a formation of centres of attraction for organic compounds it is possible to dissolve appreciably higher proportions of sparingly water-soluble monomers in micellar solutions than in water alone. Monomers which are essentially non-polar in character may be expected to dissolve only inside the hydrocarbon portion of the micelle. Compounds with polar groups can at least partially be accommodated in the water phase or at the micelle surface so that the requirements for the micelle dimensions are less critical.

9.2.3.1.1 Surfactants in emulsion polymerisation: the mechanism. The role played by surfactants in emulsion polymerisation[17,18] can best be demonstrated in a simple recipe involving only water, surfactant, monomer(s) and polymerisation initiator. When a sparingly water-soluble monomer is stirred into an aqueous surfactant solution it will be broken into droplets of varying size and give a more or less stable emulsion depending on the choice and quantity of surfactant present. If the surfactant concentration exceeds the CMC, some of the monomer will be solubilised in micelles. On the addition of a polymerisation initiator, and this almost invariably means a substance capable of producing free radicals, polymerisation will begin as soon as the initiator has been activated either thermally or chemically to yield free radicals.[19]

The process of emulsion polymerisation begins when the free radicals derived from the, usually water-soluble, polymerisation initiator enter the monomer-saturated micelles where they find a sufficient number of solubilised molecules to start a rapid chain reaction.[19] Each polymer radical first exhausts the monomer contained in the micelle and then captures additional supplies from 50 or more other micelles before the chain reaction is terminated. Some of the depleted micelles then break up and the released emulsifier molecules are adsorbed at the surface of the newly formed primary polymer particles.[18] The remainder are replenished by diffusion from the emulsified monomer droplets which act essentially as reservoirs.

The formation of fresh polymer particles continues until all the emulsifier originally contained in the micelles is adsorbed at the larger polymer/water interface and the surfactant concentration has dropped below the CMC.[18] A new and different situation is thus created for the polymerisation which would come to a halt if it could only proceed inside surfactant micelles. The growing polymer particle is, however, still very similar in general characteristics to a true (spherical) micelle. The chief difference is the composition and size of its lipophilic centre. On the outside there are still the layers of orientated surfactant molecules with their hydrophilic groups pointed towards the external water phase while the lipophilic groups are now associated, not with each other, but with the dispersed internal polymer phase. The conditions for monomer solubilisation are in fact greatly improved. Quite apart from the fact that the polymer particles have a far greater affinity for monomer molecules than the micelles ever had, there is now a considerable interfacial area available where interface solubilisation, akin to intramicellar solubilisation, can take place. It is this interface solubility of the monomers which is chiefly concerned in the second phase polymerisation, i.e. after the disappearance of the micelles. The loci of polymerisation have then shifted from the micelles to the polymer particles to which all the emulsified monomer is gradually transferred.

Now consider the case of polymerisation of monomers with an appreciable water solubility and their copolymerisation with less water soluble comonomers. Monomers such as vinyl acetate or methyl acrylate have sufficient water solubility to permit their rapid polymerisation in the water phase even in the absence or after the disappearance of surfactant micelles. New polymer particles can be formed as long as the monomer concentration in the water phase remains high enough. In most cases, there is strong monomer/polymer affinity so that more and more monomer will be extracted from the water phase. As polymer concentrations increase, polymerisation in the water phase will finally cease and with it the formation of new particles. Henceforth, conversion will proceed at the surface of the polymer/monomer particles in much the same way as in the case of water-insoluble monomers after the disappearance of the surfactant micelle.

9.2.3.1.2 Stabilisation of polymer particles. At the end of the emulsion polymerisation polymer particles will exist in the dispersed water phase with surfactant molecules adsorbed on the particle surface. It is usually considered that it is the hydrophobic portion of the surfactant that is adsorbed onto the surface while the hydrophilic portion goes into the water phase. The role of the surfactant is now to keep the system stable by preventing coalescence of the polymer particles in the dispersion. If particle coalescence is not prevented the settling of the coagulated particles can take place giving a non-redispersible sediment.

For anionic surfactants stabilisation of the polymer particles is achieved by the negative charge on the adsorbed surfactant molecules repelling neighbouring particles which are similarly charged. Stabilisation by this means is termed 'electrostatic stabilisation'. In effect, an energy barrier to particle coalescence is set up by charge–charge repulsion between particles. This energy barrier can be reduced by the addition of electrolyte, e.g. salts.

Nonionic surfactants by definition carry no charge and with these surfactants stabilisation is mainly by volume reaction. Again the hydrophobic portion is adsorbed on the polymer surface while the hydrophilic portion, usually long ethylene oxide chains, extends into the water phase. Because of the volume occupied by these ethylene oxide chains the polymer particles cannot easily approach each other, i.e. there is an energy barrier to coalescence due to the spatial presence of the adsorbed surfactant. Stabilisation by this means is termed 'steric stabilisation'. The energy barrier to coalescence can be reduced by reducing the proportion of the ethylene oxide chains, e.g. either by salt addition or by heating the latex.

It should be noted that instead of stabilisation by surfactants the stabilisation of polymer particles may also be achieved by long chain water-soluble polymers. These are usually termed 'colloids' and typical examples are polyvinyl alcohol,[20,21] hydroxyethyl cellulose and starch. Stabilisation is again by steric means. The main functions of colloids are:

- to give increased viscosity to the latex and to influence the structure and rheological properties, and
- to influence the application properties of the latex, e.g. give increased bond strength in adhesives.

Polyvinyl alcohol is supplied as a white free-flowing powder. Although the term 'polyvinyl alcohol' is used it is generally a copolymer of vinyl alcohol and vinyl acetate.[20,21] The degree of hydrolysis represents the percentage number of moles of vinyl alcohol to the total number of moles present. Typical degrees of hydrolysis encountered in emulsion polymerisation are 80–100%. Quite different emulsion polymer properties and application results will be obtained using different hydrolysis grades of polyvinyl alcohol. The viscosity effects produced by polyvinyl alcohol will also depend very much on the molecular weight (MW), since higher molecular weight species produce higher viscosity latices.[22]

Polyvinyl alcohol is rarely used in paint systems as irreversible structures may be obtained with the gelling agent. For paints the structure may be achieved by using colloidal agents, such as hydroxyethyl cellulose.

9.2.3.2 Monomers. The polymers existing in emulsion form are produced by addition polymerisation of one or more unsaturated monomers, the commonest of which are listed in Table 9.3.

Table 9.3 Monomers

Monomer	Monomer name	Approx. MW
$HOOC.CH:CH_2$	Acrylic acid	72
$N{:}C.CH.CH_2$	Acrylonitrile	53
$C_4H_9O.OC.CH:CH_2$	Butyl acrylate	128
$C_4H_5O.OC.CH:CH_2$	Ethyl acrylate	100
$CH_2:CH_2$	Ethylene	28
$CH_3O.OC.C(CH_3):CH_2$	Methyl methacrylate	100
$HOCH_2.NH.OC.CH:CH_2$	*N*-methylol acrylamide	102
$C_6H_5.CH:CH_2$	Styrene	120
$CH_3.COO.CH:CH_2$	Vinyl acetate	86
$Cl.CH:CH_2$	Vinyl chloride	62.5
$CH_2–CH.OH:O–C_9H_{19}$	Vinyl versatate	

If a single monomer is reacted, the polymer obtained is termed a homopolymer. Many of the above monomers, however, can also react with each other, forming copolymers; varying the proportions of these comonomers affords an excellent means of controlling polymer properties.

Not all the monomers listed in Table 9.3 react together satisfactorily: thus vinyl acetate and styrene, ethylene and styrene, are examples of unfavourable pairs. Moreover, monomers without terminal unsaturation, e.g. the fumarates, and the related maleates, do not generally form homopolymers although readily forming copolymers with, for example, vinyl acetate.

The properties of the polymers are governed mainly by the monomer(s) on which they are based. All are thermoplastic, that is to say they soften progressively as their temperature is increased, but regain their original condition when cooled. Except when special reactive monomers are included, there is no cross-linking reaction, as occurs in thermosetting polymers, for example melamine–formaldehyde or urea–formaldehyde systems.

The hardness and softness of the polymer at a given temperature is predominantly determined by the monomer(s) employed. In polymers of identical make-up, differences in molecular weight are found to have a secondary effect, with higher molecular weight tending to reduce thermoplasticity slightly and lead to greater toughness and better resistance to solvents and water. Examples of monomers yielding hard generally brittle homopolymers requiring external plasticisers include: vinyl acetate, vinyl chloride, styrene, methyl methacrylate and acrylonitrile. Examples of monomers yielding soft internally plasticised homopolymers include: butyl acrylate, 2-ethylhexyl acrylate and the vinyl esters. Ethylene also acts as a plasticising comonomer with vinyl acetate and vinyl chloride but by a different mechanism.

All the polymers are fully saturated, i.e. they contain no residual double bonding. In consequence they are not prone to ageing effects of embrittlement and discoloration.

All monomers impart certain characteristic properties to polymers of which they form a part, but some have a particularly marked effect and can bring about significant changes in polymer behaviour even when present as only a few percent of the total monomer mix. As examples, acrylic, methacrylic acids and other unsaturated acids are used to give alkali-soluble copolymers in conjunction with methacrylic acid and acrylic esters. *N*-methyl acrylamide and other polyfunctional monomers produce reactive copolymers capable of secondary polymerisation or cross linking by etherification or ionic bonding, usually after the watery portion of the emulsion has dried off.

9.2.3.3 Initiators. Initiators produce free radicals which act on the monomer to form the polymer in aqueous conditions. Initiation may be thermal using a single chemical type or a redox system may be used.[19] Redox systems are suitable for polymerisation at low temperatures. Some common initiator systems are listed in Table 9.4.

These redox chemicals may also be added at the end of the polymerisation process, as a finishing off stage (FOS), in order to get maximum polymerisation and to reduce residual unreacted monomer.

Other additives may be found in a polymer emulsion system which contribute to its total composition. Examples of these additives are:

- Buffering agents – for pH control, e.g. sodium bicarbonate
- Antifoams – for preventing foaming
- Salts, e.g. sodium sulphate
- Metal catalysts, e.g. ferric chloride
- Preservatives – for antimicrobial protection from spoilage.

9.2.4 Manufacture of synthetic polymer emulsions

Emulsion polymerisation is carried out in a reactor where temperatures may reach 90°C during the process. Thermally initiated emulsion processes are manufactured at temperatures between 70–90°C. In redox initiated

Table 9.4 Initiators for emulsion polymerisation

Oxidising agents	Reducing agents
Potassium persulphate	Sodium formaldehyde sulphoxylate
Sodium persulphate	Sodium metabisulphite
Ammonium persulphate	Sodium thiosulphite
t-Butyl hydroperoxide	Ascorbic acid
Cumene hydroperoxide	
Hydrogen peroxide	

processes temperatures of between 40–50°C are usually the norm. It is worthy of note that in the authors' experience microbial contamination has not been found in emulsion samples taken directly from the reactor after polymerisation. In many instances, the temperatures and chemical environment during which polymerisation is carried out are sufficient to destroy many microbial contaminants arising from raw materials.

From the reactor the emulsion is pumped through a network of pipework to the blenders, degassing tanks, an open filtration system and finally to the bulk storage systems. During this time the emulsion is cooling and possibly could become contaminated by microorganisms. This is the reason for adding a chemical biocide in order to preserve the emulsion against spoilage. Biocides, though required in very small amounts, are often expensive and complex chemicals that are very labile to the hostile chemical environment of the polymerisation process and therefore their addition must be carefully controlled and monitored.[23,24] The biocide is therefore added into the blender at a time determined when the temperature is low enough (<50°C) to prevent thermal degradation and when the free radicals have been consumed in the process.[24] This time is determined by a simple test to monitor the redox state of the emulsion.[25,26] Under these conditions the biocide has the best chance of survival to allow it to perform its function as a long term preservative against microbial spoilage of the emulsion. The chemical stability of the biocide and the accuracy of dosing of the biocide into the emulsion is determined by high performance liquid chromatography.[27] This is a quality control test method carried out to confirm if the emulsion is adequately preserved and is a standard specification along with other specifications of solids content, pH, viscosity, particle size distribution, residual monomer levels and so on. Another reason for checking the level of the biocide is for regulatory purposes where different emulsions depending on their end use have limits on their preservative type and content. Examples of these regulations include: FDA and BGA regulations for direct and indirect food contact in food packaging applications for adhesives; BGA regulations for non-woven applications where binders are used in sensitive areas, e.g. feminine hygiene products, medical wipes and disposable nappies/diapers; and last but by no means least, areas where only food grade biocides can be incorporated, e.g. the adhesive used in cigarette manufacture for the seams in the paper and in the wax coatings for certain cheeses.

9.3 Adhesive formulations

9.3.1 *Applications and uses of adhesives*

To demonstrate the key properties required for an adhesive, consider the choice of an adhesive in two cases, a broken china mug to be repaired and a wooden wine rack to be assembled. For the wine rack, the likely adhesive choice is a standard wood glue, commonly one based on aqueous synthetic polymer emulsions; a small 'blob' of glue would be placed in each of the joints, the joint made and the final assembly left to dry for 24 h before subjecting it to stress, i.e. filling it with bottles. For the china mug, a cyanoacrylate adhesive is chosen; its use requires the cleaning of both surfaces to be bonded, followed by the application of a small amount of adhesive to one or both surfaces and then holding these close together for a short time while the adhesive dries.

The choice of these two adhesives is not by accident since they are effective in different ways. The wood glue, which contains a polymer dispersed in water, acts by using water as a carrier to facilitate penetration of the wood fibres by the polymer. Then, slowly, the water dries leaving only polymer which holds the joint together since it is intrinsically entangled with the wood fibres. Important adhesive characteristics are: the viscosity, in that the adhesive must be thin enough to penetrate the wood but not so thin that it penetrates one side of the joint completely before assembly or that it drips readily; the solids content, whereby water is necessary as a carrier but too much water slows the drying and may distort the joint; the set speed which is different for different emulsion types; and the bond strength which is a requirement determined by the stress which the adhesive receives when the wine rack is full.

This wood adhesive may perform well in this application, but it would not be suitable for bonding a mug designed to hold liquids and so be impervious to water; no mechanical adhesion could be obtained by transport of the polymer into the china. Without transport, other mechanisms of adhesion would be ineffective with this adhesive/substrate combination since the low intermolecular forces would be insufficient for the demanding end use (e.g. hot tea and cleaning in dishwashers) and again chemical adhesion is low since there are no chemical groups which can react between the polymer and the china surface.

The cyanoacrylate, on the other hand, acts in a totally different way with no carrier involved though the adhesives are usually of low inherent viscosity. Although little penetration is seen, a good bond can be obtained since a chemical polymerisation occurs after application forming strong bonds between the adhesive and itself and between the adhesive and the surface of the mug. The final bond is much stronger than with the woodworking emulsions due to the strength of this reacted polymer. So why not use this to make the wine rack? One reason is the higher cost in use, another is the irreversibility if mistakes are made aligning the joints.

These two adhesives comprise two of the four main types of adhesive, namely aqueous-based (i.e. solution and dispersion) and reactive adhesives. The third adhesive type consists of hot melt types and the fourth is based on solvents.

9.3.2 Range and chemical types of adhesive

9.3.2.1 Hot melt adhesives. Hot melt adhesives are formulated to give a material which is solid at room temperature but which softens on heating. At application temperatures, commonly in the region of 150 to 170°C, the hot melt is liquid enough to act as its own carrier to allow the polymer to penetrate into the substrate. These adhesives are generally faster setting than aqueous-based adhesives since less energy change is required to cool them to form a reasonable bond than is required to drive off water to set aqueous-based adhesives. Their disadvantage often lies in their heat sensitivity, since a bond subjected to high temperatures may fall apart. These adhesives will not be mentioned further in this chapter since preservatives are rarely required for hot melt adhesives because the fundamental requirements for microbial growth, nutrients, moisture and oxygen are not present in either the adhesive or the adhesive film in the finished assembly. In addition, the high application temperatures would kill any microorganisms that may be present.

9.3.2.2 Solvent-based and reactive adhesives. Solvent-based adhesives use solvent as a carrier to transport dissolved polymer into the substrates. Their history and efficacy is well documented but increasingly environmental pressures are causing these 'older' formulations to be replaced by aqueous-based materials. In some cases these environmental pressures may be misplaced since the risks of environmental damage are simply redirected from the air (solvent evaporation) to the ground (waste water). Key differences from aqueous-based adhesives are that they are often of a lower molecular weight, higher tack, lower solids content and a faster set speed. Again, these types of adhesive are much less prone to microbial attack than the aqueous-based adhesives.

In addition to the cyanoacrylates mentioned above, other types of reactive adhesives include the anaerobics and epoxies which can be cured by heat or by two part reactions with amines or polyurethanes which cure by reaction of isocyanate groups with polyols and silicones. Preservation is, as above, often not an issue with these adhesives.

9.3.2.3 Aqueous-based adhesives. The various types of aqueous-based adhesives are the most conducive to microbial growth. They can be classified in many ways with the simplest classification dependent on their solubility. The solution type of adhesives are based on starch, dextrine, casein or polyvinyl alcohol and the disperson type adhesives are based on synthetic polymer emulsions ranging from vinyl acetate systems to rubber latex or polyurethane emulsions.

9.3.2.3.1 Solution adhesives. Starch, a biopolymer, from various sources including potato, maize and tapioca is one of the oldest raw materials[28] for adhesive manufacture and in its various guises or modifications is still commonly used as an adhesive.[29] Starch is a natural polymer of repeating glucose units which are linked by α-glycosidic bonds to give linear (amylose) or branched (amylopectin) chains and the relative amount of each of these chains determines in part the starch properties. In the highly branched amylopectin chains the starch polymers are packed less densely than in the straight chain amylose starch polymers and have a greater stability, greater water solubility and lower strength.

Starch itself is processed to provide discrete granules of 5–40 μm in size, but to be useful as an adhesive these particles have to be dissolved by heating a slurry of the starch with water to a 'gel point' where the granule will be swollen and, if subjected to strong agitation, can rupture. Native starches can be treated in this way but give only a low solids content, a high viscosity and form an unstable paste, and as such it is generally the modified starches that are used in the manufacture of starch pastes. The simplest modifications that are employed involve acid conversion (when the starch slurry is treated with a mineral acid) to give a fluidity starch or chemical oxidation (when the starch slurry is treated with oxidising agents), but again the adhesives produced from these after gelatinisation are fairly low stability pastes and they are used for simple applications like paper bag manufacture.

More importantly, chemical modifications of starch are employed to widen the scope of these adhesives using the chemical reactivity of the highly reactive hydroxyl groups on the glucose ring. Wet phase modifications are performed by adding reagents to the aqueous starch slurry under controlled reaction conditions. Cross-linking starches with, for example, epichlorohydrin, esterifying with acetic anhydride or etherifying with propylene oxide, all give starches with greater viscosity stability than native or fluidity starches and this increases the range of adhesive end uses. For example, wallpaper pastes are often made from chemically modified starches.

An alternative to cooking starches to give a final adhesive is to drum dry them after partial gelatinisation during the conversion process, which gives rise to cold water-soluble starches which may then be prepared *in situ* without the need for further cooking. Dry conversion of the starches is achieved by roasting with an acid catalyst to give dextrines. Two classes are produced but many variations are possible due to the complexity of the various starches and the various parameters of the process which can be independently controlled. White dextrines are less converted than the yellow dextrines and have slower set speeds, lower tack and lower viscosity stability. Conversely, yellow dextrines can be used at very high solids content and are often treated with borax to give high tack products used in paper converting, for example tube winding.

Jelly gums are made by cooking starch in strong alkali to give thick, tacky adhesives for bottle labelling applications.

Other natural sources of solution or semi-colloidal adhesives are the caseins which use milk-derived products to give pale, high tack, high viscosity and high adhesion products. These properties and the strongly temperature-dependent viscosity profile of caseins also makes them suitable for this bottle labelling application.

Like starch, cellulose is also a biopolymer and has a similar molecular origin, again it is based on glucose units but with the units linked by β-1,4-(D)-glycosidic bonds as opposed to the α-glycosidic bonds in starch polymers. Cellulose polymers are also used as adhesives in simple applications such as tissue and towel lamination.

Other biopolymers, namely polyhydroxybutyrate produced from microbial metabolic transformations are increasingly being used in the field of adhesive applications. A commercial process for the production of polyhydroxybutyrate by microbial transformations with the microorganisms has been patented by ZENECA BioProducts in Billingham, Cleveland, UK and called Biopol. Biopolymers, being natural products, are becoming more interesting to industry due to an increased environmental awareness and needs for recyclability of packing materials.

Natural rubbers also provide a source of adhesive raw materials. Natural rubber latex adhesives may be difficult to handle due to the high pH (often ammonia-based) required for stability but show great advantages in the ability of latex to bond well to itself but little to other substrates. This makes it an ideal choice for self-seal applications such as in envelope manufacture.

Solution adhesives do not all come from natural sources. Polyvinyl alcohol[20,21] is used to make adhesives used in applications such as multi-ply board lamination, as well as its use as a formulation tool for emulsion-based adhesives. The polyvinyl alcohol is prepared by hydrolysis of polyvinyl acetate and it is supplied as a dry powder which must be dissolved prior to use.

9.3.2.3.2 Dispersion adhesives. Applications for solution adhesives from starch or other sources are often limited by poor adhesive properties such as a low adhesion to certain materials, for example plastics, a relatively poor viscosity stability often coupled with a low solids level and a relatively slow drying speed (since a great proportion of the water must be lost to allow film forming). Synthetic polymer emulsions were developed to overcome these shortcomings and a description of these and their chemical constitution and manufacture have already been discussed.

The great advantages of using synthetic emulsion polymers in adhesives lies in their ability to set quickly due to the hydrophobic nature of the polymer, their relatively high molecular weight and their facility for manufacture at high solids level (40–50%) at a stable, low viscosity. The thermoplastic polymers also have a good adhesion onto plastics.

Several types of polymer are used. The polyvinyl acetate homopolymer (PVA) from vinyl acetate is a fairly hard polymer due to restriction of rotation in the molecular chain caused by the proximity of bulky acetate groups and the use of homopolymer emulsions usually requires formulation to soften or swell the polymer particles by the addition of plasticisers. An alternative route to a softer emulsion comes from the use of copolymer emulsions from the copolymerisation of vinyl acetate and ethylene monomers. The relatively low steric volume of the ethylene links allows greater freedom for chain movement giving a polymer which is less hard when dry and the use of softening agents can be avoided. This effect can be measured by looking at the glass transition temperature of these two polymers when dry, that of the copolymer is around 30°C lower. The drawback of using these vinyl acetate/ethylene (VA/E) copolymers is their higher cost which arises from the capital investment required for their manufacture but this is often offset by the advantages of their higher adhesion. In general, PVA is suitable for use in bonding involving paper, cardboard or wood and VA/E is also suitable for these materials and for more difficult substrates, e.g. treated paper or plastics.

Other copolymers commonly used are between vinyl acetate and dibutyl maleate or between vinyl acetate and vinyl acrylate. The latter gives acrylic-type emulsions which are characterised by very low glass transition temperatures and are seen to be permanently tacky or pressure sensitive. Adhesives based on these are commonly used for self-adhesive labels. Terpolymer emulsions are also used, e.g. vinyl acetate, ethylene and vinyl acrylate.

More recently polyurethane emulsions have gained importance and these show the advantages of high adhesion and high heat resistance (especially if cross linked *in situ* with isocyanates) though they have restricted compatibility with some other emulsions and can be both shear and pH sensitive. The manufacturing process is very different from those above since the polymerisation is essentially done in a non-aqueous organic phase which is then dispersed in water and the residual organic phase stripped off by vacuum.

9.3.3 The key properties and tests on adhesives

Some of the requirements for adhesives used in the home have been considered. When industrial applications are involved the requirements become considerably more varied and demanding. Adhesives may be applied by spray systems allowing the deposit of several kilos of adhesive each minute or by systems of rollers, this requires not just a specific viscosity but also a specific rheology (shear dependence of viscosity). The end-use requirements may also be stringent since although the adhesive may be a small part of an assembly and contribute only a small part to its cost, the consequences of adhesive failure could be severe with the whole assembly being rendered useless.

To summarise, some of the key characteristics required of aqueous-based adhesives are described below.

9.3.3.1 Viscosity and rheology. This determines the ability to penetrate into surfaces and the suitability for different types of application equipment. Viscosity is generally measured on a Brookfield viscometer and rheology on various types of rheometer. Pseudoplastic adhesives have lower viscosities as shear is increased, dilatant adhesives have higher viscosities as shear increases and thixotropic adhesives show a time-dependent behaviour.

9.3.3.2 Solids. This is a measure of the amount of material left after drying, i.e. after water loss.

9.3.3.3 Set speed. This quantifies how quickly a given adhesive forms a workable bond under given conditions. The bond must not develop before the assembled pieces are in their final position but must set quickly enough to withstand the stresses in any subsequent manufacturing operation.

9.3.3.4 Bond strength. The final bond strength of an assembly is determined by the weakest of three types of interaction, the adhesion of the adhesive onto the substrate, the internal strength of the adhesive and the internal strength of the substrate itself. The finished assembly may be subjected to many stresses in its lifetime and the bond strength must be high enough to withstand these both initially and after ageing.

9.3.3.5 pH. Specific pH ranges may be required, for example to prevent attack of the adhesive on aluminium machine parts.

9.3.3.6 Adhesion. Adhesives may be designed for specific adhesion to difficult substrates. The adhesive strength can be quantified by measuring the force required to remove adhesive strips from a substrate on a tensiometer.

9.3.3.7 Wet tack. This qualifies the amount of grab/hold an adhesive has before drying. It is important in holding assemblies together temporarily, especially in high speed operations.

9.3.3.8 Redispersibility. The properties of redispersibility are often important in allowing an adhesive to run cleanly on a machine, since buildup of dried adhesive can cause non-uniform adhesive coverage. This property is also becoming increasingly important in the search for recyclable packaging materials since in paper packaging, it may be the adhesive which is the most difficult part of the package to recycle.

9.3.3.9 Odour. A low odour can be important, especially in adhesives which are used for food packaging but also for the comfort of operators in large scale industrial processes.

9.3.3.10 Stability. Adhesives are required to remain stable and suitable for use, often for one year from their date of manufacture. This places requirements on their ability to maintain all of the above properties over time and not to settle out of phase. In order to do this, the adhesives must be capable of resisting extremes of storage temperature and, of course, attack from microorganisms.

9.3.4 Chemical nature and formulation of aqueous-based adhesives and their raw materials

An aqueous adhesive is designed to have specific values for the properties described in section 9.3.3, with the actual values determined by the adhesive end use and its method of application. Each adhesive is formulated with various raw materials to produce these properties and these raw materials contribute to the susceptibility of the product to microbial attack.

In addition to an emulsion or a blend of emulsions, a product may contain the following components.

9.3.4.1 Plasticiser. Plasticisers such as dibutyl phthalate, triacetin or diethylene glycol dibenzoate are added at up to 15%. They usually come in the form of non-volatile organic liquids and act by swelling emulsion particles, increasing the viscosity of the wet adhesive and the softness of the dry film. They are particularly important in the formulation of PVA homopolymers, without them these polymers are usually too brittle. Copolymers can be formulated without plasticiser.

9.3.4.2 Humectant. Up to 5% glycerine, urea, sugar or glycols can be added to emulsion adhesives. These hygroscopic materials reduce the speed at which the adhesive dries due to their affinity for water. They also control surface drying or skinning on the emulsion surface and they can be used to make the dry film very susceptible to picking up moisture, hence their use in adhesives for high speed machines.

9.3.4.3 Polyvinyl alcohol. Polyvinyl alcohols are water-soluble polymers that are added to emulsions to increase open time, tack and redispersibility and to control the viscosity. They are supplied as dry powders and need to be cooked in water to dissolve, this can be done *in situ* before adding the emulsions or as a separate premix. Both partially and fully hydrolysed polymers are used.

9.3.4.4 Tackifiers. Boric acid is often added to emulsions used in fast packaging applications since it cross links any polyvinyl alcohol present and significantly augments the tack. Tackifying resins from rosin or hydrocarbon sources can also be added for this purpose.

9.3.4.5 Thickeners. Small quantities of materials such as acrylics, cellulosics or polyurethanes are added to give a required viscosity or rheology or to control solids by allowing various dilution factors for a constant end viscosity.

9.3.4.6 Fillers. Calcium carbonates, calcium sulphates, clay or starch can be added at up to 50% to emulsion products. As well as enabling cost control in the application due to their gap filling properties, fillers help to give hard films and can affect rheology. Titanium dioxide can be used to control adhesive colour. Processing of filled emulsions requires strong mechanical agitation to ensure uniform dispersion.

9.3.4.7 Defoamers. Foam control agents or surfactants are added as antifoam to prevent foam formation or as a specific defoamer to destroy foam both during adhesive manufacture and during the final adhesive use. A foamed adhesive will give a poor, non-homogeneous film.

9.3.4.8 Surfactants. Wetting agents such as surfactants are also added to modify the surface tension of the final adhesive and then control its ability to wet out on certain surfaces. This is particularly important in the realm of pressure-sensitive adhesives and when bonding to plastics.

9.3.4.9 Preservatives. These are used to prevent microbial attack.

9.4 Microbial spoilage of polymer emulsions and adhesives

Having considered the uses, applications and chemistry of different types of polymer emulsions and adhesives it is clearly indicated that their aqueous-based formulations provide, in the water phase, all the essential ingredients required for microorganisms to reproduce and flourish.[30–33] Although the limited amount of unreacted residual monomer present will give some antimicrobial protection this is becoming less important as levels are reduced due to increasing legislation and awareness of environmental pollution. The trend is towards manufacturing and supplying emulsions free of organic emissions by employing methods to 'scrub out' these components so as to gain market share for more 'environmentally friendly' products. Conversely, certain monomers will actually aid in the spoilage of the emulsion itself. This statement, however, contradicts a widely held belief in the polymer industry that a residual level of monomer will act as a microbicide.[33] It is very much a question of levels. Microbial attack of vinyl chloride has been shown to occur[34] and still further it is known that vinyl acetate breaks down to give acetaldehyde and acetate[35] and both these metabolities are microbial nutrients at the levels found in typical polymer emulsions, where free vinyl acetate levels are of the order of 0.1%. At higher levels, as may have been found some years ago, acetaldehyde would have been at preservative concentrations. Free acrylate monomers may be available to some types of bacteria especially Pseudomonads. The precise mechanism by which they are utilised is not clearly demonstrated but probably occurs by hydrolysis of the monomer to acrylate and the alcohol. In the past, the residual monomer was regarded as an essential component of the emulsion system to afford some degree of antimicrobial protection, although this is only adequate for normal storage of emulsions in closed containers. For prolonged storage, or in vessels that are not well closed, the protective action from the residual monomer will slowly evaporate, so additional preservative will need to be added. Moreover, the protective effects from the residual monomer are not sufficient to withstand the additional populations of microorganisms likely to be introduced with pigments, fillers etc., or withstand heightened microbial activity in the presence of additional colloids.

The stabilising colloids are also a good source of microbial nutrient. Some surfactants contain long chain fatty acids and are in consequence fairly easily degraded. Substituted celluloses which are commonly used as stabilisers may be attacked by microorganisms, the amount of nutrient available from this source depending on the type and degree of substitution. Celluloses may also provide nutrients from their catalytic chemical breakdown throughout the reaction. The degradation of cellulosic materials by a variety of microorganisms has been reported abundantly in the scientific literature over the past 100 years as reported in two recent reviews on the subject.[36,37]

The polymer component itself is virtually unaffected by microorganisms, but the degradative effect on the water phase can lead to spoilage of the emulsion manifested by the production of a number of undesirable effects.

9.4.1 *Consequences and effects of microbial spoilage*

If microbial spoilage does occur, the consequences can be severe since it can cause deterioration in both the polymer emulsion and adhesive properties and potentially render the formulated product unsuitable for its intended purpose. The effects of microbiological spoilage of these aqueous-based synthetic products may often go unnoticed until it is too late to take remedial action. As such, if substrates are contaminated and the microorganisms and their byproducts, e.g. extracellular enzymes, are not inactivated, or at least their growth and activity inhibited heavily, they are able to decompose the product's ingredients, a consequence which manufacturers or consumers will realise when one or more of the following unacceptable symptoms are noticed.[30–32]

9.4.1.1 Effects on products.

1. Viscosity loss of the emulsion and of formulated products by breakdown of colloids and surfactant-stabilising systems of emulsions by microbial attack or by extracellular enzymes. Since the viscosity and rheology are fundamental to the product's applicability characteristics, a contaminated product can be unusable. Another consequence of viscosity instability is that phasing may be induced leading to a non-homogeneous product with thinner liquid on the top and thicker material on the bottom. This phenomenon is often termed 'syneresis' and is the result of settling and destabilisation of the product after effects to its rheological and viscocity properties.
2. An unpleasant odour or foul smells, e.g. sulphurous 'bad egg' odours produced by the sulphate-reducing bacteria which under anaerobic conditions can utilise oxygen from sulphates leading to hydrogen sulphide production. This type of odour production is often combined with a noticeable blackening in the colour of the product. In itself this odour can be unpleasant for the user but potentially more serious is the possibility that the odour will reappear if a dried adhesive film is remoistened. A contaminated adhesive used for food packaging could lead to product rejection by the final consumer. Other unpleasant rancid or musty odours are derived from various microbial by-products such as butyric acid and 1,6-dimethyl-3-methoxypyrazine.
3. Colour changes and discoloration as a result of the formation of insoluble sulphides such as iron sulphide produced by sulphate bacteria. Other microorganisms, particularly the pink yeasts such *Rhodotorula rubra* and *Sporobolomyces roseus*, and pigmented mould fungi can produce other discolorations. This is often visualised in the appearance of an oily yellow film on the product's surface.
4. Gas formation, often accompanied by unpleasant odours, produced by fermentative bacteria from the breakdown of colloidal thickeners, e.g. cellulose to glucose which is then fermented to yield acid and carbon dioxide. Such gas production is not usually noticed during production but can cause distortion and even splitting or 'lid popping' of containers.

5. Surface growth of the aerobic mould microfungi on the surface of products leading to surface disfiguration. This is particularly true where moulds grow on the surface giving a 'green carpet' on the surface of a product. In these cases, testing can reveal that only the surface is contaminated and the bulk of the product below the surface is free from microorganisms.
6. pH changes, e.g. lowering of pH due to metabolic activity of microbial cells by fermentative metabolic processes breaking down colloidal components of the product. This acid production can lead to product destabilisation as well as possible corrosion to storage vessels and containers.
7. Changes in properties affect the use of the formulated product due to microbial contamination. The resulting destabilisation and viscosity loss of the product as well as attack on the polymer particles can lead to a reduction in the average molecular weight of the polymer. In this context the bond strength of a contaminated adhesive would be expected to be less than that of a corresponding non-contaminated adhesive. In practice, this is not often seen since if the deterioration has progressed to this stage, the contamination is likely to be detected before the adhesive is applied.
8. Enzyme production, e.g. cellulases and amylases produced by bacteria and mould fungi may lead to viscosity loss in cellulosic and starch based products. Cellulase enzymes are effective at concentrations as low as 10^{-5} enzyme units per ml in breaking down long chain cellulosic molecules into shorter oligomers, thereby destroying their viscosity regulating ability and will thus cause a dramatic viscosity change of formulated products.[38]
9. Corrosion from metabolic byproducts and acid production.

Once a polymer emulsion or adhesive has been used, contamination can still take place especially in high moisture environments. To prevent this, dry film preservatives can be incorporated into the formulated product prior to application, e.g. ceramic tile adhesives and in water-based emulsion interior paints and exterior masonry paints.[5,13,39–41] However, dry film preservatives are not generally necessary in most aqueous-based adhesives and polymer emulsions.

In general, for end uses where a polymer emulsion or adhesive will be in contact with an environment conducive to the propagation of microbial infestation, a contaminated product cannot be ignored as a possible source of further contamination. All the effects detailed above resulting from microbial spoilage of polymer emulsions and adhesives must be fully investigated in terms of product quality complaints because they can lead to significant economic loss to the manufacturer if products reach the consumer in a deteriorated condition.

9.4.1.2 Effects on the environment.

- Appearance of sour and putrid odours.
- Health risks.

These effects are less apparent to the manufacturer than to the end user of the polymer emulsion, but they must not be overlooked. Both the foul odours, particularly from hydrogen sulphide, and the actual presence of microbial spores can have a serious effect on man. Hydrogen sulphide levels in contaminated products have never been shown to reach toxic concentrations but even very small amounts can render a product unsaleable. In both these examples there are strict guidelines on occupational exposure. In the case of microbial spores, overexposure can lead to respiratory disorders and asthmatic symptoms.

9.4.1.3 Effect on production plants.

- Blocking of pipes,
- Foam formation,
- Filtration problems,
- Corrosion of plant vessels and machinery.

These effects are more physical and are often associated with the result of a microbiological spoilage outbreak in a manufacturing plant. They can range from simple situations that are relatively easy to eradicate with the use of sanitation measures combined with the use of effective biocides, to more complex situations if spoilage is allowed to go unchecked for prolonged periods. Then significant economic loss to the company is realised, ranging from that caused by long periods of downtime whilst the problem is resolved to the need for major capital investments to replace pipework and tanks or even corroded metalwork in the plant.

9.4.1.4 Effects on production. The effect on product quality can severely damage a manufacturer's reputation in the market to the extent that they may lose the business in part or totally, or even suffer high financial claims for the end-user's losses which may be settled in or out of court. All this can lead to interrupted work in production and additional time which must be put in to rectify the quality incident.

9.4.2 Types of spoilage microorganism encountered

The majority of bacteria, yeasts and mould microfungi commonly living in the environment can be found in samples of polymer emulsions, as well as in formulated products thereof. Microorganisms are extremely versatile in their modes of nutrition, many organisms have very simple requirements needing only an organic carbon source and inorganic source of nitrogen, phosphorus and sulphur. Some individual genera are capable of utilising a wide variety of compounds, e.g. *Pseudomonas* spp. which can grow in many hundreds or organic compounds ranging from simple compounds like acetate to complex organic chemicals used as disinfectants in hospitals, e.g. phenols and organohalogen compounds often used as ingredients in the formulation of many biocides.

Various chemical, physical and environmental factors will determine the types of microorganism that can be isolated from different synthetic polymer emulsion and adhesive systems. For example, Evertsen in 1988[31] noted that products with a pH of less than 4–5 were rarely susceptible to microbial attack although slight contamination was occasionally found in acidic plasticised vinyl acetate homopolymers. Evertsen also found that alkaline acrylic and styrene acrylic types of polymers were particularly susceptible to bacteria. In contrast to these findings the authors (unpublished data) and Elsom in 1988[30] found that acidic polymers, including

Table 9.5 Types of microorganism isolated from polymer emulsions

Microorganism type	Genera and/or species
Fungi and moulds	*Geotrichum candidum* *Aspergillus niger* *Aspergillus terreus* *Penicillium ochrochloron* *Fusarium solani* *Trichoderma viride* *Cladosporum herbarum* *Alternaria alternata*
Yeasts	*Saccharomyces cerevisiae* *Candida albicans* *Rhodotorula rubra* *Pichia* spp. *Torulopsis* spp.
Bacteria	*Pseudomonas aeruginosa* *Pseudomonas cepaciae* *Pseudomonas fluorescens* *Pseudomonas paucilomobilis* *Pseudomonas putida* *Pseudomonas stutzeri* *Aeromonas hydrophila* *Escherichia coli* *Alcaligenes faecalis* *Arthrobacter* spp. *Achromobacter* spp. *Citrobacter fruendii* *Bacillus subtilis* *Ochrobactrum anthropi* *Micrococcus luteus* *Xanthomonas maltophilia* *Proteus vulgaris*
Anaerobic bacteria	*Delsulphovibrio desulphuricans*

homopolymers of vinyl acetate and copolymers of vinyl acetate and ethylene that are stabilised with polyvinyl alcohol and/or starch or cellulose colloidal systems, in the pH range 3.5–6.5, were particularly susceptible to yeast and mould fungi. It is postulated that the colloid system has an influence on the susceptibility of the product and in the case of the copolymers of vinyl acetate and ethylene it may be that ethylene has a growth-enhancing effect.[42] Table 9.5 lists some of the major contaminating microorganisms that have been encountered, isolated and identified by the authors from spoilt samples of polymer emulsions as well as formulated products.

9.4.2.1 Moulds and microfungi. In the authors' experience the principal organism isolated is *Geotrichum candidum*. This organism is very important in the biodeterioration of polymer emulsions and has been isolated in many cases of spoiled product. The reason for its prevalence is due to its ability to spread rapidly and disperse throughout the substrate matrix by the production of arthrospores or fragmenting hyphae. This arthrospore-forming process of reproduction has often led some microbiologists to misclassify *Geotrichum* as a yeast since the colony characteristics on agar plates make it appear like a powdery yeast-like species. *Geotrichum candidum* is a known degrader of polyvinyl alcohol and cellulose and starch-based colloids and its degradative powers have also been reported in the polyvinyl acetate system.[43–51]

Other moulds and fungal organisms have also been isolated but these are more likely to be general spoilage organisms growing on components of the system. In 1983 Jakubowski *et al.*[8] found many mould contaminants comprising most of the well-known biodeteriogenic species such as *Aspergillus*, *Penicillium*, *Fusarium*, *Geotrichum*, *Scopulariopsis*, etc. These mould species are more often found on the surface of the emulsion in storage vessels at the air/surface interface. This is seen as a surface layer of mould growth, or small coloured spots, and arises when non-volatile preservatives are employed in the emulsion giving no surface protection. As such condensate from the lid of the containers will drop onto the emulsion's surface and being unpreserved will thus become rapidly contaminated from spores in the environment. Moulds and fungi, however, are unlikely to have a major role in the polymer degradation itself. However, their control is essential in order to maintain a well-preserved product of extended shelf life.

9.4.2.2 Yeasts. Various types of yeast are regularly isolated from polymer emulsions. The yeast types found and isolated are often fermentative and distributed throughout the emulsion. The common types of yeast include the genera of *Saccharomyces*, *Rhodotorula* and *Torula* and these have been noted by the authors as well as being reported by

Jakubowski *et al.*[8,52] who has also isolated *Sporobolomyces* on a number of occasions.

9.4.2.3 Bacteria. Bacteria may be found in specific types of polymer emulsions often of high pH, e.g. acrylic and styrene acrylic types reported by Evertsen.[31] After isolating Gram-negative Pseudomonads from every contaminated low viscosity paint examined, Winters and Guidetti[53] and Oppermann and Goll[54] concluded that this group of bacteria is involved in the great majority of instances of loss of viscosity of emulsion/dispersion paints. This conclusion was supported by Carter[55] who stated that the Gram-negative, non-sporing microorganisms such as those belonging to the genera *Pseudomonas*, *Proteus*, *Enterobacter*, *Escherichia* and *Flavobacterium* are most frequently associated with degradation, whilst Gram-positive microorganisms are regarded as secondary invaders. However, Buono *et al.*[56] studied the cellulolytic activity of several strains of Gram-negative and Gram-positive bacteria and found the Gram-positive microorganisms (*Bacillus* spp.) to be more active than the Gram-negative microorganisms. Furthermore, in more recent work (unpublished) the authors have isolated other less common species as well as those previously found by other workers on a number of occasions (see Table 9.5).

In the authors' experience the most common spoilage bacteria isolated from polymer emulsions are non-lactose fermenting Gram-negative rods. The most predominant genera are the pseudomonads, which are organisms with very powerful biodeteriogenic capacities. Other genera are also isolated from time to time in specific emulsion types, including occasionally Gram-positive organisms.

It is probable that, as strains of both Gram-negative and Gram-positive bacteria are capable of polymer degradation, the type of microorganism responsible for the degradation of particular polymers will be dependent upon the polymer constituents and upon the accessibility of each microorganism to the polymer.[43–45] In either case, it is imperative that the polymer manufacturer uses a biocide which is active against both Gram-negative and Gram-positive bacteria.

9.4.2.4 Anaerobic sulphate-reducing bacteria. Anaerobic sulphate-reducing bacteria (SRBs) have also been found in polymer emulsion systems by the authors and many other workers including Elsom[30] and Gillatt.[57] These microorganisms tend to occur in systems that have been previously contaminated with numerous aerobic species and are generally regarded as the last organisms to arise in microbial ecological succession. SRBs tend to occur at the bottom of bulk storage vessels/tanks and cause blackening of the emulsion and are responsible for foul sulphurous odours due to the terminal electron acceptor in their respiratory chain being hydrogen sulphide.

Table 9.6 Sources of microbial contamination

Source	Control
Raw materials (including process water) Environment Manufacturing equipment Operators Packaging materials	Manufacturer can control
In use	Manufacturer cannot control

Oppermann and Goll[54] also investigated the incidence of anaerobic microorganisms and in their study of seventy contaminated samples found a wide variety of species of which *Bacteroides*, *Clostridium*, *Desulphovibrio* and *Bifidobacterium* were the most common.

9.4.3 Sources of microbial contamination

As stated previously polymer emulsions when first produced are sterile. It can therefore be assumed that spoilage microorganisms must gain access subsequent to the completion of the polymerisation reaction. When a major outbreak of contamination does occur, initial responses tend to centre around methods of wiping out the immediate problem. Identification of the source of contamination is considered as an afterthought, if at all.

The sources of contamination are limited and generally within the control of the manufacturer (Table 9.6). Realistic action taken to identify these sources, quantify the potential risk from each source and reduce or eliminate the degree of contamination at that source can save a great deal of time and expense later.

9.4.3.1 Aerial contaminants and the environment. This is often overlooked because microorganisms cannot be seen with the naked eye. It is accepted that the human race lives in equilibrium with a wide spectrum of microorganisms in the environment. A dusty atmosphere from building operations, animal pollution from farms, stables etc., proximity to orchards, gardens, etc. leading to yeast and mould contamination, are typical hazards for any factory making aqueous-based products that are vulnerable to microbiological spoilage.

Polymer emulsion and adhesive manufacturing plants tend to be open to the atmosphere, in many locations, due to the nature of the chemicals used. The open filtration and sieving processes are often the stage where most microbial contamination can occur. For example, Diehl[10,11] reported that air may contain 500–2000 microorganisms per m^3 which may contain spores of some of the polymer emulsion spoilage species.

9.4.3.2 Water. The process water is unlikely to be a primary source of microbial contamination, as any microorganisms present tend to be destroyed in the reactor, however, some workers have reported that certain spore-forming microorganisms and some yeasts can survive the polymerisation process and these have been isolated from fresh polymer emulsions soon after production. It is therefore necessary to ensure that process water is treated before its use since water may come from a number of sources including wells, boreholes or even rivers. Chlorination alone may not be totally effective to treat the water, especially if organic matter is present, when residual contamination may be left. As suggested by Duddridge[58] even supposedly good quality town tap water will contain bacteria, for example *Pseudomonas* species at up to 1000 cells per ml. In order to improve the process water quality, an ultraviolet treatment system for the water has been installed in one of the Vinamul polymer emulsion production plants. This process is checked on a monthly basis to ensure it is functioning properly to produce sterile water.

Assuming adequate control measures are taken to ensure the supply water is sterile the most likely source of microorganisms from water would be from untreated residual wash water if it is left in mixing vessels and hoses, etc. which could lead to contamination.

9.4.3.3 Post-added raw materials. Any new materials added at the end of the manufacturing process can be contaminated with microorganisms, e.g. addition of thickening agents such as polyvinyl alcohol. Nichols[12] suggested that raw materials were probably the single biggest potential contamination source. Malcolm[6] arbitrarily classed raw materials into three groups; non-aqueous synthetics, aqueous synthetics and natural.

Some of the powdered raw materials, especially those originating from natural sources, such as extenders/fillers can often be contaminated with dormant spores of bacteria and fungi which can germinate once an aqueous environment is provided.

Liquid raw materials such as defoamers, surfactants, starch solutions (Liquid Natural Polymer, LNP) and hydroxyethyl cellulose solutions may themselves be susceptible to microbial attack and, unless carefully manufactured and protected with biocides, can also introduce contamination into the product.

9.4.3.4 Manufacturing procedures. Increased attention to quality control, and the introduction of ISO 9000 quality systems, has led to more detailed examination of the raw materials used. However, even if raw materials are free from infection before use, the method of manufacture itself may result in contamination of the end product. The addition of biocide to a product as soon as practicably possible after the polymerisation process is therefore recommended. Ideally, it should be added to the product as soon as it leaves the reactor on pump out to the blender. Incorporation of the biocide can be made with the dilution water required to bring the product into its solids content specification. It must also be noted that if thickening solutions of colloids are employed post-manufacture these also should be protected with a biocide.

A number of other areas in the manufacturing procedures are also important sources of microbial contamination and these should be noted. These are detailed further below.

9.4.3.5 Pipelines. It is not uncommon that excessively long and complicated runs of transfer pipelines with sharp bends and deadspots are present in manufacturing plants which contain accumulated product, often diluted with wash water. Microbiological contamination of the dilute emulsion can occur rapidly in such pipelines which can then be a source of inoculum for fresh product that is pumped through the system. Furthermore, flexible hoses used for transfer of product can, if improperly cleaned and stored, also become contaminated, especially if diluted product, resulting from ineffective cleaning, accumulates.

9.4.3.6 Mixing vessels and bulk storage tanks. Dirty vessels and bulk storage tanks, at either the manufacturer or the customer, may be an important source of microbial contaminants.

Open hatches for the passage of air during the filling and emptying of storage tanks will also allow aerial inoculation of the product with contaminating microorganisms from the environment. This can easily be resolved by fitting an air filter suitable to permit the passage of air to prevent a vacuum from building up and filter out particulate dusts and microorganisms.

Condensation will wash contaminating microorganisms onto the surface of the bulk phase of the product where, especially if the vessel remains unstirred for some time, dilution of the surface layer occurs along with some degree of syneresis of the product and hence dilution of the biocide allowing profuse microbial growth to occur. When mixing recommences, a high microbial loading will enter the bulk phase and may 'overwhelm' a biocide.

Empty vessels and storage tanks should be cleaned down at regular intervals but the residual wash water, if left in them, is effectively diluted product and will rapidly become contaminated, even in just a few hours in a warm environment. If the residual wash water is left in the vessel or tank it will provide a heavy source of contamination for the next batch produced. The use of an effective biocide or disinfectant can be employed to prevent this situation from arising.

In addition to regular cleaning and sterilising of vessels and storage tanks other measures can be taken to prevent problems in such vessels. These can include fitting a stirrer to ensure the product is well mixed at the surface and in the bulk, or if this option is too expensive a simple device fitted to spray a dilute biocide or disinfectant onto the surface of the product at regular intervals can be employed. Either of these options will minimise microbial contamination in vessels and storage vessels.

9.4.3.7 Road tankers. Road tankers are often not cleaned properly between loads and therefore may be an important source of microbial contaminants. Strict guidelines should be rigorously adhered to to ensure tankers are properly cleaned between loads and as with storage tanks the residual wash water should be treated with a biocide.

9.4.3.8 Product storage containers: drums, mausers and buckets. If a sterile emulsion is filled into a sterile, impervious container, microbial growth will not occur.

First the container itself may not be biologically clean. If stored in unsuitable conditions, microorganisms may be present before the product is filled into it. Plastic containers will often attract microbial spores and dusts electrostatically and these can lead to contamination of the product when it is filled into the container. When the container is filled and the lid is applied, condensation may irrigate the lid and form a pool of diluted, underprotected product on the product's surface in which contaminants can grow. It is therefore important that biocide is used that has broad spectrum activity against bacteria, yeast and the mould fungi as well as some degree of headspace protection and is stable in the product for the duration of the product's life.

9.4.3.8.1 Case study: a polymer emulsion spoilage scenario. A typical scenario for microbial contamination of polymer emulsions exists due to the fact that an emulsion will spend most of its time in a storage system, whether it is a bulk tank, road tanker, drums, etc., and it is here that the most likely sources of infection will arise. When a storage tank has been in use for some time with repeated filling and emptying then polymer inevitably builds up on the walls of the tank. Water condensing on the polymer film will wash out any preservative leaving it open to microbial contamination, which enters with air that is drawn into the tank on emptying. In the absence of preservative the microorganisms may multiply. The rate of formation of the polymer film and its subsequent infection is influenced by the type of emulsion, the type of storage system, the nature of the tank lining and the ambient temperature. It is, however, a fairly slow process taking months rather than weeks to occur. Once an infection becomes established in the polymer film then microorganisms can potentially be washed continually onto the surface of the contents of the tank by condensation. In an unstirred tank, this may give rise to a thin layer of infected water on the surface of the emulsion. On emptying the tank it may be hosed down, leaving a pool of diluted microorganisms at the bottom of the tank. As those organisms washed off the polymer film on the tank wall are those which are capable of growth on the constituents of the emulsion then rapid growth should occur in this diluted emulsion. On receipt of the next delivery polymer emulsion is introduced into the tank with its infected material. Thus, a high level of contamination can occur.

Another means by which microorganisms may be introduced is by way of a 'wet breather' system which is fitted to some storage tanks to prevent the contents from skinning over. There are a number of arrangements whereby air entering the tank is first drawn over wet surfaces. This type of system will also act as a filter removing dust from the atmosphere. The dust filtered out may provide nutrients and the wet breather may become heavily fouled with microorganisms. Air that is drawn through a fouled wet breather may contain large numbers of droplets containing significant numbers of microorganisms.

9.5 Prevention and control of microbial spoilage in polymer emulsions and adhesives

In the previous sections of this chapter the emphasis was on the susceptibility of synthetic polymer emulsions and formulated adhesives, the ease at which they can be spoiled by microorganisms and some of the effects of contamination both in the emulsion and the formulated products. Therefore to prevent microbial contamination of aqueous-based products and so retain their beneficial properties, it is essential that biodeterioration is prevented. This is effected in two ways: by controlling the number of microorganisms which may come into contact with the product and by the use of an adequate level of a suitable preservative. It is also very important to ensure that the preservative is stable in the product for the time frame over which protection is required. These criteria equally apply to the end user of the emulsion system, e.g. the paint and adhesive manufacturer for without their cooperation the useful life of the emulsion will be greatly reduced.

Preservatives are designed to protect clean products from occasional microbial contamination. At typical use levels, preservatives are not able to withstand continual repeated challenges of high levels of microorganisms. In such situations, the preservative can be consumed and the product will no longer be protected from subsequent microbial contamination.

Unlike formaldehyde, most other preservatives have very low vapour pressures which result in little or no preservative present in the storage tank headspaces. As a result, microorganisms introduced into the headspace of a storage tank can often proliferate in the condensation on the walls and ceiling of the tank and on the agitator shaft. When new material is added to the tank, all of the preservative can be consumed in eradicating these microorganisms.

In addition, the greater the range of microorganisms exposed to a preservative system, the greater the likelihood of encountering microorganisms able to survive in the presence of the preservative. Thus, continued poor storage and handling conditions can lead to development of a microbial population which is not controlled by the preservative system, particularly when an inadequate level of preservative is being used. In situations where large microbial populations are exposed to inadequate levels of preservative, some microorganisms may respond to the preservative and alter their susceptibility. Development of a resistant, or more precisely tolerant, microbial population is a common problem for some preservatives such as formaldehyde. However, due to the complex chemical nature of preservative systems true resistance is an uncommon event.

Microbial contamination in storage tanks can be greatly reduced or eliminated if the following procedures are implemented in conjunction with the use of adequate levels of a suitable preservative.

1. Supply a microbe-free atmosphere to the tank headspace. After thoroughly cleaning and sanitising the storage tanks, options such as the following can be initiated:
   (a) Use nitrogen or air purified with ultraviolet radiation or submicron filters, or use clean compressed air.
   (b) Keep a small positive pressure (a couple of inches of water) of the clean atmosphere in the tank, or keep a low continuous flow through the tank. The advantage of a low continuous flow is a significant reduction in condensation. If using a low continuous flow, ensure that the flow sweeps the entire headspace and that no other air enters the tank.
2. Monitor the microbial quality of raw materials and products, particularly water.

   The microbiological quality of water supplies varies considerably. Bore holes and wells usually contain relatively high levels of a range of organisms including *Pseudomonas* spp. Thus it is important to maintain water quality by the adequate use of chlorine, or other biocides or even by physical agents such as ultraviolet sterilisation systems and more recently by the use of ozone. Deionised water which is stored, generally needs to be chlorinated to prevent microbial growth. Maintaining a residual chlorine level of 2 ppm in water storage tanks is an excellent means of controlling microbial growth.

   Other heavily contaminated raw materials should also be treated prior to introducing them into the manufacturing environment, especially those materials that are aqueous-based products, e.g. surfactants, defoamers, emulsifiers, thickeners and colloids etc.

Typical microbiological tests for quality control can range from agar-plate techniques using specific microbiological media or dipslides to rapid methods such as impedance measurements. The newer rapid methods allow for screening of products in less than 24 h. Systems such as the Bactometer have been used for a number of years in the food[59] industry and are increasingly employed in the detergents,[60] personal products, cosmetics[60] and toiletries[61] and pharmaceutical industries.[62,63] Vinamul currently use such a procedure and results have shown that impedance methodology is equally applicable to the emulsion and related industries, e.g. paints, adhesives, etc.[52,64,65]

3. Ensure that microorganisms are not introduced into the storage tanks when new shipments, batches, etc., are pumped into the tanks:
   (a) Flush and sanitise all the lines and hoses frequently and do not allow product to stagnate in pipes.
   (b) When possible, eliminate bends in piping and hoses, valves, lips, ridges and lines that do not drain (<5% slope). These all provide excellent environments for bacteria to gain a foothold and establish biofilms. Biofilms are protective layers produced by bacteria and are difficult to remove.
   (c) Ensure that ambient air is not inadvertently pumped into storage tanks when new additions are made to tanks.
   (d) Ensure microorganisms are not introduced into storage tanks by level indicator air sources. If level indicators bubble ambient air through the tank, install a submicron filter in the air line or switch to purified compressed air.

9.5.1 *Manufacturing plant hygiene, cleaning and sanitisation methods*

In order to prevent contamination problems, a suitable sanitisation programme is extremely important.[66–68] All good emulsion producers and end users will have such a regime in place. Typically, this would involve a cleaning and sanitisation programme to clean regularly the storage tanks, pipework etc. with high pressure water jetting apparatus to remove polymer residues on surfaces combined with caustic washing and possibly sterilisation with dilute sodium hypochlorite (e.g. 1000 ppm available chlorine). Such a regime carried out three to four times a year at the emulsion producer and the end-users' plants will be adequate to reduce the risk of microbial contamination. Additionally, this sanitisation programme should be carried out in the event of an infection prior to refilling with fresh product. To prevent skinning at the emulsions surface as well as surface contamination at the air interface the most appropriate design of bulk tank is one with a stirrer. This greatly improves storage times and quality of the product.

The following procedure is commonly used for cleaning and sanitisation of storage tanks.

9.5.1.1. Cleaning. Drain the lines and flush well with water. If possible, combine with mechanical cleaning, such as water pressure surges, or putting a 'pig' through the line, to aid in the removal of any biofilm that may have formed.

9.5.1.2 Sanitisation.

9.5.1.2.1 Halogenation. Fill the tank with 0.02–0.20% sodium or calcium hypochlorite, or similar bromine-release compounds which pose less of a threat than the organochlorine effluents after hypochlorite treatment. Allow the solution to stand in the tank for 24 h. Flush all lines well with 0.02% of the sanitising solution. Rinse the tank and lines well with microbially clean water.

Alternatively, tanks can be thoroughly swabbed with a 5% hypochlorite solution, allowed to stand for 24 h, then rinsed well with microbially clean water. This method reduces the volume of waste water.

9.5.1.2.2 Steam. Stainless steel tanks and lines can be steam cleaned. When using steam, ensure all parts of the system being cleaned reach 75–80°C for 15–30 min. One way to monitor the steam sanitisation process is to place temperature stickers on equipment throughout the system. Stickers that do not change colour indicate areas which are not sufficiently sanitised.

9.5.2 Preservation of polymer emulsions and adhesives with antimicrobial agents

It is of course also important that an emulsion and adhesive is adequately preserved and to this end a preservative is added during the manufacturing process.[69,70] The type and level of the preservative system used may vary with the type of emulsion, because as previously stated due to differences in available nutrients, pH and chemistry of the emulsion matrix some are more difficult to preserve than others. Generally, the emulsion manufacturers aim to protect and preserve their products against microbial spoilage under reasonably good storage conditions for up to 12 months. It is not the aim to add sufficient preservative to provide protection for any finished product arising from it. New impending European legislation, the Biocidal Products Directive due in 1997,[71] will ensure that the amount of preservative used is adequate for protection of the product it is contained within without being in excessive quantities. This will be determined by appropriate efficacy tests for the preservative in that product. If a product was to be supplied with a higher level of a preservative, it would be regarded by the Directive as a biocidal product itself. This would mean it would have to be listed with toxicity data, environmental fate and biodegradability data as well as efficacy data in the annexe of the Directive.[71]

In order to choose the right preservative from a wide range of biocides available that are based on many different active ingredient chemicals and stabilisation systems a wide variety of factors must be considered. Some of these factors are listed here.

- Broad spectrum antimicrobial activity
- Stable over a wide pH range
- Stable at elevated temperatures (to 50°C)
- Stable to various chemical agents
- Aqueous-based solutions or dispersions
- Compatible with a wide range of polymer types
- Low environmental impact
- Low toxicity, e.g. skin irritancy
- Ease of handling and safe to factory workers
- Relevant regulatory approvals, e.g. FDA, BGA, EPA, EEC
- Cost effective

Though cost effectiveness is a factor, less emphasis is placed on this attribute now than 5–10 years ago. Toxicity has taken on an increasing importance in recent years as have the environmental implications of the waste disposal of the emulsion. Related to this is the issue of safety of handling and the use of diluted solutions is preferred. Also dilute solutions are easier to pump and meter into products. Solid biocides pose handling difficulties. A dispersion of a solid biocide is easier to handle if a solution cannot be prepared.[72–74] In the case of compatibility with manufacturing processes some preservatives appear to give good results in the laboratory but when used on the plant scale have failed to perform. Careful consideration of compatibility of preservatives with the emulsion in question must be given and appropriate tests carried out to verify it. It is also necessary to look for colouration effects, viscosity changes, gelling, grit formation, etc. Some biocides fail in the plant due to the presence of excess chemical species that are antagonistic to most preservatives, e.g. redox chemicals. The stability of the preservative to these sorts of chemical must therefore be understood.[24] Lastly, suitable efficacy tests must be carried out in order to ensure that the biocide is adequately biostatic or biocidal for its intended end use.

9.5.3 *Chemistry of active ingredients as antimicrobial agents*

Some of the different active ingredient chemicals that have been employed as preservatives over the years are summarised here.[74]

- Heavy metals, e.g. Hg, As, Sn
- Phenolic compounds, e.g. halogenated cresols
- Aldehydes, e.g. formaldehyde, glutaraldehyde
- Oxoazolines, e.g. Amical
- 1,2-Dibromo-2,4-cyanobutanes, e.g. Tektamer
- Thiocyanates, e.g. Tolcide
- Triazines, e.g. Dazomet
- *N*,*S*-heterocycles: various substituted isothiazolinones and benziso-thiazolinones, e.g. Kathon, Proxel and Promexal

One of the major groups of biocides used for aqueous-based product preservation are the substituted isothiazolinones. These classes of anti-microbial chemicals have been known for more than 15 years. Both the 1,2-benzisothiazolin-3-one ('Proxel' types) and the mixtures of the two substituted isothiazolinones, 5-chloro-2-methyl-4-isothiazolin-3-one and 2-methyl-4-isothiazolin-3-one ('Kathon' and 'Acticide' types) have been employed for the longest as aqueous-based product preservatives. Both have advantages and disadvantages, depending upon the system, and these have to be determined when evaluating and selecting the right biocide for the system. Two other substitutions are the 2-*n*-octyl-isothiazolin-3-one and the 4,5-dichloro-2-*n*-octyl-isothiazolin-3-one, which have primarily been employed as dry-film preservatives in surface coatings.

By their chemical nature, biocides are toxic to the microbial cell by targeting the microbial cell wall cytoplasmic membrane, cytoplasm and various organelles or combinations of these cellular constituents. The effect of biocides on several cell functions and their relative importance may vary depending on the value of that function to the species of microorganism challenged. Altered sensitivity towards biocides may reflect change in accessibility of targets within cells and the way it affects that target. These changes in sensitivity may be brought about by changes in the physiology of the organism or the environment (such as the formulation or process).

9.5.4 *Physical and chemical factors influencing the choice of preservative strategies*

The choice of preservative must be taken knowing the nature of the raw materials in the polymer emulsion and adhesive formulation and the chemical environment they create. Some of the factors which must be considered are:

9.5.4.1 Chemical constituents. The choice and level of preservative required is partly determined by the raw materials in a formulation since they may be contaminated when received or particularly prone to spoilage. A guideline to susceptibility can be written to predict the extent of protection required:

- Most susceptible. Fillers such as china clay and calcium carbonate (strong differences can be seen due to the origins of the filler); natural materials such as starch; dextrines.
- Susceptible. Cellulose derivatives; polyvinyl alcohol solutions.
- Less susceptible. Emulsion bases, these are generally attacked through their stabilising system rather than inherent attack on the polymer. With these we can produce an approximate susceptibility series starting with the most susceptible.
    - PvOH-stabilised PVA
    - PvOH-stabilised VA/E
    - Cellulosic-stabilised VA/E
    - Rubber emulsions (SBR, natural and polyurethane)
    - Surfactant-stabilised VA/E or acrylic
    - Styrene acrylate
    - Acrylamide, *N*-methyl acrylamide, *N*-butyl methyl acrylamide
- Least susceptible. Concentrated or solid surfactants; defoamers.

With emulsion bases, the residual monomer levels are important since these free monomers can themselves have an antimicrobial effect. The current trend of using low VOC (volatile organic component) emulsions for adhesive manufacture has to be run hand-in-hand with preservative testing. This is particularly important in Scandinavian countries where the move to reduce VOC levels is being accompanied by moves to restrict allowed preservative levels. In some cases, the techniques used in emulsion polymerisation to reduce the VOC level can destroy the preservative.[75,76]

9.5.4.2 pH. Preservatives are active over specific pH ranges and may prove completely ineffectual or decompose outside of these. Generally polymer emulsions and adhesives are prone to yeast and mould growth at pH of 3–7 and to bacteria at pH 6–8. Over pH 9 and under pH 2, adhesives are less prone to attack. Some of the acid-catalysed cross-linking adhesives used at very low pH as woodworking adhesives may not require preservative addition at all. Standard VA/E and PVA emulsions are generally in the pH range 4–6 so a wide range of preservatives can be used. For alkali-converted starches and casein adhesives where the pH is over 8.5

the chlorinated isothiazolinones (e.g. 'Kathon') would be ineffective and the benzisothiazolinones (e.g. 'Proxel') or the dibromocyanobutanes (e.g. 'Tektamer'), or the cationic biguanide hydrochlorides (e.g. 'Vantocil') are used instead.

Care with pH levels must also be taken when adding pre-dissolved preservatives. Sodium benzoate and the sodium salts of *p*-(4)-hydroxybenzoic esters all give alkaline pre-mixes which may cause coagulation if added directly to acidic emulsions. Drift of pH with time should also be monitored for these.

9.5.4.3 Solids. Dilution may also play an important role in preservation. Very low solids products have shown a particular sensitivity to attack and this can be exacerbated if regulatory restrictions fix allowable levels based on the dry component of the adhesive and not the wet state.

9.5.4.4 Climate. Polymer emulsions and adhesives destined for export to warmer countries can be particularly susceptible to growth due to condensation of dilute films onto the product's surface. These low solids films are readily attacked by bacteria.

9.5.4.5 Compatibility. Compatibility problems can be observed in addition to simply acid/base interactions and should be verified before finalising an adhesive formulation. This can be seen with polyurethane emulsions where addition of some biocides can cause coagulation or with caseins where formaldehyde or formaldehyde-release preservatives can cause gelatinisation.

9.5.4.6 Processing. Processes involving 'cook-ups' require the preservative to be heat stable or to be post-added after cooling. Formaldehyde-release preservatives are generally unstable over 50°C.

9.5.4.7 Redox state. Excess oxidising agents occasionally used in adhesives, e.g. hypochlorite, can have deleterious effects on certain preservatives. The same is true for reducing agents such as sodium metabisulphite which is used to maintain a light colour in starch-based adhesives as well as being used as initiators and end-stripping chemicals in polymer emulsions.

9.5.4.8 Nucleophiles. Heterocyclic preservatives can be decomposed by nucleophilic attack rendering them inactive. Primary, secondary and even tertiary amines can deactivate some isothiazolinone preservatives (see chapter 2). Sulphur nucleophiles like mercaptans show the same effect.

9.5.4.9 Physical form. For aqueous-based products, preservatives in liquid form are easier to incorporate than powders. Liquids can also be dispensed using automatic metering systems to reduce the amount of operator exposure to the preservatives during manufacture.

9.5.4.10 Solubility. Given the importance of preservation during processing and storage for polymer emulsions and adhesives, the preservatives used should be soluble in the aqueous phase.

9.5.4.11 Versatility. A broad spectrum of antimicrobial activity is required.

9.5.4.12 Cost. In some cases the preservative can contribute a relatively high proportion of the raw material cost of a formulation. The high cost of some preservatives can be offset by their potency which may allow them to be used in low doses.

9.5.5 *Test methodology for the selection of biocides and analytical methods for their quantification*

In order to make recommendations for the use of a chosen or selected biocide for the preservation of polymer emulsions and formulated products the first criterion that must be satisfied is that of its microbiological efficacy, i.e. how effective and good it is at giving prolonged protection to the product formulation in terms of preventing microbial infection. In order to test for the microbiological efficacy of biocides in product formulations two specific testing procedures should be employed.

9.5.5.1 Microbiological challenge testing for biocide efficacy. The microbiological effectiveness of a biocide is tested by repeated sequential challenges with a specific acclimatised mixed microbial inoculum.[51,64,65,77]

At present there are no standard methods in existence for determining the biocidal activity of products intended for the preservation of polymer emulsions and adhesives. However, there are a number of international standards such as ISO, CEN, ASTM, BSI, DIN, AFNor, as well as many industry associations and groups. However, it is more often the case at present that in-house methods are preferred. These in-house methods can be adapted to manufacturer's own products and use spoilage microorganisms that are highly specific to the industry in question. There is of course a disadvantage with such testing in that it is not very reproducible and as such comparisons are often not easy if at all possible.

The International Biodeterioration Research Group (IBRG) was originally established under the auspices of the Organisation for Economic Cooperation and Development (OECD). It is an international group of scientists from industry, academic institutes, national testing institutes and organisations and government and privately funded test laboratories. One aim of the IBRG is to develop standard test methods to determine the antimicrobial activity of microbicides in a range of products that can be acceptable to the EU for the Biocidal Products Directive requirements on efficacy testing of active substances.[71] One IBRG project is currently to develop a standardised method for testing biocides for use in polymer emulsions[78] that can equally be applied to adhesives and related formulations with some minor adaptations.

The group is currently developing a standardised method for biocides for polymer emulsions that will give statistically valid results when comparing different biocides in different polymer formulations. Areas of interest include the microorganisms in the challenge inoculum, how to adapt them to growth in the product formulation, inoculum density and whether a specific diluent should be employed, pH of the substrate and whether ageing of the polymer is important. Some workers go further by questioning whether it would be possible to develop a chemically defined medium that is representative of a typical polymer matrix (a very complex medium) which could be used as a model system. In this way, biocide efficacy for such products can be evaluated under standard laboratory parameters thus generating results that are statistically viable without the need to work in the complex matrix itself. The ultimate aim of the IBRG is to have a suitable test method available more quickly, in time for the EU Directive, which both biocide producers and users alike can use to test active ingredients and thus generate meaningful comparisons of relative effectiveness and efficacy.

Once a satisfactory biocide has been found which gives the best microbiological protection in the product formulation, the next stage in the assessment of the antimicrobial compound is to ensure its stability in the specific polymer emulsion and/or adhesive system.

9.5.5.2 Biocide active ingredient compatibility with the formulation. Some relatively simple tests can be performed in order to check the biocide's chemical compatibility in the polymer or adhesive matrix to ensure there are no undesirable side reactions such as colour changes, odours, gelation, precipitation and sedimentation. These tests essentially involve dosing the biocide into the product at various levels, below and above the recommended dose, and then storing the samples for prolonged periods of time at low and elevated temperatures and observing any side effects. Other factors such as pH compatibility, viscosity loss and chemical interactions are also evaluated.

9.5.5.3 Biocide active ingredient chemical stability in the product formulation. Testing the chemical stability of a biocide in a polymer matrix or adhesive formulation can be achieved by the use of quantitative analytical methods to monitor the level of active ingredient. This will demonstrate if the preservative is stable over the product's lifetime, as well as discovering if the formulation is antagonistic to the biocide.[79] This form of procedure is necessary and critical and must be adopted to assess the potential of a biocide for preservation of polymer systems where it is known that there may be antagonistic components to the active ingredients present, e.g. redox chemical-containing systems.

Thus, it is important that analytical procedures for the active ingredient of the biocide are available. This will also be a requirement for the EU Biocidal Products Directive positive list of active substances. Furthermore, methods are increasingly required in order to carry out quality control (QC) checks such as:

- QC checking of dosing accuracy at recommended set level,
- QC checking of biocide stability in the polymer emulsion or adhesive batch in question.
- QC checking of biocide active ingredient content with suppliers' certificate of analysis,
- Marketing purposes where biocide type and level need to be stated on safety data sheets where there are regulatory limits for their inclusion.

9.5.6 Regulatory issues governing the choice and selection of antimicrobial agents and biocides

The amount and level of preservative which can be used in a synthetic polymer emulsion or an adhesive formulation can be subjected to regulatory control. Various legislation exists and is applied according to requirements laid down by the specific nature of the application and the country where this is to be used. The form of the law is often a positive list of acceptable substances. In addition, for some applications many adhesive customers have their own guidelines on acceptable preservatives and these may be more stringent than local regulations.

We can consider some of the legislation relevant to emulsion polymers and adhesives.

9.5.6.1 American Food and Drug Administration (FDA) legislation. This governs all the ingredients which are permitted in adhesives for certain end uses and although designed for the USA it is also widely used and recognised in Europe and elsewhere. Three levels are important for adhesive manufacture with different degrees of severity or applicability.

GRAS status (generally recognised as safe) is given to ingredients, including preservatives which can be used as food additives. These preservatives are used for example in the preservation of fruit juices and they include some of the less potent preservatives such as sodium benzoate and potassium sorbate. These can also be used in adhesives but show low inhibitory powers in many cases.

The two main areas which concern adhesive manufacturers in the FDA are the direct and the indirect food contact regulations.

If there is a possibility of contact between the dry adhesive film and the food then Title 21 Part 176 must be consulted since Title 21 Part 176 (Indirect Food Additives: Paper and Paperboard Components) contains two sections relevant to coatings which can be extended to adhesives. These sections differ according to the type of food which may be in contact with the coated board. Section 176.170 names components allowed for contact with aqueous and fatty food and Section 176.180 names those allowed for contact with dry food. Materials with GRAS status automatically conform.

Title 21 Part 175 Section 105 (Indirect Food Additives: Adhesives and Components of Coatings) names the ingredients which are allowed for adhesives where the adhesive is separated from the food by a functional barrier. This regulation is relevant to packaging adhesives for food, for example a 'cornflakes' packet or box as long as the dry adhesive film can not touch the food, i.e. the 'cornflakes'. Again, materials with GRAS status are automatically included here.

9.5.6.2 German BGA legislation. In Germany, substances which are used in food are subject to restrictions decreed in the BGA legislation. This is published by Carl Heymanns Verlag KG under the title 'Kunstoffe im Lebensmittelverkehr' and contains various chapters detailing allowed components for various materials. Restrictions on the ingredients allowed for use in adhesives are governed by chapter XIV. As well as polymers, surfactants and other raw materials, the acceptable preservatives are listed.

Some subparts of this legislation restrict the use of preservatives even further. In the tobacco industry, the DTV regulations (Verordnung über Tabakerzeugnisse Tabakverordnung) retain essentially only food grade preservatives as acceptable for use in adhesives for tipping and seaming applications. These restrictions are echoed in the tobacco legislation of other European countries.

Generally, in Europe, conformance to the BGA or the indirect FDA statements are sufficient for adhesives for many applications and these are often used as substitute guidelines where there is no directly relevant legislation, as in the tissue and towel market where, although food contact is not involved, adhesives are often required to conform to the same standards.

Though not always governed directly by legislation, the choice of preservatives for a system is equally guided by a desire for systems showing low impact on the environment and also a low toxicity or tendency to cause irritation or sensitisation to both workers handling the preservatives during adhesive manufacture and during end-use processing of the adhesive.

9.5.6.3 EEC biocidal products legislation. During the time when the EC Council for European Member States prepared the Directive 91/414/EEC concerning the placing on the market of plant protection products,[71] it was felt that another Directive should be drafted dealing specifically with other dangerous substances and preparations in the field of non-agricultural pesticides. After taking advice from experts the EC decided that such products should be grouped together under the heading of biologically active products for use in non-agricultural applications. It was then found that the nature and the applications of these products made them very different from pesticides, and so the term 'biocides' was selected. These biocidal products can be used both on microscopic organisms, such as viruses, bacteria, moulds and yeasts, and on larger organisms, such as insects, small rodents (rats, mice, etc.), molluscs, algae and even some birds. The Directive thus became known as 'The Council Directive concerning the placing of biocidal products on the market'.[71]

The organisms detailed above can harm human and animal health, damage natural or manufactured products and attack materials, such as plastics, masonry, wood and paint. They can also cause pollution in sanitary and air-conditioning systems, and can weaken, if not undermine dykes.

A summary of the types of biocidal products in this current EU Directive legislative process includes the following wide range of biologically active substances: rodenticides, molluscicides, insecticides, acaricides, fumigants, disinfectants for swimming pools and food applications, antimicrobial active ingredients such as the general biocides, sanitary biocides, air-conditioning biocides; wood preservatives; textile preservatives; artwork preservatives; masonry preservatives; consumer-product preservative; and other specialist biocides; and antifouling paints and other products where a biological functionality is claimed.

In essence this Directive will have many implications in the adhesive and polymer emulsions and related industries in that it will govern and regulate what can be used, at what level and how it is employed. The Directive will authorise the actual placing of active substances on the market that have proven technical efficacy in specific applications at the prescribed dosage without being harmful to human and animal health or having other harmful effects, directly or indirectly (e.g. through drinking water, food or feed). The Directive will also require the substances that are authorised to have substantial toxicological and environmental fate dossiers.

Once a product has been given authorisation it will be filed with the following information: the applicant (name and address), chemical identity, physical and chemical properties, methods of detection and identification, effectiveness against target organisms, toxicological profile for man and animals including metabolism, ecotoxicological profile including environmental fate and behaviour, measures necessary to protect man, animals and the environment and classification and labelling. Each of these categories is subdivided into many other classifications to substantiate the filing of appropriate data, but will not be discussed further in this chapter.

In conclusion, this Directive will limit and control the use of antimicrobial preservatives which can be used in both adhesives and synthetic polymer emulsions and related products to the benefit of human and animal health and as such will assist in the choice of the right product for a specific application. With standard test methods also available for efficacy testing the best preservative systems for adhesives and polymers should be able to be selected more easily from a list of acceptable active substances. No known harmful and toxic ingredients will be included in the Directives' annexes, for example mercury-based materials.

9.6 Preservation strategies and related issuesšaffecting polymer emulsions and adhesives for the next millennium

The following points summarise the requirements or attributes for biocides that will be employed for the preservation of polymer emulsions and adhesives in the future. With increasing public awareness of environmental issues and waste disposal, toxicological issues, regulatory approval requirements and the advent of the Biocidal Products Directive biocide manufacturers will have to become more environmentally aware and supply of biocides will be very much market driven.

The following issues are some of those that will increasingly need to be addressed by reputable biocide manufacturers:

1. Lower environmental impact: no VOCs, no solvent, no halogen,
2. Lower toxicological effects: *Daphnia magna* test for 'EcoLabel'.
3. Low skin sensitisation: local lymph node assay,
4. Biodegradable below use concentrations: waste effluent treatment,
5. Ease of handling and safe to factory workers,
6. Chemically stable and compatible with polymer system,
7. Regulatory approvals: EPA, EU, FDA, BGA,
8. Use in low cost-effective doses.

9.6.1 Safety, health and environmental issues

9.6.1.1 Ease and safety of application. The biocide user, as part of his own manufacturing process, must incorporate the biocide into the product to be protected. The ease with which the biocide can be added and incorporated into that product is an important factor.[72,73]

Liquid formulations, either solutions or dispersions with low viscosity, are the easiest to handle as they can be stored in holding tanks and then automatically pumped and metered into the final product. Mixing, to ensure uniform distribution of the preservative throughout the whole batch of the product, is also easier with a liquid. Solids, on the other hand, are more difficult to use. Free-flowing powders and granules are relatively easy to add, but still require manual rather than pumped addition. Precise dosing is also difficult, requiring either the weighing of a solid, or the addition of a whole number of sacks or drums of preservative. Addition of wet or sticky solids is labour intensive and messy.

Safety in use is an important consideration for any biocide user. Because biocides are toxic to the microbial cell, there is potential to be toxic to other cells, including mammalian cells. There is therefore a fine balance between efficacious concentrations of the active ingredient of the biocides in products and processes, and safety of those concentrations to humans and the environment.[80] In the concentrated forms in which biocides are available on the market, the toxic potential of the product is extenuated and eye and skin irritation and sensitisation do occur.[80]

9.6.1.2 Toxicological and environmental features. The toxicological and environmental features of biocides are becoming increasingly important. This encompasses not only the active ingredients, but also the additives used in formulation, and containers and packaging used in transport.

9.7 Concluding remarks

In conclusion to this chapter, the following questions can be posed, which can only be answered with time and continued testing of new molecules and novel antimicrobial systems in many different formulations.

1. What will be the active ingredients for future antimicrobial agents and biocide systems?
2. Do these active ingredients meet the needs of the industry?
3. Are these active ingredients cost effective?
4. Are these active ingredients safe from an environmental and toxicological point of view?
5. What does the future hold for the biocide industry and preservation of aqueous- and surfactant-based formulated products and are there any other developments yet to be discovered?

It must be recognised that the biocide is the only raw material added to polymer emulsions and adhesive formulations with the specific intent to kill living organisms and therefore it will have some degree of toxicity. The future biocide is likely to exhibit a fine balance between the high toxicity to micoorganisms with low toxicity toward higher organisms.

Given the required properties for a biocidal preservative development we are unlikely to see the emergence of a truly universal system. Modern trends have shown that the way forward is in the production and formulation of a number of active agents that together may fulfil the requirements of the ideal biocide. It is possible that the ideal biocide does exist. Perhaps it is just waiting to be discovered!

References

1. Jakubowski, J.A. and Rowe, A.R. (1970) Key to superior mildewicide-resistant paints. *Paint Varnish Production*, 1–4.
2. Ferguson, J.F. (1970) Microbiology of paint films: II. *Paint Technol.*, **33** (7), 23–29.
3. Jakubowski, J.A. (1971) Paint mildewicide evaluation in the laboratory parallel field performance. *Am. Paint J.*, **9**, 56–75.
4. Miller, W.G. (1973) Incidence of microbiological contamination of emulsion paint during manufacture. *JOCCA*, **56** (7), 307.
5. Ludwig, L.E. (1974) Formulation of mildew-resistant coatings. *J. Coat. Technol.*, **46** (594), 31–39.
6. Malcolm, S.A. (1980) Microbiological monitoring and quality control. *Mfg. Chem.*, 55–56.
7. Lederer, S.J., Jakubowski, J.A. and Birnbaum, H.A. (1982) An effective preservative for
adhesives with reduced health risks. *Adhesive Age*, 28–32.
8. Jakubowski, J.A., Gyuris, J. and Simpson, S.L. (1983) Microbiology of modern coating systems. *J. Coat. Technol.*, **55** (707), 49.
9. Forshaw, S. (1985) The assessment of microbiological product stability – pitfalls and problem solving. *Mfg. Chem.*, 47.
10. Diehl, K.-H. (1986) Biocides in aqueous coating systems and our environment. Lecture notes from *Preservation for the 1990s*, Windmill Park Hotel, Stratford-on-Avon, UK, organised by Schulke and Mayr, 20 Oct.
11. Diehl, K.-H. (1986) Biocides in aqueous coating systems and our environment. *Farbe and Lack*, **92** (3), 193–195.
12. Nichols, C.J. (1989) *Microbial Spoilage of Surfactant-based Products*. Paper presented at Chemspec 1989, Manchester, UK.
13. Springle, W.R. (1990) Prevention of organic growth in buildings. *Polymers Paint Colour J.*, **180** (4254), 92–93.
14. Gillatt, J.W. (1991) The microbiological spoilage of emulsion paints during manufacture and its prevention. *JOCCA*, **9**, 324–328.
15. Springle, W.R. (1992) Surface coating microbiology. Lecture notes from *Preservation for the 1990s*, Windmill Park Hotel, Stratford-on-Avon, UK, organised by Schulke and Mayr, 20 Oct.
16. Farmer, D.B. (unpublished) *Emulsion Polymers (Basic Theory)*. Vinamul Limited internal report.
17. Bondy, C. (1966) The role of surfactants in emulsion polymerisation and emulsion paints. *JOCCA*, **49**, 1045.
18. Dunn, A.S. (1971) The role of the emulsifier in emulsion polymerisation. *Chem. Indus.*, 1406.
19. Elgood, B. and Gilbekian, E.V. (1973) The emulsion polymerisation of vinyl acetate by redox initiation. *Br. Polym. J.*, **5**, 249.
20. Finch, C.A. (ed.) (1973) *Polyvinyl Alcohol Properties and Applications*, J. Wiley, London.
21. Finch, C.A. (ed.) (1992) *Polyvinyl Alcohol*, J. Wiley, London.
22. Farmer, D.B. (1992) Polyvinyl alcohol in emulsion polymerisation. In *Polyvinyl Alcohol*, 2nd edn, Finch, C.A. (ed.), J. Wiley, London.
23. Reeve, P. (1987) Raw material compatibility of Kathon CG. Notes from *Preservation*

Towards 1990 and Beyond, Ladbroke Hotel, Warwick, organised by Rohm and Haas 11/12 June.
24. Conquer, L. (1993) Looking at polymer emulsion reagents. *Perf. Chems.*, **6/7**, 21–24.
25. Vinamul Microbiology Test Method, VMTM/11, Detection of Reducing Activity in Polymer Emulsions, proprietary information, unpublished.
26. Vinamul Microbiology Test Method, VMTM/12, Detection of Oxidising Activity in Polymer Emulsions, proprietary information, unpublished.
27. Vinamul Microbiology Test Method, VMTM/07, Analysis of Preservative Active Ingredients by HPLC, proprietary information, unpublished.
28. Rosenbaum, J.L. (ed.) (1945) *Industrial Cold Adhesives*, C. Griffin, London.
29. Davison, R.L. (ed.) (1962) *Water-Soluble Resins*. Reinhold.
30. Elsom, S.J. (1988) The biodeterioration of polymer emulsions. Paper presented at *International Biodeterioration and Biodegradation Research Group Meeting* in Paris, France. Paper number IBRG/P89/04A. The International Biodeterioration and Biodegradation Research Group Secretariat, Building Research Establishment, Watford, UK.
31. Evertsen, P. (1988) The biodeterioration of polymer emulsions/dispersions. Paper presented at *International Biodeterioration and Biodegradation Research Group Meeting* in Paris, France. Paper number IBRG/P89/04B. The International Biodeterioration and Biodegradation Research Group Secretariat, Building Research Establishment, Watford, UK.
32. Cresswell, M.A. (1993) Preservation of polymer emulsions: Selection and requirements of biocides for the millennium. *Proceedings of the Biodeterioration Society Symposium*, on Biocides for Aqueous-based Systems, Stratford-on-Avon, UK.
33. Cheesman, G.C.N. and Conquer, L. (1979) Some aspects of microbiology of synthetic resin emulsions. Paint Research Association seminar on *Paint Microbiology*, Excelsior Hotel, 18/19 June, pp. 1–4.
34. Nelson, Y.M. and Jewell, W.J. (1993) Vinyl chloride biodegradation with methanotrophicattached films. *J. Environ. Eng.*, **119**(5), 890–907.
35. Nieder, M., Sunarko, B. and Meyer, O. (1990) Degradation of vinyl acetate by soil, sewage, sludge and the newly isolated aerobic bacterium V2. *Appl. Environ. Microbiol.*, **5**(10), 3023–3028.
36. Doelle, H.W. (1984) Microbial cultures in the utilisation of cellulosic materials. *Biotech. Adv.*, **2**, 1–19.
37. Enari, T.M. (1983) Microbial cellulases: In *Microbial Enzymes and Biotechnology* (ed. W.M. Fogarty), Appl. Sci. Publishers, London.
38. Gillat, J.W. (1992) Bacterial and fungal spoilage of water-borne formulations during production and storage. Remedial and preventive measures. *Surface Coatings Int.* **10**, 387–392.
39. Paulus, W. (1992) Microbicides for material protection. *Farg ah Lack Scandinavia*, **9**, 161–166.
40. Paulus, W. (ed.) (1992) *Microbicides for Material Protection*, Chapman and Hall, London.
41. Miller, G.A. and Lovegrove, T. (1980) 3-(2*H*)-Isothiazolinone: A new class of antifouling toxicant. *J. Coat. Technol.*, **52** (661), 69–72.
42. Cresswell, M.A. (unpublished) Vinamul Technical Report on the *Ethylene-enhancing effect on Microbial Colonisation of Polymer Emulsions*.
43. Hesketh, A.J., Cresswell, M.A. and Gowland, P.C. (1995) Microbial biodegradation of polyvinyl acetate (PVA) emulsions. *Biodeterioration & Biodegradation*, **9** (eds. Bouscher, A. and Edyvean, R.G.J.), Inst. Chem. Engrs. In press.
44. Hesketh, A.J., Cresswell, M.A., Gowland, P. and Gowland, P.C. (1995) Detection of polyvinyl acetate-degrading microorganisms using an enriched nutrient agar medium. *J. Environ. Polymer Degradation*. In press.
45. Hesketh, A.J., Cresswell, M.A., Gowland, P. and Gowland, P.C. (1994) Approaches to studying the biodegradation in emulsion polymers. Paper presented at Int. Biodet. and Biodegr. Research Group Meeting in Braunschweig, Germany. Paper number IBRG/P94/13. The International Biodeterioration and Biodegradation Research Group Secretariat, Building Research Establishment, Watford, UK

46. Heyn, C., Petersen, K. and Krumbein, W.E. (1995) Investigations on the microbial degradation of synthetic polymers used in the conservation and restoration of art objects. *Biodeterioration & Biodegradation*, **9** (eds. Bouscher, A. and Edyvean, R.G.I.), Int. Chem. Engrs. In press.
47. Kay, M.J., McCabe, R.W. and Morton, L.H.G. (1993) Chemical and physical degradation occurring in polyester polyurethane during biodegradation. *Int. Biodeterioration & Biodegradation*, **31**, 209–225.
48. Trejo, G. (1988) Fungal degradation of polyvinyl acetate. *Ecotox. & Environ. Safety*, **16**, 25–35.
49. Griffin, G.J.L. and Mivetchi, H. Biodegradation of ethylene/vinyl acetate copolymers. *Proc. Biodet. Soc.*, Sept., 8 pages.
50. Pahren, H.R. and Bloodgood, D.E. (1961) Biological oxidation of vinyl compounds. *J. Water Pollut. Contr. Fedr.*, **3**, 233–238.
51. Smith, L.H. and Cresswell, M.A. (1994) Biodegradation of textile nonwoven binders: Theories and applications for using the Bactometer. Vinamul Internal report, TP1426, Unpublished.
52. Jakubowski, J.A., Gyuris, J. and Simpson, S.L. (1982) Microbiological spoilage of latex emulsions: Causes and prevention. *J. Coat. Technol.*, **54** (595), 39–44.
53. Winters, H. and Guidetti, G. (1974) Growth of a typical paint bacterial isolate in aqueous emulsion paint. *J. Coat. Technol.*, **46** (594), 69–72.
54. Oppermann, R.A. and Goll, M. (1984) Presence and effects of anaerobic bacteria in water based paints. I. *J. Coat. Technol.*, **56** (712), 51–56.
55. Carter, G. (1982) The preservation of latex emulsions. *Proceedings Plastics and Rubber Institute Conference* on Emulsion Polymers, paper 12, 1–9.
56. Buono, F., Stewart, W.J. and Friefeld, M. (1973) Evaluation of latex preservatives. *J. Coat. Technol.*, **45** (577), 43–53.
57. Gillatt, J.W. (1990) The biodeterioration of polymer emulsions and its prevention with biocides. *Internat. Biodeterioration Biodegr.*, **26**, 205–216.
58. Duddridge, J.E. (1987) Problems associated with microbiologically contaminated mains water. Notes from *Preservation Towards 1990 and Beyond*, Ladbroke Hotel, Warwick, organised by Rohm and Haas 11/12 June.
59. Holah, J.T., Higgs, C., Robinson, S., Worthington, D. and Spenceley, H. (1990) A conductance-based surface disinfection test for food hygiene. *Letts. Appl. Microbiol.*, **11**, 255–259.
60. Kahn, P. and Firstenberg-Eden, R. (1984) A new cosmetic sterility test. *Soaps/Cosmetics/Chemical Specialities*, May, 4 pages.
61. Whittingham, L. (1988) *Use of the Bactometer for Cosmetics and Toiletries.* Vitek Internal Paper, 4 pages.
62. Booth, P.E. (1987) Evaluation of the Bactometer for microbial bioburden testing of challenged fire products. In *Bactomatic's User's Seminar.* Proc. 2nd Annual Advanced Impedance Workshop, Las Vegas, Nevada, 15–16 June. Princeton, New Jersey: Bactomatic Inc.
63. Connolly, P., Bloomfield, S.F. and Denyer, S.P. (1994) The use of impedance for preservative efficacy testing of pharmaceuticals and cosmetics. *J. Appl. Bacteriol.*, **7**, 68–74.
64. Provan, F. and Cresswell, M.A (1994) Bactometer–Impedance method for the rapid detection of microbial activity. Development and standardisation of procedures for polymer emulsion testings. Vinamul Internal report, TP1421, Unpublished.
65. Jacques, P.A. and McLaurin, M.C. (1993) An efficient impedimetric procedure to demonstrate bacterial contamination in water-based coatings and their raw materials. *J. Coatings Technol.*, **8**, 1–5.
66. Rigarlsford, J. (1987) Plant hygiene and disinfection. Notes from *Preservation Towards 1990 and Beyond*, Ladbroke Hotel, Warwick, organised by Rohm and Haas 11/12 June.
67. Siegert, W. (1994) Production of microbiologically faultless cosmetics. *Euro-Cosmetics*, **7**, 45–48.
68. Siegert, W. (1993) Production hygiene in the manufacture of water-based coating materials. *Farbe & Lack, Scandinavia*, **9**.

69. Wood, W.B. (1982) Prevention of microbial spoilage of latex paint. *J. Waterborne Coat.*, **111**, 2.
70. Springle, W.R. (1990) *Guide to Preservatives for Water-Based Coatings.* Paint Research Association report.
71. EU (1993) Proposal for a Council Directive concerning the placing of biocidal on the market. *O.J. of the E.C.*, **C**, (239/3).
72. Payne,, J. (1992) Formulation of industrial biocides. *Spec. Chem.*, 485–489.
73. Godfrey, D. (1993) Blending for the future. *Cosmetics and Toiletries Manufacture Worldwide*, **4/5**, 149–151.
74. Gillat, J.W. (1993) The ideal biocide for the protection water-based formulations. *Spec. Chem.*, **18**(6), 336–342.
75. Conquer, L. (1991) Environmental needs shape biocides future. *Perf. Chems.*, **6/7**, 31–34.
76. Conquer, L. (1993) Volatile organic compounds (VOC) and the biocide industry. *EuroCoat*, **9**, 592–597.
77. DePasquale, D., Kahn, P., and Firstenberg-Eden, R. (1985) Impedimetric determination of antimicrobial preservative efficacy testing. An automated challenge test. Presented at the Annual Meeting of the American Society for Microbiology, Las Vegas, Nevada.
78. Gillatt, J.W. (1994) Development of a method for evaluating biocidal components in aqueous polymer emulsions – I: Establishment of a recommended mixed microbial inoculum. Paper presented at *International Biodeterioration and Biodegradation Research Group Meeting* in Helsinki, Finland. Paper number IBRG/P94/13. The International Biodeterioration and Biodegradation Research Group Secretariat, Building Research Establishment, Watford, UK.
79. Parkin, M. (1992) Evaluating biocides using analytical methods. *Spec. Chem.*, 390–392.
80. DeGroot, A.C. (1991) Choosing preservatives: Dermatoallergenic considerations. *Cosmet. Toiletries*, **106** (2), 37–38.

Further reading

Corbett, R. and Wood, P. (1994) Novel biocide design using an inorganic composite. *Spec. Chem.*, 223–228.

Barman, B.N. and Preston, H.G. (1992) The effect of pH on the degradation of isothiazolinone biocides. *Tribiology Int.*, **25**(4), 281–287.

Krzeminski, S.F., Brackett, C.F. and Fisher, J.D. (1975) Fate of microbial 3-isothiazolinone compounds in the environment: Modes and rates of dissipation. *Agricult. and Food Chem.*, **23**(6), 1060–1064.

Krzeminski, S.F., Brackett, C.F. and Fisher, J.D. (1975) Fate of microbial 3-isothiazolinone compounds in the environment. Products of degradation. *Agricult. and Food Chem.*, **23**(6), 1064–1074.

Hopton, J.W. and Hill, E.C. (eds) (1987) *Industrial Microbiological Testing*, Society for Applied Bacteriology Technical Series No. 23, Blackwell Scientific Publications, London.

Board, R.G., Allwood, M.C. and Banks, J.G. (eds) (1986) *Preservatives in the Food, Pharmaceutical and Environmental Industries*, Society for Applied Bacteriology Technical Series No. 22, Blackwell Scientific Publications, London.

Collins, C.H., Allwood, M.C., Bloomfield, S.F. and Skinner, F.A. (eds) (1981) *Disinfectants: Their Use and Evaluation of Effectiveness*, Society for Applied Bacteriology Technical Series No. 17, Blackwell Scientific Publications, London.

Gould, G.W. and Correy, J.E.L. (eds) (1980) *Microbial Survival in Extreme Environments*, Society for Applied Bacteriology Technical Series No. 15, Blackwell Scientific Publications, London.

Board, R.G. and Lovelock, D.W. (eds) (1975) *Some Methods for Microbiological Assay*, Society for Applied Bacteriology Technical Series No. 8, Blackwell Scientific Publications, London.

Board, R.G. and Lovelock, D.W. (eds) (1973) *Sampling – Microbiological Monitoring of Environments*, Society for Applied Bacteriology Technical Series No. 7, Blackwell Scientific Publications, London.

10 Preservation of inorganic systems

P.M. PRICHARD and J. MARTIN

10.1 Introduction

Microbiological contamination is prevalent in inorganic mineral processing where the mineral is either processed or sold as an aqueous suspension. Severe contamination can seriously affect processing and product quality. Because processing inorganic minerals generally requires suspension of the mineral in aqueous media, microbiological growth must be controlled.

Mineral refining requires discrete, individual particles. Refining includes such aqueous processes as degritting, dispersion, blunging, classification, flotation, selective flocculation, delamination, screening, chemical modification and magnetic separation.

Aqueous suspensions of inorganic minerals provide a suitable environment for the growth of microorganisms. Temperature and pH conditions during processing/shipping are often physiologically ideal for microbiological growth. Aerobic organisms are supported by oxygen that is introduced through mixing, pumping, flotation, etc. Processing also requires the addition of any number of chemicals, many of which can serve as nutrients and, quite often, sources of microbiological contamination. Areas that are not circulated tend to become anaerobic, supporting the growth of anaerobic microorganisms.

The manufacturing process utilizes apparatus which are rarely cleaned, much less sterilized. Therefore, the primary source of microbiological contamination may be the processing equipment.

The advantages of processing minerals as aqueous suspensions (slurry) are many. Particle size separation can be achieved with centrifuges, hydrocyclones and settling tanks. Slurry is passed through electromagnets for the removal of paramagnetic materials. Chemical processing, such as reductive bleaching with sodium hydrosulfite, oxidation of organic substances by ozonation, and removal of impurities by flotation, requires aqueous media. The removal of residue is not only more efficient and effective in aqueous media, but much smaller residue fractions can also be removed. Product uniformity can be easily obtained through mixing of dispersed, aqueous suspensions. Sampling is much more representative. Transportation and blending (blending may be needed at the beginning or end of the process) are greatly facilitated. The environmental issue of dust is eliminated.

There are several advantages of shipping slurry to customers. Unloading and storage are easier; the effect of moisture on dry discharge is eliminated. The cost of drying is eliminated. The homogeneous suspension of dried minerals in water (makedown) is very energy and chemical intensive. The customer typically does not have a makedown facility.

There are also distinct disadvantages in shipping slurry to customers. There is the cost of transporting water; therefore, it is necessary to achieve the highest solids possible. This is possible only with the use of costly dispersants. Mixing (using energy and apparatus) is required. There can be settling, gelation, corrosion and microbiological growth, all of which can diminish the quality of the slurry. Dispersants and suspension aids are used to keep the mineral particles in suspension; all of these greatly enhance microbiological growth.

It is important to use water of good microbiological quality for processing. A strong oxidant, such as chlorine, is normally used to treat water; however, the amount of residual oxidant is usually grossly inadequate to control the levels of microorganisms that will be encountered during processing. The control of microbiological growth in aqueous suspensions of inorganic minerals during processing and in final product is typically accomplished with the use of biocides.

10.1.1 Kaolin and calcium carbonate

Kaolin is a naturally occurring mineral which contains, and is processed with, large amounts of water. The economic kaolin deposits in Georgia were formed by the erosion of feldspar and their subsequent transportation by water to the coast of Georgia during the Cretaceous period. Because of their sedimentary formation, the kaolin deposits contain varying amounts and types of both nutrients and microorganisms. Both kaolin and water are potential sources of microorganisms.

Calcium carbonate used for ground calcium carbonate (GCC) is in the form of marble. It is a naturally occurring, metamorphic mineral which is essentially devoid of nutrients, water and microorganisms. However, it is also processed with large amounts of water and is therefore subject to microbiological contamination from water, processing apparatus and added chemicals.

10.1.2 Titanium dioxide

TiO_2 is used as a white pigment. It is typically produced from ore that contains ilmenite ($FeTiO_3$) and ferric oxide (Fe_2O_3). The ore is digested with an aqueous sulfuric acid solution to produce an aqueous solution of titanyl sulfate and ferrous sulfate. Iron sulfate is removed. Water is added to hydrolyse the titanyl sulfate to H_2TiO_3, which is insoluble and precipitates, and H_2SO_4. The precipitate is heated to drive off water, leaving TiO_2. The use of water for processing and the introduction of sulfate makes the control of aerobic and anaerobic organisms necessary.

10.1.3 Processing kaolin and calcium carbonate

10.1.3.1 Kaolin. Kaolin is typically processed by 'blunging' (mixing) the crude clay with water containing first, dispersants such as sodium polyacrylate or sodium hexametaphosphate and second pH modifiers such as ammonium hydroxide, sodium carbonate or sodium hydroxide. Blunging is followed by degritting, where sand is removed by gravity sedimentation, classification (particle size selection), electromagnetic separation, reductive bleaching with sodium hydrosulfite and water/salts removal by filtration. Blunging, degritting, classification and electromagnetic separation are done at neutral pH. Reductive bleaching and filtration take place at very low pH, inhibiting the growth of bacteria. Following filtration, the suspension is made pH neutral again with the addition of sodium hydroxide, ammonium hydroxide or sodium carbonate. Additional dispersants are also added.

The filter product is pumped to holding tanks, after which solids are increased by various methods. It may be pumped to spray driers for drying, or mixing facilities where spray dried clay is added to increase solids content, or evaporators where water is removed under decreased pressure and increased temperature. The spray drying process kills the majority of bacteria; however, spores survive and become active again when the spray dried material is resuspended in water. The evaporative process also kills the majority of bacteria, but some survive and must be controlled with biocides. Nutrients remain plentiful after the spray drying and evaporative processes.

In addition to these processing steps, some clays are processed using flotation techniques to remove impurities and improve brightness. The flotation process uses additional chemicals which can serve as nutrients/contaminants; also, temperature, pH and oxygen are generally optimal for microbiological growth. Air is bubbled through the suspension to 'force' the less dense impurity complexes to the top where they are removed as froth.

Settling of coarser products during transport is prevented by the addition of suspension aids such as xanthan gum, alginate and carboxymethyl cellulose. Unfortunately, these carbon-based complex materials are excellent nutrients and frequently microbiological contaminants. The use of nutrient-type suspension aids makes preservation more difficult.

Some kaolin deposits are grey in color because they contain organics; these deposits do not respond to the usual processing. Oxidation is required to remove the organics. This is generally accomplished by either ozonation or calcination. Obviously, the use of ozone or calcination for processing has a positive effect on microbiological control. Microorganisms are killed and food sources are oxidized. However, ozone is very short term and microbiological control is absent on further contamination. Calcination takes place at extremely high temperatures (800–1000°C) which oxidize all organic material including microorganisms. When the calcined clay is made into aqueous suspension and transported, a suspension agent is typically used to prevent settling. The suspension agents provide sufficient nutrients to support rapid microbiological growth and some suspension aids may be contaminated with microorganisms. Some suspension aids, such as Kelzan, are produced using microbiological processes. Once again, by suspending an otherwise sterile mineral in water and adding chemicals which serve as nutrients, the dilemma of controlling microbiological growth is created.

10.1.3.2 Calcium carbonate. Ground calcium carbonate is typically processed by 'autogenous' grinding which generates biocidal temperatures. Like kaolin, processing includes centrifugation, screening and sometimes flotation. Coarser products include the use of 'settling bowls' at ambient temperature which, in the absence of biocide, contribute to the level of microbiological contamination. Like kaolin, settling of coarser final products is prevented by the addition of suspension aids. On the other hand, the finer products are 'sand ground' near the end of the process. Therefore, biocidal temperatures are achieved at the beginning and end of the process for the finer products. Once the slurry cools, however, microbial activity will increase in the absence of biocide.

Precipitated calcium carbonate (PCC) is produced in two steps: first 'slaking' (addition of water to calcium oxide to form calcium hydroxide) and second addition of carbon dioxide to the calcium hydroxide to form calcium carbonate. Calcium oxide is produced from calcium carbonate at very high temperatures and the 'slaking' reaction is highly exothermic. Therefore, the starting material, calcium oxide, and the slaking reaction are not conducive to microbiological contamination or growth. The carbonation reaction also takes place at biocidal temperatures.

Generally, PCC is used immediately after production. Extended storage of PCC suspensions will require biocide treatment. The use of flue gas as a source of carbon dioxide can introduce unwanted sulfur compounds which, as nutrients for anaerobic microorganisms, can cause odor and discoloration if the suspension is allowed to become anaerobic.

10.1.4 Uses

10.1.4.1 Kaolin. Kaolin is sold as bulk and slurry material. It has many industrial applications, primarily in the paper industry. It is also used in rubber, catalysts, fiberglass, insulation, paint and ceramics. In the paper industry kaolin is used as a filler, i.e. it is added to the fiber furnish prior to sheet formation to improve light scatter (opacity), paper smoothness and brightness. It is also used as a coating pigment to improve the gloss and printing properties of paper.

10.1.4.2 Calcium carbonate. Ground calcium carbonate is sold as both a bulk and slurry material. It is a filler in paper, rubber, paint and plastics. GCC is also used in paper coating. Precipitated calcium carbonate is sold as a slurry material and is generally used as a filler in paper manufacturing.

10.1.4.3 Titanium dioxide. TiO_2 is an excellent pigment with excellent light scattering properties due to its very high refractive index. TiO_2 is used in paper and print manufacturing. Its short supply and high price have forced the paper manufacturer to look at other pigments such as kaolin for use as TiO_2 extenders and replacements.

10.2 Contamination sources

Microbiological contamination includes both aerobic and anaerobic microorganisms; it is excessive and ubiquitous. Contamination is found in the crude mineral, water, air, blungers, pipes, centrifuges, pumps, flotation cells, magnets, storage tanks and other processing apparatus. Pockets of anaerobic microorganisms will be present where oxygen is depleted, such as tank, blunger and flotation cell bottoms. A primary source of contamination is plant apparatus that has not been cleaned.

10.2.1 Types of organisms found in kaolin aqueous manufacturing processes

10.2.1.1 Kaolin. As expected, microbiological surveys have shown many types of microorganisms to be prevalent in our kaolin manufacturing facility. These include *Alcaligenes faecalis*, *Pseudomonas diminuta*, *Erwinia carnegieana*, *Yersinia*, *Arthrobacter protophormiae*, *Micrococcus luteus*, *Acinetobacter baumannii*, *Aeromonas hydrophila*, *Serratia marcescens*, *Arthrobacter globiformis*, *Pseudomonas syringae*, *Bacillus mycoides*, *Xanthobacter flavus*, *Ochrobactrum anthropi*, *Acinetobacter haemolyticus*, *Pasturella multocida*, *Phyllobacterium rubiacearum* and *Pseudomonas aeruginosa*.

10.3 Consequences of microbiological contamination

If microorganisms are not controlled, the consequences of severe contamination can range from slightly detrimental to catastrophic for processing, final product and the paper making/coating process itself. In the absence of biocide, aerobic microorganisms, such as *Pseudomonas aeruginosa*, proliferate rapidly in kaolin and carbonate slurry and deplete oxygen to the point where anaerobic bacteria such as the sulfate-reducing bacteria *Desulfovibrio desulfuricans* can grow.

10.3.1 Kaolin

During kaolin processing, several sulfur-containing compounds are used, including sulfuric acid, sodium hydrosulfite and alum. The sulfate-reducing bacteria use sulfur instead of oxygen as a terminal electron acceptor and form hydrogen sulfide. The H_2S reacts with iron hydroxide that has been formed by the reaction of sodium hydrosulfite with iron oxide to form the black compound, iron sulfide. Hence, kaolin slurry that has become anaerobic has the typical rotten egg odor (H_2S) and the level of iron sulfide is sufficient to turn the clay grey. Reclamation of kaolin slurry that has become anaerobic and chemically altered by the reaction of H_2S with iron to form FeS must be accomplished with a strong oxidant, such as H_2O_2. The quality of kaolin slurry that has become anaerobic can not be recovered with the addition of biocides alone.

Calcination of kaolin removes sulfur. Even though calcined kaolin slurry will not become discolored due to the lack of sulfur, it will become highly contaminated with aerobic microorganisms and become anaerobic if microbiological growth is not controlled.

10.3.2 Calcium carbonate

Ground calcium carbonate contains very little sulfur. If aerobes are not controlled, the slurry will still become anaerobic and if sulfate-reducing bacteria are present, enough sulfur is present to detect the odor of H_2S. Iron content is minimal and significant discoloration is not observed.

Precipitated calcium carbonate also contains very little sulfur because calcium carbonate has been heated at high temperatures (800–1000°C) to drive off the CO_2. However, the use of flue gas that contains sulfur will introduce sufficient sulfur that, if the PCC suspension is allowed to become anaerobic in the presence of sulfate-reducing bacteria, H_2S will be formed. Although the level of iron is minimal, the FeS formed is obvious because of its color contrast to the very high whiteness of PCC.

10.3.3 General

In addition to foul odor and discoloration, viscosity and pH can also be adversely affected. The excretion of organic acids by the bacteria lowers the pH. Viscosity is adversely affected by both a lowering of pH and microbiological degradation of the dispersant, for example sodium polyacrylic acid. Degradation of dispersant can lead to settling and unloading difficulties.

10.3.4 Paper

The introduction of slurry that is highly contaminated with microorganisms into the customer's storage tank can lead to very serious problems with fillers and coatings. These include paper breaks, holes, fish eyes, poor runnability, excessive microbiological contamination in food grade paper, loss of strength, loss of brightness and loss of gloss. High aerobic contamination will ultimately contaminate the customer's system and, if left unchecked, will result in an anaerobic condition. All machine additives must be microbiologically acceptable for optimum performance. The use of the proper biocides for both head box and coating color sterilization/ preservation will be discussed later.

10.4 Sampling

The collection of a representative slurry sample is, of course, very important when testing a slurry sample for any quality parameter. In order for the sample to be truly representative, the suspension must be as homogeneous as possible and the sample either taken from the suspension itself or from an uncontaminated valve. Homogeneity requires good mixing.

Compounding the difficulty of collecting a representative slurry sample is the problem of obtaining a sample that will be representative of the microbiological contamination. Not only is homogeneity required, contamination must be avoided. Although most quality parameters are static, microbiological quality is dynamic. If one contaminates a sample for microbiological testing with only a minimal amount of microbiological contaminant, the data can be completely erroneous.

Prerequisites for microbiological sampling are clean, sterile containers and careful handling of the sample to avoid contamination. Perhaps the most difficult part is obtaining the sample itself. Assuming homogeneity, the best sample is taken with a sterile container introduced directly into the top of the suspension. However, this can be very time consuming and the sterile container holding apparatus must be washed after each sample is taken. The use of strong oxidants to wash/decontaminate the apparatus risks introducing oxidants into the next area of sampling which may alter the microbiological population.

Samples taken from side valves on static tanks or other containers containing mineral suspensions cannot be assumed to be representative. The valves themselves can also create erroneous results. These valves plug easily, contain contaminated material and the flow from them is typically insufficient to 'clean out' the valve. The dead space will tend to become contaminated and contaminate the clay at the entrance of the valve which, during mixing, can contaminate the entire tank.

Samples taken from large bottom valves are also suspect because they are typically below the level of the mixing paddles, are not aerated and become contaminated with anaerobic microorganisms. A suspension that is not mixed and contains insufficient biocide to control microbiological growth will become anaerobic. This phenomenon is observed wherever 'dead spaces' occur.

If time permits, the sample should be taken from the suspension itself without any intermediate pipes, valves, etc. to contaminate the sample. If this is not possible, then obtaining the sample from a moving stream is acceptable. A stainless steel valve on the outlet side of the pump in a recirculating/unloading system provides sufficient quantity of material and stream velocity to clear the valve of most residual material. The valve should be thoroughly flushed before taking the sample.

10.5 Equipment design

Typically a system has to be dealt with that was not designed to facilitate microbiological control: flat-bottom tanks that do not drain properly, lack of recirculation system, poor mixing, inadequate sampling valves, tank composition, inability to drain slurry from pipes/loading lines, etc. No mineral suspension processing/storage facility should be built without the appropriate design for cleaning, recirculation, mixing, sampling and chemical addition.

10.6 Exponential growth

Once the work force understands the deleterious effects of excess microbiological contamination, perhaps one of the most difficult facts about microbial growth to communicate to them is the rate at which microorganisms multiply. As was discussed earlier, unlike other quality parameters which tend to be static, microbiological growth is extraordinarily dynamic. The growth rate is exponential with a doubling time every 20 min under optimum conditions.

The following method is effective in conveying this growth phenomenon to operators. Assuming that bacteria can double every 20 min, and starting with only one bacterium per ml, operators are asked to calculate the number of bacteria that will accumulate in 1 ml of slurry during an eight-hour work shift. Since there are 24 twenty-minute increments in 8 h, they are asked to enter '1' into their calculator and multiply by two 24 times in succession. When the operators see that the answer is greater than 10 million, they begin to realize just how rapidly bacteria can grow and how important it is to prevent microbiological growth. Of course, the suspension typically contains >1 bacterium per ml prior to adding biocide, which of course exacerbates the problem. What may have seemed like minor contamination can become catastrophic in a matter of hours.

10.7 Microbiological test methods

The level of microbiological contamination must be monitored continuously to ensure control is effective. The primary goal in mineral suspension preservation is to control the aerobic microorganisms, thus preventing the suspension from becoming anaerobic. Techniques are used to measure and monitor the levels of aerobic organisms while biocides are used to prevent their growth.

The two techniques typically used in the mineral industry are standard plates and 'dip slides'. Although the dip slide method is generally acceptable, the standard plate method is more accurate, especially for low levels of contamination. However, at moderate to high contamination levels, the two methods are reasonably correlatable. In a quality control (QC) environment where there are large numbers of tests and time is of the essence, the dip slide method is preferred.

To carry out effectively a meaningful microbiological test requires an uncontaminated, representative sample and a test duration of 72 h. It is extremely important to collect samples in a clean sterile container. Also, the samples should not be contaminated by residual material in valves, hoses, etc.; these should be kept clean and flushed thoroughly before catching the sample.

The test procedure requires a 24 h incubation at the optimal growing temperature (this provides sufficient time for the microorganisms to react with the biocide), dilution and plating of the sample and further incubation for 48 h to observe microbiological growth.

This time requirement is certainly a limiting factor in most production situations. 'Just-in-time shipments', scheduling, and rail space typically require test results to be provided in a matter of hours, certainly less than 24 h. By the time the microbiological test results have been obtained, the slurry shipment is usually on its way to the customer. A microbiological test of slurry shipments, therefore, becomes more of a secondary check than test for shipment approval. The quantitation of residual biocide (which has been shown to be effective and stable in a system) is a more expeditious test for shipment approval. This will be discussed later.

10.7.1 Standard plate count procedure

The slurry sample is serially diluted and an aliquot of each dilution is placed in or on suitable culture media. Each colony is assumed to have grown from one viable unit, which may be one organism or a group of many. This standard method is described in most microbiology text books.

One method is the use of tryptose glucose extract (TGE) agar. Dilutions are made with sterile Butterfield buffer.

10.7.2 Dip slide method

Dip slides are available from several manufacturers and are of various types, depending on the medium and microbiological population to be examined. Each side of the paddle will be coated with a specific medium containing nutrients optimum for the microbiological species to be examined. The 'paddle' may have different media on each side such that two types of microorganisms, such as bacteria and fungi, can be examined simultaneously.

10.7.2.1 Equipment needed.

1. 1 ml Sterile syringes or sterile pipettes – syringes are preferred to eliminate the hazard of mouth pipetting.
2. Tubes containing 9 ml of sterile buffer solution – (Butterfield–monopotassium phosphate).
3. Dip slides with Tryp–Soy agar impregnated with TTC – (Triphenyltetrazolium chloride) on one side and Sabouraud (SAB) dextrose and malt extract agar on the other. TTC, colorless at neutral pH, is changed to the red-colored formozan by the acids excreted by most species of bacteria such that bacterial colonies appear as red dots.

   The SAB dextrose side is adjusted slightly acidic to favor the growth of fungi/yeast/mold but suppress bacterial growth. This side is also impregnated with chloramphenicol, an antibiotic which inhibits the growth of prokaryotic (anucleated) cells but not eukaryotic (nucleated) cells such as fungi. Thus, bacterial growth is inhibited on the 'fungus' side which allows one to differentiate between bacteria and fungus/yeast/mold contamination. Numerous types of dip slides are available from several companies.
4. Incubator typically set at 35°C.

10.7.2.2 Procedure. (Be very careful not to contaminate mineral suspension and test equipment):

1. Obtain 1 ml of slurry using a syringe or pipette. Ensure that there are no bubbles present.
2. Introduce the 1 ml of slurry into the buffer tube with 9 ml Butterfield buffer to give a 1 to 10 dilution. Dilution is required because of the opaqueness of the mineral suspension and to dilute out any biocide present. Undiluted slurry will prevent observation of the bacterial/fungal colonies. Of course the dilution factor must be taken into consideration when the assessment of colonies is made.
3. Shake the slurry sample–buffer mixture until homogenous (10–20 s).
4. Remove the cap from the dip slide container (being very careful not to contaminate the dip slide) and pour the diluted sample into the container, replace dip slide, tighten and *gently* agitate the container such that the agar surface comes into contact with the slurry for a minimum of 5 s. Do not agitate too vigorously as the agar can be dislocated from the paddle. Remove the dip slide and pour excess slurry from the container over the dip slide; let drain.
5. Replace the dip slide into its container and label accordingly. incubate for 24 h at 35°C.
6. After 24 h, count the colonies (red dots) present. If there are no colonies, incubate for a further 24 h and count. Sometimes the bacteria grow slowly and, quite often, the bacteria colonies are present as very tiny dots but have not excreted sufficient acid to turn the TTC red.
7. Bacteria will appear on the non-chloramphenicol-treated side of the dip slide as colonies that look like red or clear dots. Fungi appear as fluffy colonies and yeast appears as ‘dry’ colonies on the chloramphenicol-treated side of the dip slide.
8. Pictorial guides are provided by the dip slide manufacturer for an estimate of bacterial/fungal population. The guides are typically photographs of dip slides that have been exposed to solutions of bacteria whose populations have been determined by the standard plate method.

10.7.3 Assessment of bacterial populations using dip slides

10.7.3.1 Counts of $<$1K organisms per ml of slurry. Slurry is considered within tolerable limits and does not require biocide treatment. Recheck every 24 h if slurry is to remain stored in tank for an extended period (longer than 1 week).

10.7.3.2 Counts of $>$1K but $<$100K organisms per ml of slurry. Slurry is considered contaminated with aerobic microorganisms but sufficient oxygen remains to inhibit the growth of anaerobic microorganisms. However, slurry must be treated with biocide immediately. Retest after 24 h and retreat with biocide as required.

10.7.3.3 Counts of >100K organisms per ml of slurry. Slurry is considered seriously contaminated and in danger of becoming anaerobic. Slurry must be treated with biocide immediately. Test for anaerobic bacteria and treat accordingly. Retest after 24 h and retreat with biocide as required. Once the contaminated slurry has been treated and utilized, the holding tank must be washed with water and sterilized before new slurry is introduced.

10.8 Anaerobic contamination

Depletion of oxygen by aerobic microorganisms in the slurry allows anaerobic microorganisms to flourish. In the case of anaerobic mineral suspensions containing sulfur, iron and sulfate-reducing bacteria, hydrogen sulfide and iron sulfide are formed.

The FDA approved biocides will not destroy the iron sulfide. If the suspension has become anaerobic, fouled with H_2S and discolored by iron sulfide, a strong oxidant such as hydrogen peroxide must be used to oxidize the iron sulfide. It should be noted that the absence of aerobic microorganisms does not necessarily preclude the existence of contamination. The suspension may have already gone through the aerobic phase, become anaerobic, and only anaerobic bacteria are viable. Obviously, discoloration and/or odor indicates that an anaerobic microbiological test for sulfate-reducing bacteria should be done.

10.8.1 Anaerobic microbial analysis

The following 'stab test' method is used to assess a population of anaerobic sulfate-reducing bacteria in a kaolin or calcium carbonate slurry sample.

10.8.1.1 Equipment needed.

1. One ml sterile syringe or sterile pipette – syringes are preferred to eliminate the hazard of mouth pipetting.
2. One tube containing 9 ml sterile buffer solution – (monopotassium phosphate).
3. One tube containing 8 ml of Peptone Iron Agar Butt.
4. One inoculation wire – (6 inch chromed 'A' wire).
5. Bunsen burner or incinerator.
6. Incubator.

10.8.1.2 Procedure.

1. Using the sterile syringe or pipette, obtain 1 ml of slurry. Ensure there are no bubbles present.
2. Place the 1 ml of slurry into the buffer tube with 9 ml of buffer solution.
3. Shake the mixture until homogeneous (10–20 s).
4. Sterilize the inoculation wire by passing it through the bunsen burner flame or into the incinerator until the wire turns red.
5. Allow the wire to cool for 5 s.
6. Insert wire into slurry and buffer mixture.
7. Remove cap from Peptone Iron Agar Butt.
8. Remove the wire from the buffer tube and immediately insert the wire into the agar tube, being careful not to touch the sides of the tube with the wire when inserting or extracting it. Make sure wire penetrates agar to the bottom of the tube. Allow the wire to remain for 5 s and remove.
9. Recap the Peptone Agar Butt.
10. Store agar tube in incubator at 35°C for 48 h.
11. Inspect the tube after the 24 and 48 h incubation period. If anaerobic sulfate-reducing bacteria are present, agar will turn black where the inoculation wire penetrated. The black color is due to iron sulfide formed from iron and hydrogen sulfide produced by the sulfate-reducing bacteria.

Note that this method is basically a qualitative method.

10.9 Microbiological control

The control of microbiological growth can be accomplished physically with frequent housekeeping (cleaning/sanitation). It can also be accomplished with radiation (UV, X-ray, gamma radiation) and chemicals such as antibiotics, biocides and strong oxidants (chlorine, ozone, chlorine dioxide and hydrogen peroxide).

10.9.1 Cleaning/housekeeping

Microbiological control cannot be properly maintained through the use of biocides alone. Equally important is the routine cleaning and sterilizing of the processing system itself. This will greatly enhance microbiological control by reducing any recontamination of the slurry as it flows through the system.

Typically, frequent cleaning and sterilization of plant equipment are very difficult because of both the continuous use and inaccessibility of equipment. However, it must be recognized that this facet of a microbiological control program is very important. The higher the contamination in the equipment the greater the risk of spoilage and the greater the biocide demand.

Equipment should be cleaned on a regular basis to eliminate the contamination that will accumulate. This includes biofilm which is very difficult to remove and not only provides a continuous source of contamination but also contributes to corrosion.

Equipment (tanks, valves, hoses, couplings, etc.) is usually cleaned with sodium hypochlorite solution. Sodium hypochlorite must be thoroughly removed by rinsing with good quality water because of its detrimental effect on biocides and viscosity. Equipment can also be cleaned with strong acids, strong bases, detergents, solvents or dispersants.

10.9.2 Radiation

The use of ultraviolet radiation for sterilization is not practical because of the inability of UV radiation to penetrate a high solids slurry and the large quantity of material. The use of X-ray or gamma radiation is also not practical to sterilize such large amounts of material and, of course, safety becomes a primary consideration. Radiation techniques, like strong oxidants, have no residual effect. Any further contamination after the radiation treatment will result in uncontrolled microbiological activity.

10.9.3 Oxidants

Hydrogen peroxide, ozone, chlorine dioxide and chlorine (chlorine gas, sodium hypochlorite) are used for sterilization. However, although strong oxidants provide good sterilization, they do not provide preservation. Strong oxidants can also destroy other important processing chemicals, such as dispersants, suspension aids and biocides, as well as oxygenate the system to support aerobic microorganisms once the biocide has been depleted. Where iron has been converted to the reduced, soluble form for extraction purposes, peroxide will oxidize it to the insoluble iron oxide.

10.10 Biocides

Chemicals are used in conjunction with housekeeping for microbiological control. Chemicals have been developed which are toxic to microorganisms and provide not only quick kill, but intermediate to long term microbiological control. These chemicals are termed biocides. Biocides that provide long term control are called biostats.

The primary function of biocides used in kaolin and carbonate slurry preservation is to control the growth of aerobic microorganisms such that sufficient oxygen is present in the slurry to inhibit the growth of anaerobic microorganisms, especially the sulfate-reducing bacteria. Obviously, the user of the kaolin/carbonate/titanium dioxide product is concerned about both aerobic and anaerobic microbiological contamination in his system. So the key is to keep the aerobes in check so that the anaerobes cannot grow. By doing this, deterioration in the quality of the slurry is prevented and the customers' pigments, storage tanks, paper machine, etc. are not contaminated.

The mechanism of action of biocides varies with the specific product. Where available, these will be discussed later.

10.10.1 FDA approval

Biocides that are used for the preservation of pigment slurry are generally those FDA approved for food contact. They meet the following approvals: CFR 176.170, CFR 176.180 and CFR 176.230. Other non-FDA approved biocides used to preserve mineral suspensions are methylene bisthiocyanate (MBT), 2-dibromonitrilopropionamide (DPNPA) and 2-bromo-2-nitropropane-1,3-diol (BNDP or Bronopol).

10.10.1.1 CFR 176.170. (Components of paper and paperboard in contact with aqueous and fatty foods). These biocides are those that may be safely used as components of the uncoated or coated food-contact surface of paper and paperboard intended for use in producing, manufacturing, packaging, processing, preparing, treating, packing, transporting or holding aqueous and fatty foods.

10.10.1.2 CFR 176.180. (Components of paper and paperboard in contact with dry food). These biocides are substances that may be safely used as components of the uncoated or coated food-contact surface of paper and paperboard intended for use in processing, manufacturing, packaging, transporting or holding dry food of the type identified in CFR 176.170.

10.10.1.3 CFR 176.230. (Special section for thione biocides). The general requirement of this section is that the thione be volatilized by heat in the drying and finishing of the paper, coated paper or paperboard.

10.10.2 Approval outside the United States

It must be ensured that biocides used in mineral suspensions that are exported comply with regulations of the country of destination. Generally, US FDA approval is sufficient. However, this is not always the case.

Typically, Europe will adhere to the German BGA guidelines. Scandinavian countries may have different regulations. Japan will generally accept US FDA approved biocides.

On the other hand, imported mineral suspensions would have to meet US FDA guidelines. Biocides such as Bronopol are accepted for mineral suspensions used in food grade packaging in Europe but not in the US. Another biocide issue is that the biocide must also meet the regulations of the country in which the paper product is used.

10.10.3 Types of biocide

FDA approved biocides are discussed below. Other biocides, depending on the end use, which can be considered for pigment slurry preservation but are not discussed here are dodecylguanidine hydrochloride (DGH), 2-bromo-2-nitropropane-1,3-diol (Bronopol) and 2-bromo-4′-hydroxyacetophenone (Busan 90).

*10.10.3.1 Tetrahydro-3,5-dimethyl-2*H*-1,3,5-thiadiazine-2-thione.* This is generally referred to as thione or dazomet. Thione is the biocide generally used in the kaolin and carbonate industry for slurry preservation. Thione hydrolyzes to several compounds including formaldehyde, methylamine, carbon disulfide, methylisothiocyanate, carbon monoxide, carbon dioxide, nitrogen oxides, sulfur oxides, disodium oxide and hydrogen sulfide.

The primary biocidal compound is methylisothiocyanate (MITC); the small amount of formaldehyde present contributes very little to the biocidal activity. Thione does not polymerize and is effective over a broad pH range. It is incompatible with strong acids, mercurials or other compounds that are not compatible with sulfur compounds, oxidizing agents, copper, carbon steel and epoxy drum liners. Color will vary from pale yellow to amber to green.

Thione is manufactured in the USA by Vinings Chemicals and Buckman and sold in three formulations: dry, 10% NaOH and aqueous dispersion. Dry thione is relatively stable but on contact with water rapidly breaks down to methylaminomethyldithiocarbamate which subsequently yields MITC and hydrogen sulfide. In acid solution, the dry product hydrolyses to form carbon disulfide.

Thione is the most cost effective FDA approved biocide. Unlike other FDA approved biocides, the maximum allowable amount that can be added is not restricted except as is stated in CFR 176.230 (b); ‘The quantity used shall not exceed the least amount reasonably required to accomplish the intended technical effect and shall not be intended to nor, in fact, accomplish any physical or technical effect in the food itself.’ The primary requirement is that the preservative is volatilized by heat in the drying and finishing of paper, coated paper and paperboard.

Thione has the following disadvantages. Some of the volatile breakdown products are malodorous, irritating and lachrimatory. Thione also emits small amounts of formaldehyde. Thione and its breakdown products are incompatible with several of the other FDA approved biocides: glutaraldehyde, Kathon and Tektamer. It is also incompatible with hydrogen peroxide which is commonly used to sterilize/recover pigment slurry.

Thione is most effective when it is added to a system containing low microbiological contamination. Thione is not as effective when added to a slurry that is in a high contamination/nutrient state.

10.10.3.2 1,2-Benzisothiazolin-3-one. This is generally referred to as BIT or its trade name, Proxel. Proxel is manufactured by Zeneca and used throughout the kaolin industry as a preservative. A very effective biocide, one of its greatest advantages is its compatibility with all the FDA approved biocides. It is incompatible with strong oxidizing agents and corrodes mild steel, aluminum, copper and other metals. It is a brown liquid with a slight odor and a pH of 12.

BIT does not polymerize and is effective over a pH range of 4 to 12. It is odorless and does not react with ammonia and amines. It has relatively low toxicity and can be used up to 475 ppm active ingredient (BIT) per wet ton. It is a good biostat, has a high percentage of active ingredient and does not emit formaldehyde.

BIT's disadvantages include cost, especially when used with a commodity-type material, and mild skin sensitization. It is also inactivated by reduction to mercaptobenzamide and BIT diamide by reducing agents such as those typically found in kaolin slurry (sodium hydrosulfite and sodium bisulfite) and hydrogen sulfide. It is also inactivated by strong oxidants such as hydrogen peroxide and sodium hypochlorite.

Proxel GXL (19% BIT in dipropylene glycol) becomes very viscous in cold weather. This should be taken into consideration when designing the chemical addition system.

10.10.3.3 Glutaraldehyde(1,5-pentanedial). This is generally referred to by its trade name, Ucarcide. Glutaraldehyde is manufactured by Union Carbide. It is a broad spectrum biocide due to its ability to cross link primary amines, such as those present in lysine residues in a microbial cell wall. This fixative action ultimately can prevent a microorganism from carrying on its metabolic functions by limiting the cell permeability. It is effective against aerobic and anaerobic organisms.

Glutaraldehyde is typically sold as a 50% active material in water. It is a colorless liquid with a strong odor and a pH of 3.7 to 4.5 and is corrosive to steel, galvanized iron, aluminum, tin and zinc.

Glutaraldehyde reacts with and is inactivated by ammonia, ammonium salts, primary amines and protein. It does not react with secondary, tertiary, quaternary or protonated amines. pH has a dramatic effect on the efficacy of glutaraldehyde. As the pH is increased, the amino groups are deprotonated and reactivity is rapid. However, above pH 9 glutaraldehyde will be inactivated by polymerization. As the pH is decreased, the amino groups become protonated and do not react with glutaraldehyde. The inactivation of glutaraldehyde with ammonia, primary amines and protein can be offset by lowering the pH.

Glutaraldehyde is stable in kaolin slurry to the presence of low concentrations of hydrogen peroxide (up to 1 lb active H_2O_2/ton). It is an excellent, rapidly acting, long term biocide in kaolin slurry at pH 7 when ammonia and primary amines are absent. The use of ammonia, primary amines or proteins excludes the use of glutaraldehyde.

Glutaraldehyde has not received widespread acceptance in the kaolin industry because of inconsistency. Particularly, it may be converted to the inactive valerolactone by a component(s) in kaolin and/or process water. This inactivation reaction is easily eliminated by pretreatment with hydrogen peroxide.

Although glutaraldehyde is an excellent, rapidly acting biocide in calcium carbonate slurry, it is not biostatic. The high pH of carbonate slurry results in inactivation of glutaraldehyde by polymerization. Glutaraldehyde must be used in combination with a stable biocide in carbonate slurry for preservation.

10.10.3.4 5-Chloro-2-methyl-4-isothiazolin-3-one and 2-methyl-4-isothiazolin-3-one. This is generally known by its trade name Kathon. Kathon is manufactured by Rohm and Haas. It is a combination of chlorinated methyl and methyl isothiazolin-3-ones at a ratio of 3.3:1. It is a broad spectrum biocide, controlling Gram-negative and Gram-positive bacteria as well as fungi and is typically sold as a water-based 1.5% active material at pH 3–5 with a copper stabilizer. It is pale yellow to green in color with a mild aromatic odor and is corrosive to mild steel, carbon steel, copper and 304 stainless. It does not polymerize.

Kathon reacts with and is inactivated by reducing agents, such as hydrosulfite, sulfite and bisulfite typically found in kaolin slurry. It is also inactivated by primary, secondary and tertiary amines, ammonia, mercaptans, sulfides, strong oxidants, and high pH typically found in carbonate slurry. These reactions are exacerbated by high temperature.

Kathon is a cost effective, quick killing biocide at low active concentrations. However, its low percent active ingredient and its maximum usage rate result in low levels of active ingredient in the slurry. Thus, the presence of minimal amounts of reducing agents or nucleophiles will result in its inactivation. It is an excellent preservative in systems where it can be stabilized.

Kathon stability in kaolin slurry can be attained by ensuring the elimination of reducing agents, primarily sulfites resulting from the use of sodium hydrosulfite for iron reduction, and the use of sodium hydroxide or sodium carbonate instead of ammonia.

Reducing agents can be removed by oxidation with hydrogen peroxide. However, because of isothiazolin's inactivation by strong oxidants such as hydrogen peroxide, hydrogen peroxide and isothiazolin have to be used sequentially. Hydrogen peroxide should be eliminated before the addition of isothiazolin.

The chlorinated methyl isothiazolin is subject to nucleophilic attack by amines. This makes this molecule unstable in systems where ammonium hydroxide/amines are used. It also makes it unstable in the presence of thione biocide and its hydrolysis products. As with all biocides, Kathon should be handled with extreme caution. Kathon, in addition to its corrosivity, is also a strong skin sensitizer.

10.10.3.5 1,2-Dibromo-2,4-dicyanobutane. This is generally known by its trade name Tektamer or DBDCB. DBDCB is a broad spectrum biocide controlling bacteria, mold, yeast and sulfate reducers. It is effective at a pH range of 2–8.5 but is incompatible with sulfites, amines and strong nucleophilic agents. It is compatible with materials such as PVC, polyethylene, polypropylene, Tygon, Teflon and glass. It is corrosive to mild steel.

10.10.3.6 Biocide blends. Several FDA approved biocide blends are available. Also, one can add biocides that are compatible with one another, as long as they are added separately. These combinations can consist of fast acting biocides and long term, but slower acting, biostats, such as a mixture of glutaraldehyde and Proxel. This allows the immediate killing of highly contaminated suspensions plus long term preservation by the biostat. Once control is established, the slower acting biostats, such as Proxel, effectively maintain control.

10.10.3.7 Alternating biocides/resistance. Although biocides tend to be fairly non-specific, some biocides can be more effective in controlling certain microorganisms than others. Microbiological competition for nutrients, oxygen, etc. is reduced by ridding the aqueous suspension of certain bacteria/yeast/fungi/mold. This allows the 'more resistant' microorganisms to flourish. The limited number of FDA approved biocides for indirect food contact coatings as well as biocide–biocide/biocide–process chemical incompatibility make it difficult to alternate biocides in most cases.

10.11 Biocide performance testing

Although biocide brochures provide excellent information concerning the performance of biocides, one must determine biocide performance for a mineral suspension 'in house'. Cost and compatibility issues must be considered.

Initially, tests are done in the laboratory with appropriate slurry samples and a wide range of biocide concentrations. A minimum of 30-day control is required. Most FDA approved biocides have maximum addition levels.

If a particular biocide is effective, one must then reinoculate ('reinsult') the suspension with additional bacteria to ensure that the biocide is sufficiently stable to control further contamination. Strong oxidants like hydrogen peroxide are short lived and, on reinoculation, a lack of control may be observed. Once satisfactory laboratory results are obtained, the biocide must be tested in the actual process.

10.12 Biocide addition systems

Biocides can be added manually or automatically. The manual addition of biocides is not recommended because of the danger inherent to these types of chemicals. If manual addition is necessary appropriate protective clothing must be worn such as chemical aprons and goggles. Ventilation must be sufficient. The addition of biocide should be controlled by mass flow meters to ensure accurate dosing regardless of slurry flow.

The effect of biocide pH on mineral slurries must be taken into consideration. Very acidic (such as Kathon or glutaraldehyde) or alkaline biocide solutions (such as thione, which is typically formulated in 10% sodium hydroxide) can flocculate a suspension. Kaolin particles, which have an overall negative charge at neutral pH, are kept separated or deflocced by maintaining the surface negative charge. The overall negative charge is also increased by using anionic dispersants. If the negative charges are neutralized, the Van der Waals attraction forces cause flocculation.

Biocides should be added at the earliest possible point in the process to control microbiological growth. Temperature, pH and process chemical compatibility must be taken into consideration when determining the point of addition. Of course, temperature will exacerbate chemical reactions. Biocides are often added to the makedown water prior to addition of pigment.

10.13 Residual biocide testing

As discussed earlier, microbiological tests are very time consuming and sterile representative sampling is crucial. A positive test may mean that the sample was contaminated. Testing for residual biocide is one way to expedite the shipment approval process. Of course, the biocide must have been shown to be effective and stable in the system. Several analytical methods are available for the determination of residual biocide in slurry. Also available are bacteria-impregnated strips to determine if there is biocide present.

The importance of a minimum amount of the right biocide to maintain long term microbiological quality as well as the high cost of biocides makes accurate biocide quantitation in final product necessary. The addition of an accurate dose of biocide does not always ensure that sufficient biocide is available. The residual biocide level depends on its stability to pH, temperature, bacteria and chemicals. Minimal biocide levels for acceptable, long term control must be determined for the mineral suspension being preserved.

10.13.1 Thione

The primary biocidal product in thione biocide is methylisothiocyanate (MITC). This compound can be detected by GC or HPLC. Formaldehyde is another byproduct which can be detected by HPLC.

10.13.2 Glutaraldehyde

Glutaraldehyde levels may be quantified with the Union Carbide 'Ucarcide antimicrobials field test kit'. Glutaraldehyde may also be quantified by gas chromatographic (GC) or high performance liquid chromatographic (HPLC) methods. Rapid tests for glutaraldehyde are also commercially available. If inconsistent control is observed, it may be due to valerolactone formation; this compound can also be quantified with HPLC or GC.

10.13.3 Benzisothiazolin-3-one

Benzisothiazoline (Proxel) levels in slurry may be quantified by high performance liquid chromatographic methods. This method is available from Zeneca. If inconsistency of control is observed, it may be due to the formation of mercaptobenzamide and/or BIT diamide; these compounds can also be quantified with HPLC.

10.13.4 Kathon

Kathon levels in slurry may be quantified by HPLC. This method is available from Rohm and Haas or Vinings. A field test using UV spectroscopy is also available from Rohm and Haas and Vinings.

10.14 Working with the external customer

A close working relationship with the external customer is required to have a successful microbiological control program. The extended process requires the supplier to work with the customer so the fitness for use quality of the slurry is maintained.

Slurry transported in rail cars or trucks will typically be at optimum temperatures for microbiological growth. Shipment samples stored at room temperature will not always indicate what actually happens to the slurry en route. Representative samples should be taken on arrival for both microbiological and biocide analysis.

Several questions have to be asked. What biocides is the customer using? Is the customer making food grade paper? What process chemicals is the customer using? Are they compatible with the biocides in the slurry? Are the process chemicals in the slurry compatible with the biocides the customer is using? Will the customer blend one supplier's slurry with another supplier's slurry? How long will the slurry be stored? Is there a housekeeping program at the customer's facility? How clean is the unloading system? Does the customer drain/clean apparatus used to unload slurry? Is there a sampling/testing program at the customer site? Is there concern about formaldehyde? Is there concern about odor? Safety must always be thoroughly discussed and closely monitored. The shipment of slurry to customers requires that not only must the slurry arrive with minimal microbiological growth, it must contain residual, compatible biocide.

10.15 Slurry made from dry material and slurry storage

Quite often, dry kaolin and/or carbonate are made into suspensions at an intermediate site or at the customer site. Chlorinated water and clean equipment should be used during slurry makedown to prevent initial contamination. Biocide should always be added during slurry makedown to control subsequent microbiological growth in the slurry. If the slurry is to be stored for an extended period (longer than one week), it should be tested regularly for microbiological growth.

11 Preservation of metalworking fluids

P.S.K. LEE

11.1 Introduction to metalworking fluids

Metalworking fluids are extensively used in a wide range of machining operations such as turning, milling, tapping, drilling, grinding, broaching, rolling, drawing and stamping.[1,2] The primary functions of metalworking fluids during these metalworking operations are to provide cooling effects on the tool and workpiece by dissipating the generated heat and to lubricate the work area by reducing the amount of frictional heat generated. Other secondary functions of metalworking fluids include waste removal, rust protection or corrosion prevention, surface part finish and tool protection during metalworking operations.

11.2 Consequences of metalworking fluids failure

Metalworking fluid plant operations are usually related to a central fluid system serving several metalworking machines or individual sumps on metalworking machines. Metalworking system sizes can therefore vary from 25 gallons to several thousand gallons of metalworking fluids use. These metalworking fluid operation systems are frequently subjected to microbial contamination which is detrimental to the working life of the metalworking fluids and is a major cause of metalworking fluids failure.[3–6]

The growth of microorganisms in metalworking fluids is usually kept in check by the addition of chemical preservatives. However, unchecked microbial growth in preservative-free metalworking fluids may result in modifications to the metalworking fluids as previously determined by chemical analytical methods.[7–9] Uncontrolled microbial growth in metalworking fluids will result in offensive odor development, decrease in pH and increase in corrosion rates due to production of organic acids or harmful microbial metabolites, darkening of metalworking fluids color that may cause surface finish blemishes, selective depletion of balanced metalworking fluid components which become utilized as microbial nutrient sources and which may cause changes in emulsion stability, and the visible presence of slimy microbial biomass that will clog up fluid lines, filters and screens in metalworking fluid systems. All these microbial activities will eventually result in the deterioration or failure of metalworking fluids.

The deterioration or failure of metalworking fluids will in turn result in the loss of plant worker productivity, the need to rework finished parts, the premature disposal and increase in dumping of metalworking fluids into the system to replenish the rapidly spoilt metalworking fluids, increase in plant operation downtime and decrease in metalworking tool life.

Most of these factors will add up to higher cost needed to operate the metalworking plant system and may not be economically beneficial. There also exists the possibility of unchecked growth of opportunistic microbial pathogens in metalworking fluids that may represent an increase in health risk to the metalworking plant workers.[10–17]

11.3 Ranges and types of metalworking fluid

There are four basic metalworking fluid types used in metal cutting and metal forming operations.[18–22] Soluble, semi-synthetic and synthetic-based metalworking fluids are three out of the four metalworking fluids that are diluted with water by end users.[23,24] These three types of metalworking fluids are often called water-based or water-miscible metalworking fluids. Water-based metalworking fluids are usually sold as water-free concentrates which can be mixed with water by end users to 1:10 or 1:40 dilutions depending on the fluid types and the metalworking applications. Straight or neat oils are not water-based metalworking fluids and are used as they are in the non-diluted form by end users.

11.3.1 Straight or neat oils

These are solely oils of petroleum (naphthenic or paraffinic mineral oils), animal, vegetable or marine in origin. These base oils may be used alone or in combination with or without additives. Additives are usually added to improve lubricity and prevent welding between metal surfaces via extreme pressure properties of the additives. These metalworking fluids offer excellent lubricating properties but have only minimal heat transfer or cooling effect during metalworking operations. Typical compositions of straight or neat oils and their applications are given below.

11.3.1.1 Base mineral oil blends. Base mineral oils are blended with a fatty oil or polar additive, e.g. lard, castor or rapeseed oil. This blended mineral oil is typically used in metalworking operations involving light duty machining of cast iron, aluminum, brass and magnesium. Other applications include fine honing and lapping metalworking operations.

11.3.1.2 Sulfured fatty mineral oils and sulfurized fatty mineral oils. Sulfur is added to the base mineral oils to provide extreme pressure lubricity. Sulfured fatty mineral oil consists of dissolving 1 to 1.5% sulfur in hot mineral base oil. The sulfur in this particular blend is loosely bound and is very reactive at low temperature metalworking operations, hence the term 'active oil'. The sulfide-forming reaction occurs at low temperature with metals like copper which can cause staining to the metal worked on. Sulfurized fatty mineral oil, however, consists of sulfur added to the fatty oil. This blended combination results in a stronger chemical bond between the fatty oil and sulfur. The resulting blend will prevent the sulfur from being active and has less tendency to stain non-ferrous metals like copper.

11.3.1.3 Chlorinated mineral oils. This mineral oil has chlorine dissolved in base oils as an additive to provide an extreme pressure lubricity property. This blend of mineral oil is typically used in metal machining operations such as turning, drilling and tapping including ferrous and non-ferrous forms of grinding.

11.3.1.4 Sulfochlorinated mineral or fatty oils. This oil consists of a combination of both chlorine and sulfur additives which can either be blended as a sulfochlorinated mineral or sulfochlorinated fatty oil. This allows both mineral and fatty oils to be available for use. Typical metalworking applications of this blend of mineral oil are in thread grinding, gear cutting and automatic screw machining of steel.

11.3.2 Soluble oils

These are essentially mineral oils with or without fatty acids that contain an emulsifier such as petroleum sulfonates, i.e. oil-based emulsifiable concentrates. These metalworking fluids are milky white dispersions of oil-in-water with the oil particle size ranging between 2 to 10 μm. Soluble oils are used mainly in low duty machining operations. The three general compositions of soluble metalworking fluid are shown in Table 11.1.[24]

11.3.3 Semi-synthetic metalworking fluids

These are fluids that contain mineral base oil at a lower percentage of 5 to 40% than that of emulsion or soluble oils at 40 to 65% mineral base oil content. Semi-synthetic metalworking fluids usually impart better lubrication than true synthetic metalworking fluids while still retaining their heat dissipating character.[25] Semi-synthetic metalwork fluids usually form translucent or transparent emulsions with oil particles size equal to 0.1 μm. This metalworking fluid type is used in high speed machining and some fine grinding metalworking operations. Water hardness has also been reported

Table 11.1 General composition of three soluble metalworking fluids

Soluble or emulsifiable oils	Functional composition
Simple	mineral base oil emulsifier coupling agent preservative
Fatty	mineral base oil emulsifier coupling agent preservative fatty acids or soaps
Extreme pressure (EP)	mineral base oil emulsifier coupling agent fatty acids or soaps preservative EP additives (chlorine, sulfur or phosphorus)

Table 11.2 General composition of a semi-synthetic metalworking fluid

Functional components	Composition by weight (%)
Mineral base oil	15.0
Emulsifier/secondary or coemulsifier	20.0
Corrosion inhibitors (multi-package)	6.0
Coupling agent	1.0
Preservative	2.0
Water	56.0
Total	100.0

to affect the lubricity of a semi-synthetic metalworking fluid.[26] A general composition of a semi-synthetic metalworking fluid is given in Table 11.2.

It was noted that an optional extreme pressure or EP package may also be included into the semi-synthetic formulation.

11.3.4 Synthetic metalworking fluids

These are water-based solutions free of mineral oils which rely on glycols and esters for lubricity.[27,28] Water-soluble additives such as corrosion inhibitors, detergents and defoamers can be present in these metalworking fluids. These metalworking fluids are typically transparent in appearance and are used mainly in high heat metalworking operations like grinding where cooling is of importance. These metalworking fluids have high detergency properties but lack the lubricating properties for machining

Table 11.3 General composition of a synthetic metalworking fluid

Functional components	Composition by weight (%)
Water	72.0
Corrosion inhibitor	10.0
Corrosion inhibitor pH buffer	4.0
Boundary lubricant	4.0
Extreme pressure lubricant	4.0
Synthetic lubricant	4.0
Preservative	2.0
Total	100.0

operations. A general composition of a synthetic metalworking fluid is as shown in Table 11.3 minus mineral base oils.

It not only helps for metalworking fluids end users to know about the various types and applications of these metalworking fluids, but the proper care, handling and disposal of these metalworking fluids are important too.[29,30]

11.4 Major surfactant types used in metalworking fluids

Surfactants are primarily used in metalworking fluids as functional components of the emulsifier system of mainly water-based metalworking fluids.[31–33] Secondary functions of surfactants in metalworking fluids include corrosion inhibiting and lubricating properties. Major types of surfactant used in the various ranges of metalworking fluid types are described here.[34–36]

11.4.1 Anionics

11.4.1.1 Petroleum sulfonates. Petroleum sulfonates can be derived from either a natural or synthetic source. There are three types of oil-soluble sodium petroleum sulfonates based on molecular weight differences, i.e. low (420–440), medium (450–470), and high (490–510). These surfactants are typically used as primary emulsifiers and secondary corrosion inhibitors in water-based metalworking fluids, especially in soluble and semi-synthetic metalworking fluid types. Petroleum sulfonates like the calcium petroleum sulfonates can also provide secondary oil soluble corrosion inhibitory functions by utilizing a film formation mechanism on metal surfaces. These are inexpensive surfactants but are typically dark in color and may contain unsulfonated hydrocarbons.

11.4.1.2 Metal carboxylate soaps. These are the sodium and potassium soaps derived from long chain fatty acids that include mixed tallow, coconut oil, tall oil or palm oil fatty acids and purified lauric fatty acids. These surfactants are effective in stabilizing oil-to-water metalworking fluid emulsions. In addition, these surfactants are stable in high pH environment of pH 10 or higher, relatively inexpensive and are used as emulsifying components in emulsifier packages for water-based metalworking fluids. Amine carboxylates such as fatty acid soaps of triethanolamine or morpholine are used in a milder alkaline environment.

11.4.1.3 Alkylsulfamido and arylsulfamido carboxylic acids. These surfactants are soluble in water as salts and are used as emulsifiers or corrosion inhibitors in metalworking fluids systems involving water-based metalworking fluid types like the semi-synthetics and synthetics.

11.4.1.4 Phosphate esters. These surfactants are complex phosphorylated molecules derived from phosphated ethoxylated phenols and long chain alcohols and may vary from free acid form to amine or alkali metals. The aromatic phosphate esters provide corrosion inhibitory, strong EP, good wetting and lubricating properties for synthetic metalworking fluids. The aliphatic phosphate esters also provide corrosion inhibitory, strong EP, good lubricating, plus emulsifying properties for water-based metalworking fluid formulations.

11.4.1.5 Sulfated triglyceride oils. These surfactants are derived from the sulfation of hydroxy group and/or the fatty acid portion of the triglyceride. Examples are sulfated castor oil, sulfated soya oil, sulfated oleic acid and sulfonated tall oil fatty acid. Used primarily as lubricants and emulsifiers in metalworking fluids.

11.4.2 Nonionics

11.4.2.1 Alkanolamides. The 2:1 amides tend to be more water soluble and are used to a greater extent in metalworking fluids than the 1:1 amides. The 2:1 amides are prepared from the reaction of 2 mol of alkanolamine (diethanolamine, monoethanolamine or monisopropanolamine) with 1 mol of free fatty acid. They contain an excess amine content of 25 to 30% with a lesser amide content of 60 to 70% than the 1:1 amides, and are usually neutralized with fatty acids to make amine soaps which can function as coemulsifiers with the amide component. An example of an alkanolamide used in metalworking fluids is the 2:1 diethanolamine (DEA) tall oil fatty acid alkanolamide used as the emulsifier, lubricant and corrosion inhibitor in soluble, synthetic and semi synthetic metalworking fluids. The 2:1 DEA lard oil soaped alkanolamide is another alkanolamide used as a secondary emulsifier with good lubricating properties in water-based metalworking fluid types. Other alkanolamines like 2-amino-2-methyl-propanol (AMP) have been considered as secondary ethanolamine replacements in light of potential health concerns about *n*-nitrosodiethanolamine contamination in metalworking fluids.[37]

11.4.2.2 Alcohol, alkyl phenol, natural oil ethoxylates. Ethoxylated alcohol may be coconut oil-derived, tallow-derived or synthetic linear chain alcohol. These surfactants are used as secondary or coemulsifiers in metalworking fluids, e.g. ethoxylated C_{12}–C_{15} primary alcohols with a typical HLB (hydrophilic-lipophilic balance) value equal to 7.8. Ethoxylated fatty alcohols are also used as mineral oil emulsifiers in soluble and semi-synthetic metalworking fluids. The ethoxylated alkyl phenol molecule contains the hydrophilic ethoxylate chains and the hydrophobic alkyl phenol portions. These surfactants are used as secondary or coemulsifiers in soluble or emulsifiable oils. Examples of these surfactants are the octyl (C_8H_{17}) phenol ethoxylates and the nonyl (C_9H_{19}) phenol ethoxylates. Ethoxylated natural oil derivatives such as the ethoxylated castor oil are the emulsifiers for fats and soaps used in the compounding of water-soluble metalworking fluids.

11.4.2.3 EO/PO block copolymers/polyglycols. EO/PO (ethylene oxide/propylene oxide) block copolymers offer a wider range of products by varying the length of both the hydrophobic (polyoxypropylene) and hydrophilic (polyoxyethylene) portions of the molecule for specific applications. An example is an EO/PO block copolymer with HLB values between 7 and 12, which is used as the emulsifying and lubricating component in water-based metalworking fluids. Molecular transposition of the previously mentioned EO/PO block copolymer structure will result in an example of a reverse EO/PO block copolymer with HLB values between 2 and 7, used to lubricate oil-soluble metalworking fluids. Tetrafunctional EO/PO block copolymers are derived from the addition of ethylene oxide and propylene oxide to ethylenediamine. These EO/PO block copolymers provides lubricating properties for water-based metalworking fluids. EO/PO esters can also be prepared from their corresponding methyl esters and EO/PO glycols for use as emulsifiers and lubricants in semi-synthetic fluids. The more hydrophobic EO/PO esters have defoaming properties and combine with alkanolamides for additional corrosion protection properties.[38]

Polyglycols are used primarily as the lubricant component in water-based metalworking fluids. An example is an unmodified random polyalkylene glycol which contributes inverse solubility and hydrodynamic lubricity properties to the metalworking fluids. This surfactant is typically used in combination with fatty acids or phosphate esters in metalworking fluids to provide good lubrication in ferrous and non-ferrous metalworking operations. Another example is a modified triethanolamine acid polyalkylene glycol mixture which is used as the sole lubricant base for water-based metalworking fluids. This surfactant is used to provide extreme pressure lubrication for heavy duty machining operations.

11.4.2.4 Long chain carboxylic acid esters. These surfactants are the sorbitan esters which are typically sorbitan fatty acid esters such as sorbitan monooleate or the ethoxylated derivative of polyoxyethylene sorbitan(5)-monooleate, and the glycerol esters such as glycerol monooleate. These two surfactant groups are primarily used as coemulsifiers in soluble metalworking fluids. The polyfunctional alcohol or polyol esters are used for compounding synthetic metalworking fluids and are the boundary lubricant component of the fluids. Polyethylene glycol or PEG esters are a wide range of surfactants with different HLB values for selective applications. They are used primarily as the emulsifiers in synthetic metalworking fluids and in some cases as the lubricant component of the fluids.

11.4.3 Cationics and amphoterics

These surfactant types are normally not used in metalworking fluid formulations. Factors that prevent the utilization of these surfactants to a greater extent than either the anionic or nonionic surfactants are mainly cost and incompatibility with typical anionics found in most metalworking fluid formulations.

Cationic surfactants used in metalworking fluids are the imidazolines surfactants. Examples of these surfactants are the oleyl, coco and soya imidazolines derived from oleic acids, coco and soya fatty acids, respectively. Imidazolines are primarily used as oil-soluble emulsifiers and corrosion inhibitors in straight or neat oil metalworking fluid formulations.

Amphoteric surfactants used in metalworking fluid are the amphoteric carboxylate salts like the tall amphopropionate derived from tall oil fatty acid imidazolines. The amphoteric surfactants are usually sodium chloride free and primarily used as oil emulsifiers in soluble oils.

11.4.4 Bioresistant surfactants

The surfactant components in metalworking fluid formulations are susceptible to microbial degradation during in-use metalworking fluid operations.[5,38] Degradation of metalworking fluid surfactants or emulsifiers by microbial action has been reported for petroleum sulfonates,[9] as well as for oil-soluble nonionic ethoxylated alcohols or water dispersable nonionic ethoxylated esters.[7] Metalworking fluid formulators have started to investigate surfactants that are more resistant to biodegradation,[39,40] but are constrained by biodegradability requirements for proper waste disposal of spent metalworking fluids.

11.5 Types of microorganism in metalworking fluids

The primary source of microbial contamination of a fresh coolant is the residual biomass left behind from the previous charge. Although cleaning the system between charges is a good practice, it may not necessarily remove all the microbial biomass in the system.[3,4,41] Other sources of microbial contamination are from the metalworking fluid concentrate itself, the water used to dilute the metalworking fluid concentrates for use, dirt entering the plant system from the air, plant workers' clothing or from the workers themselves through poor personal hygiene, general poor plant hygiene, metal stocks and hydraulic oil leakage.[4,25,42,43] In order for the microorganisms to survive in the metalworking fluid system, nutrient requirements for microorganisms are often obtained from the various components present in the metalworking fluids formulation itself.[38,44–46] These nutrients can be organic in nature such as mineral oil base stocks, fatty acids soaps, fatty oils, synthetic esters, phosphate esters, glycols, amines or inorganic mineral salts. Three common components of metalworking fluids such as naphthenic petroleum-based mineral oil, petroleum sulfonates and fatty acids (linoleic and oleic acids) were evaluated as sources of carbon for microbial growth. The study indicated that the fatty acids component of metalworking fluids was the primary source of carbon for bacterial growth.[47]

Other primary factors that contribute to the growth and proliferation of microorganisms in the metalworking fluid system are microbial growth conditions of pH between 9.2 and 9.5, optimal bulk temperatures of metalworking fluids in the system for microbial growth and oxygen availability for certain microorganisms. Secondary factors that influence microbial growth in metalworking fluid systems include the effects of iron fines,[48] oil-to-water ratio[49] and water hardness.[50]

There are usually three types of microorganisms associated with metalworking fluids deterioration. They are the aerobic bacteria, anaerobic bacteria and fungi (mold and yeast). The presence of biofilm has also been reported in water-based metalworking fluids which is different from the typical planktonic or free-floating microorganisms associated with an aqueous system.[51] The biofilm present in a metalworking system is made up of bacteria and/or fungi species and may account for residual biomass that recontaminates fresh additions of metalworking fluid into the operating system. Biofilm presence in the metalworking fluid system may also be responsible for the regrowth of the microbial population following preservative treatment for microbial control of the system.[52]

11.5.1 Aerobic bacteria

These microorganisms include the commonly found *Pseudomonas* species which are the prevalent spoilage agents of cutting fluids.[4,38,53–57] Other aerobic bacteria found in metalworking fluids are the *Enterobacter* species, *Klebsiella* species, *Escherichia* species and *Proteus* species.[4] The *Acinetobacter* bacterial species have also been indicated as a primary bacterial colonizer in metalworking fluids. Although not the dominant species, its presence may maintain the other bacterial species in contaminated metalworking fluids.[47] These microorganisms are highly oxidative and adapt very well to a wide variety of metalworking fluids. Their growth and proliferation lead to emulsion destabilization, loss of metalworking fluids lubricating properties and increase in corrosive properties of metalworking fluids used in the plant system. A microbial ecological study of a metalworking fluid system has indicated that the levels of fungi and anaerobic or sulfate-reducing bacteria are quite dependent on the initial levels of aerobic bacteria.[58]

11.5.2 Anaerobic bacteria

These microorganisms are essentially sulfate reducers such as *Desulfovibrio desulfuricans* which reduce sulfate to sulfide.[59] Anaerobic conditions can exist in stagnant areas within the metalworking fluid system for anaerobic bacterial growth. The mechanism of anaerobic growth stems from an interrelationship with the aerobic bacteria,[60,61] where the aerobic bacteria lower the redox potential of the metalworking fluids and synthesize nutrients for themselves. The presence of iron swarf or metal fines may contribute to the increase in sulfate reducing bacteria growth.[48,62] Anaerobic bacteria which generate sulfide by the reduction of sulfur in petroleum sulfonates, organic polysulfides and sulfurized oil/fats instead of sulfate, are better known as sulfide-generating bacteria.[45] Growth of sulfate-reducing and/or sulfate-generating bacteria usually causes metalworking fluids to blacken or grey, offensive odor to form and corrosion to metalworking tools and the workplace.

11.5.3 Fungi

These microorganisms grow well in synthetic metalworking fluids and have been implicated in the spoilage of metalworking fluids.[44,63–65] Most commonly found fungi in metalworking fluids include the *Fusarium* species, *Cephalosporium* species and *Candida* species. Fungi can grow well when the preservatives added to metalworking fluids are more active against bacteria and would then clear the way for the growth of fungi.[51,66–68] Fungi found in metalworking fluid systems are mainly slime formers and can physically plug system pipelines and thereby disrupt the flow of metalworking fluids. Fungi can be found free living in the water phase and at the same time be found metabolizing the ingredients that migrate from the oily phase of metalworking fluids. Due to susceptibility of metalworking fluids to fungal contamination and uncontrolled fungal growth, metalworking fluids are commonly treated with preservatives active against fungal contamination.

11.6 Types of preservative used in metalworking fluids

Metalworking preservatives are used to prevent the biodeterioration of the metalworking fluid functions and extend the life of the metalworking fluids used in metalworking operations.[4,69,70] Metalworking fluid preservatives are incorporated as a functional component of metalworking fluid concentrates by metalworking fluid formulators, or added by metalworking fluids end users as regular tankside additions to the diluted metalworking fluids.

Factors to keep in mind when selecting a preservative for a specific metalworking fluid application are toxicology profile, environmental friendliness, biocidal effectiveness, chemical compatibility and physical stability.

A favorable and extensive toxicity profile is a must. Preservative users should be informed of potential health hazards associated with the preservative for safe handling, storage and use of the product.[41,70,71]

To prevent contamination of the environment, safe disposal procedures must be obtained from the preservative manufacturer. This information should be made available to the preservative users and in compliance with all relevant disposal laws and regulations to prevent contamination of the environment.

Preservative biocidal effectiveness should be broad spectrum against a wide variety of bacteria and fungi commonly found in biodeteriorated or spoilt metalworking fluids at recommended end-use dosages. The biocidal effectiveness of preservatives used in a metalworking fluid system may be impacted by the presence of biofilm[52,72] and microbial resistance to the preservative treatments.[73–75]

The preservative should also be chemically compatible with the other components of the metalworking fluid concentrates,[76] or with other constituents or byproducts present in the in-use metalwork fluid system, such as filtering agents,[77] metals[78] and hydraulic fluids.[79]

Physical stability of the preservative should include the pH stability in an alkaline environment of pH 8.5 to 9.5,[80] thermal stability,[81] corrosivity effects on worked metals[82] and oil-in-water solubility where the relative volume of oil-to-water influences the equilibrium concentration of the preservative in the aqueous phase during end-use diluted metalworking applications.[49]

The Federal Insecticide, Fungicide and Rodenticide Act (FIFRA) requires the United States Environment Protection Agency to regulate legally the pesticide registration of a stated application use of a preservative in the United States including preservatives used in metalworking fluids applications.[83] Registration of metalworking fluid preservative use may also apply to other countries where pesticide registration is also required by law.[84]. Although there are a number of metalworking fluid preservatives presently registered for use in metalworking fluid applications, the bulk of the metalworking fluid control is accomplished through a limited number of preservative manufacturers. The various chemical types used as metalworking fluid preservatives include the formaldehyde releasers, e.g. triazines and hexamines,[66,67,85,86] phenolic derivatives,[87] thiazoles,[88–90] alkane derivatives,[91,92] morpholines,[93] glutaraldehyde[94,95] and others.[96]

A sampling of metalworking fluid preservatives registered for use in the United States is described below based on manufacturer, trade name, active ingredient/s and applications.

11.6.1 ANGUS Chemical Company

11.6.1.1 Bioban® CS-1135 (EPA Reg.No.48301–8). Contains 4,4-dimethyloxazolidine as 74% active ingredient (AI) and 3,4,4-trimethyloxazolidine as 2.5% AI. This water-based preservative is used in soluble oils, semi-synthetics and synthetic metalworking fluid concentrates and tankside additions. Recommended end-use level in final diluted metalworking fluid is 0.10% w/w of product.

11.6.1.2 Bioban® N-95 (EPA Reg.No.1100–82–48301). Contains 5-hydroxymethoxymethylaza-3,7-dioxabicyclo(3.3.0)octane as 24.5% AI, 5-hydroxymethyl-1-aza-3,7-dioxabicyclo(3.3.0)octane as 7.7% AI and 5-hydroxypoly[methyleneoxy(74%C-2, 21%C-3, 4%C-4, 1%C-5)]methyl-1-aza-3,7-dioxabiocyclo(3.3.0)octane as 7.8% AI. Recommended end-use level in final diluted metalworking fluid is 0.1 to 0.3% w/w of the product.

11.6.1.3 Bioban® P-1487 (EPA Reg.No.48301–7). Contains 4-(2-nitrobutyl)morpholine as 70% AI and 4,4′-(2-ethyl-2-nitrotrimethylene)-dimorpholine as 20% AI. This water-based preservative is used in soluble oils, semi-synthetics and synthetic metalworking fluid concentrates and tankside additions. Recommended end-use level in final diluted metalworking fluids is an initial dose of 0.1% active ingredient followed by a maintenance dose of 0.01% w/w active ingredient on a weekly basis.

11.6.1.4 Tris Nitro® (EPA Reg.No.48301–11). Contains tris(hydroxymethyl)nitromethane as 50% AI. Used mainly with metalworking fluids as tankside additions. Recommended end-use level in final diluted metalworking fluids is 0.05% w/w of the product.

11.6.2 Buckman Laboratories, Inc.

11.6.2.1 Busan® 30WB (EPA Reg.No.1448–55). Contains 2-(thiocyanomethylthio)benzothiazole as 30% AI. This is a water-based preservative used in water-diluted soluble oils, semi-synthetic and synthetic metalworking fluids especially as tankside additions. Recommended end-use level in final diluted metalworking fluid is 0.01% w/w of product.

11.6.2.2 Busan® 77 (EPA Reg.No.1448–242). Contains poly[oxyethylene(dimethyliminio)ethylene(dimethylimino)ethylene dichloride] as 60% AI. This preservative is used in water-based synthetic and oil-based metalworking fluids with emulsified nonionic surfactants. However, it is not recommended for use in straight oils or metalworking fluids that contain anionic surfactants in their formulation. Recommended end-use level in final diluted metalworking fluids is 0.01 to 0.1% w/w of product.

11.6.2.3 Busan® 1030 (EPA Reg.No.1448–55). Contains 2-(thiocyanomethylthio)benzothiazole as 30% AI. This is a solvent-based preservative used in soluble oils, semi-synthetic and synthetic metalworking fluid concentrates. Recommended end-use level in final diluted metalworking fluids is 0.005 to 0.01% w/w of product.

11.6.2.4 Busan® 1060 (EPA Reg.No.1448–243). Contains hexahydro-1,3,5-tris(2-hydroxyethyl)-*s*-triazine as 78.5% AI. This preservative is stable and water soluble. Used in soluble oils, semi-synthetic and synthetic metalworking fluid concentrates and tankside additions. Recommended end-use level in final diluted metalworking fluid is 0.10 to 0.15% w/w of product.

11.6.3 The Dow Chemical Company

11.6.3.1 Dowicide® A (EPA Reg.No.464–78). Contains sodium salt of *O*-phenylphenol as 97% AI. This water-soluble preservative is used in soluble oil, semi-synthetic and synthetic metalworking fluid concentrates as well as tankside additions. Recommended end-use level in final diluted metalworking fluid is 0.005 to 0.2% w/w of product.

11.6.3.2 Dowicide® 1 (EPA Reg.No.464–70). Contains *O*-phenylphenol as 99% AI. This is mainly an oil-soluble preservative used in soluble oils, semi-synthetic and synthetic metalworking fluid concentrates as well as tankside additions. Recommended end-use level in final diluted metalworking fluid is 0.05 to 0.15% w/w of product.

11.6.3.3 Dowicil® 75 (EPA Reg.No.464–403). Contains 1-(3-chloroallyl)-3,5,7-tria-1-azoniaadamantane chloride as 67.5% AI combined with an inert sodium bicarbonate stabilizer. Used in water-diluted soluble oils, semi-synthetic and synthetic metalworking fluids especially as tankside additions. Recommended end-use level in final diluted metalworking fluid is 0.015 to 0.25% w/w of product.

11.6.4 Olin Corporation

11.6.4.1 Triadine® 3 (EPA Reg.No.1258–1071). Contains hexahydro-1,3,5-tris(2-hydroxyethyl)-*s*-triazine as 78.5% AI. This preservative is stable and water soluble. Used in soluble oil, semi-synthetic and synthetic metalworking fluid concentrates and tankside additions. Recommended end-use level in final diluted metalworking fluid is 0.10 to 0.15% w/w of product.

11.6.4.2 Triadine® 10 (EPA Reg.No.1258–990). Preservative combination of hexahydro-1,3,5-tris(2-hydroxyethyl)-*s*-triazine as 63.6% AI and sodium 2-pyridinethiol-1-oxide as 6.4% AI. This preservative is used in soluble oil, semi-synthetic and synthetic metalworking fluid concentrates and tankside additions. Maximum end-use level in final diluted metalworking fluid is 0.2% w/w of product. Typical end-use level in diluted metalworking fluid is 0.07% w/w in synthetic fluids or 0.1% w/w in oil-containing fluids.

11.6.4.3 Triadine® 20 (EPA Reg.No.1258–1205). Preservative combination of hexahydro 1,3,5-tris(2-hydroxyethyl)-*s*-triazine as 71.4% AI and sodium 2-pyridinethiol-1-oxide as 3.64% AI. This preservative is used in soluble oil, semi-synthetic and synthetic metalworking fluid concentrates and tankside additions. Recommended end-use level in final diluted metalworking fluid is 0.15 to 0.20% w/w of product.

11.6.4.4 Sodium Omadine® 10% aqueous solution (EPA Reg.No. 1258–1213). Contains sodium 2-pyridinethiol-1-oxide as 10% AI. This preservative is used in soluble oil, semi-synthetic and synthetic metalworking fluid concentrates and tankside additions. Its lower use-concentration allows for easier tankside additions of smaller sumps. Maximum end-use level in final diluted metalworking fluid is 0.5% w/w of product.

11.6.4.5 Sodium Omadine® 40% aqueous solution (EPA Reg.No. 1258–843). Contains sodium 2-pyridinethiol-1-oxide as 40% AI. This preservative is used in soluble oil, semi-synthetic and synthetic metalworking fluid concentrated and tankside additions. Maximum end-use level in final diluted metalworking fluid is 0.125% w/w of product. Recommended end-use level in final diluted metalworking fluid is 0.016% to 0.025% w/w of product.

11.6.5 Rohm and Haas Company

11.6.5.1 Kathon® 886MW (EPA Reg.No.707–129). Contains 5-chloro-2-methyl-4-isothiazolin-3-one as 8.6% AI and 2-methyl-4-isothiazolin-3-one as 2.6% AI. Used in soluble oil, semi-synthetic and synthetic metalworking fluid concentrates and diluted tankside additions. Recommended end-use level in final diluted metalworking fluids is an initial dose of 10 to 21.8 ppm active ingredient followed by a maintenance dose of 5 ppm active ingredient every 4 weeks or 5–21.8 ppm active ingredient every 8–12 weeks.

11.6.5.2 Kathon® 893MW (EPA Reg.No.707–195). Contains 2-*n*-octyl-4-isothiazolin-3-one as 45% AI. Used in soluble oil, semi-synthetic and synthetic metalworking fluid concentrates and diluted tankside additions. Recommended end-use level in final diluted metalworking fluid is an initial dose of 25–75 ppm active ingredient followed by a maintenance dose of 5–30 ppm active ingredient every four weeks.

11.6.5.3 Kathon® MWC (EPA Reg.No.707–220). Contains 5-chloro-2-methyl-4-isothiazolin-3-one as 2.69% AI and 2-methyl-4-isothiazolin-3-one as 0.8% AI. Used in soluble oil, semi-synthetic and synthetic metalworking fluid concentrates and diluted tankside additions. Recommended end-use level in final diluted metalworking fluids is an initial dose of 4.9–20.5 ppm active ingredient followed by a maintenance dose 4.9–20.5 ppm active ingredient every 1 to 3 weeks.

11.6.6 R.T. Vanderbilt Company, Inc.

11.6.6.1 Vancide® TH (EPA Reg.No.1965–55). Contains hexahydro-1,3,5-triethyl-*s*-triazine as 95% AI. This preservative is stable as well as water soluble and oil soluble. Used in soluble oil, semi-synthetic and synthetic metalworking fluid concentrates and diluted tankside additions. Recommended end-use level in final diluted metalworking fluid is 0.04 to 0.15% of product.

11.6.6.2 Vancide® 51 (EPA Reg.No.1965–8). Preservative combination of sodium dimethyldithiocarbamate as 27.6% AI and sodium 2-mercaptobenzothiazole as 2.4% AI. This preservative is used in water-diluted soluble oils, semi-synthetic and synthetic metalworking fluids especially tankside additions. Recommended end-use level in final diluted metalworking fluid is 0.5 to 0.1% w/w of the product.

11.6.7 Stepan Company

11.6.7.1 Onyxide® 200 (EPA Reg.No.1839–99). Contains hexahydro-1,3,5-tris(2-hydroxyethyl)-*s*-triazine as 78.5% AI. This preservative is stable and water soluble. Used in soluble oil, semi-synthetic and synthetic metalworking fluid concentrates and tankside additions. Recommended end-use level in final diluted metalworking fluid is 0.10 to 0.15% w/w of product.

11.6.8 Union Carbide Corporation

11.6.8.1 UCONEX® Antimicrobial 345 (EPA Reg.No.10352–28). Aqueous solution of glutaraldehyde as 45% AI. Used in water-diluted soluble oil, semi-synthetic and synthetic metalworking fluids especially tankside additions. Recommended end-use level in final diluted metalworking fluid is an initial dose of 220 to 670 ppm followed by maintenance dose of 90 to 440 ppm (daily to weekly) of the product.

11.6.9 US Professional Laboratories (L & F Products Group)

11.6.9.1 Grotan® (EPA Reg.No.365–76). Contains hexahydro-1,3,5-tris(2-hydroxethyl)-*s*-triazine as 78.5% AI. This preservative is stable and water soluble. Used in soluble oil, semi-synthetic and synthetic metalworking fluid concentrates and tankside additions. Recommended end-use level in final diluted metalworking fluid is 0.10 to 0.15% of product.

11.6.10 ZENECA Biocides

11.6.10.1 Proxel® MW200 Preservative (EPA Reg.No.10182–30). Contains a solution of 1,2-benzoisothiazolin-3-one as 17% AI. This preservative is stable and water soluble. Used in soluble oil, semi-synthetic and synthetic metalworking fluid concentrates and tankside additions. Recommended end-use level of the product in the final diluted metalworking fluid is 0.025 to 0.100% w/w for soluble oils, 0.050 to 0.125% w/w for semi-synthetics and 0.075 to 0.150% w/w for synthetics.

11.6.10.2 Proxel® MW300 Preservative (EPA Reg.No.10182–1). Contains an aqueous solution of 1,2-benzoisothiazolin-3-one as 30% AI. This preservative is stable and water soluble. Used in soluble oil, semi-synthetic and synthetic metalworking fluid concentrates and tankside additions. Recommended end-use level of the product in the final diluted metalworking fluid is 0.02 to 0.07% w/w for soluble oils, 0.03 to 0.08% w/w for semi-synthetics and 0.05 to 0.10% w/w for synthetics.

11.6.11 Potentiation of preservatives

Potentiation of the biocidal effectiveness of metalworking preservatives was reported with the use of corrosion inhibitors such as alkanolamines,[97–100] metal chelating agents such as ethylenediaminetetraacetic acid (EDTA),[101–103] copper citrate[104–106] and even in combination with other preservatives.[89]

However, there is no universally potent preservative; each user has to evaluate and select the ideal preservative in terms of whether it works or not in a formulated concentrate or in-use diluted metalworking fluid operation, with the help of standard testing methods.

11.7 Testing protocols used in metalworking fluid preservative selection

Many test protocols have been developed to evaluate and select various types of preservative for use in metalworking fluid systems.[107–109] There are usually two testing phases involved in the selection of the right preservative for a particular metalworking fluid plant system. The first phase involves preliminary laboratory screening of the test preservative for efficacy to control microbial growth in a laboratory simulated metalworking plant system. The second phase is the confirmatory field testing of the selected preservative for efficacy to control microbial growth in an actual metalworking plant operation. In some instances, both phases can be done in parallel to each other in order to select the right preservative for microbial growth control for a metalworking fluid system.

To facilitate the standardization of preservative testing and selection, the microbial inoculum used for the testing has to be predetermined. The microbial inoculum selection can either be obtained by isolating the primary microorganisms indigenous to the contaminated metalworking fluid system of interest,[110,111] or be a defined microbial inoculum developed specifically for metalworking fluid preservative testing.[112–114] The defined microbial inoculum can survive and grow in a large number of metalworking fluids and usually consists of three bacterial species of *Pseudomonas aeruginosa* (dominant metalworking fluid biodeteriorant), *Proteus mirabilis* (produces strong fecal odor), *Klebsiella pneumoniae* (slime former), including two fungal species of *Fusarium* and *Cephalosporium*. The microbial count of the bacterial inoculum is approximately 10^9 colony forming units or cfu ml^{-1} for each bacterial species and approximately 10^5 cfu ml^{-1} for each fungal species. Other factors to be considered in determining whether to adopt a particular test method are the general acceptance of that particular test method by metalworking fluid end users, the ability to simulate metalworking plant operation use, reliability and statistical reproducibility, ease of conduct, and possibility of completion within a reasonable amount of time.

11.7.1 Laboratory testing of metalworking fluid preservatives

A number of laboratory tests for evaluating and selecting metalworking fluid preservatives have been previously covered.[109] These include the recirculating test methods that have aerobic/anaerobic cycles, repeated microbial inoculations of the test systems, aeration and circulation of metalworking fluids through iron chips,[108,115–117] and the aerated bottle test methods,[4,112] upon which the current American Society for Testing and Material (ASTM) testing protocols for preservative selection for use with water-based metalworking fluids were based. Steps were originally initiated by ASTM to develop these standard test protocols for antimicrobial agents which included testing of metalworking fluid preservatives.[118] This resulted in the development of the two-week bioresistance test (ASTM# D 396–92) and the six-week bioresistance test (ASTM# E 686–91) standard methods for testing.

The two-week bioresistance test is a standard test method for evaluating the microbial resistance of water-based metalworking fluids.[119] The level of bioresistance of the water-based metalworking fluids is evaluated by microbiologically challenging the fluids with an inoculum prepared from actual biodeteriorated metalworking fluid obtained from the end-user's site. This method allows the end users to determine whether a particular metalworking fluid is susceptible or not to microbial spoilage, and if so, to select a preservative for microbial control.

The six-week bioresistance test is a standard test method for evaluating the selecting preservatives for use in water-based metalworking fluids.[120] The efficacy of any antibacterial and antifungal preservatives at use concentrations are evaluated in controlling the microorganisms responsible for the biodeterioration of metalworking fluids. This test method can also be used to evaluate the efficacy of preservatives which are directly formulated into the metalworking fluid concentrates. The test runs for six weeks with multiple microbial challenge either isolated from the spoilt metalworking fluid or a defined inoculum. The metalworking fluids are then evaluated for physical changes and pH changes, including bacterial and fungal microbial counts. The metalworking fluids either pass or fail the preservative selection test based on these criteria.

11.7.2 Field testing of metalworking preservatives

Field testing allows for metalworking fluid users to assess the actual interactions of prior microbial growth and its byproducts with the actual preservative used in a metalworking plant operation.[121] Efforts were attempted in the past to correlate laboratory testing results with actual field evaluation testing of the selected preservative of choice for microbial control in the plant system being evaluated.[122–124] The field test confirms the efficacy of the selected preservative from the laboratory screen to protect the metalworking fluid system against uncontrolled microbial contamination and eventual spoilage of the metalworking fluids. Once a preservative has been selected to treat successfully a particular metalworking fluid system, it is important for the end users to know when to add more preservative for continuous control of microbial contamination. This can be accomplished by routine monitoring and record keeping of microbial levels in the metalworking system using specific microbial assays for determining microbial growth and proliferation levels in metalworking fluids.

11.7.3 Monitoring for microbial contamination in metalworking fluid systems

Early microbial assays for monitoring the microbial growth and proliferation levels of metalworking fluid systems included the use of the Warburg apparatus, 'red-spot' test, methylene blue reduction test and the 'corn meal' test methods.[107] The Warburg apparatus was based on the measurement of oxygen consumption of the actively growing microorganisms in metalworking fluids.[38] The red-spot test is based on the ability of actively growing microorganisms to reduce a tetrazolium salt to form a red colored formazan.[125] The methylene blue test is also based on dye reduction chemistry where methylene blue is reduced by the actively growing microorganisms in the metalworking fluids to a colorless leuco base.[115,126] The corn meal test measures the microbiological stability of metalworking fluids for at least 21 days without going 'sour'. This prepared metalworking fluid mixture would be evaluated at periodic intervals for slime formation, stability and odor production.[127] These early microbial assays did not provide enough information concerning the identity or types of spoilage microorganisms present in the contaminated metalworking system and the microbial counts obtained as the 'actual' number of microorganisms were not accurate or consistent, and were marginally quantitative.[107]

More practical microbial assays were developed to determine the level of microbial load in the metalworking fluid system. These were based on the enumeration of microorganisms by viability testing, microbial metabolic activity measurements and microbial cell components measurements.[128]

Viability assay is based on the enumeration of microbial numbers present in the metalworking fluid system. This can be done through standard plate counts or by five tube dilution-to-extinction microbial assays.[4,110] The 'dip stick' or 'dip slide' method is a simplified plate count method developed for the on-site estimation of specific microorganisms found in the metalworking fluid system.[108,129,130] The microbial assay consists of a plastic paddle or filter paper coated with an appropriate microbial recovery medium and/or indicator, dipped into the metalworking fluid sample, returned to its container and incubated appropriately. The microorganisms left behind on the dip slide can grow and produce visible colonies. Instead of counting microbial numbers in cfu ml^{-1}, the observer compares the results obtained to a standard colonies density chart (ranging in order of magnitude from 10^3 to 10^7 cfu ml^{-1} for total aerobic bacterial numbers). The results obtained are then measured according to the corresponding order of magnitude of colonies density in cfu ml^{-1}. An increase of more than two orders of magnitude, i.e. from 10^4 to 10^6 cfu ml^{-1}, indicates that microbial growth is out of control and the metalworking fluid system requires appropriate additions of preservative for control. This method can provide specific microbial isolates for inoculum preparation and is relatively easy to do but limited to time requirements of 24 h or one week for results to develop and for safe disposal considerations. These microbial assays are available to detect total aerobic bacteria and fungi or combinations of both bacteria/fungi. The detection of sulfate-reducing bacteria utilizes specially prepared media tubes. This involves a reaction by the sulfate-reducing bacteria with iron present in the tube to form a visible black iron sulfide precipitate.[131]

Metabolic activity assay measures the overall metabolic activity of the microorganisms present in the metalworking fluid system. A modified biochemical oxygen demand (BOD) assay specifically measures the metabolic rate at which the microbial population is consuming oxygen within the metalworking fluid system.[128] Loss of more than half of the dissolved oxygen in a test sample left standing for four hours indicates significant microbial contamination. The microbial assay is relatively rapid and measures the overall impact of the microbial population in the system but it requires more technical skill and purchase of expensive equipment.

Microbial cell component assays involve the selective measurement of specific constituents present in microbial cells growing in the metalworking fluid system. Specific microbial assays were developed for measuring cell components such as adenosine triphosphate (ATP) and catalase. The ATP microbial assay utilizes indirect measurement of ATP concentration with the luciferase enzyme and luciferin to form ATP–luciferin–luciferase complex that is extremely reactive and emits a photon of light. Detection of this emitted light can be measured using a photometer and the value obtained is proportional to the amount of ATP per unit of microbial biomass present in the metalworking system.[132] The ATP assay has not been successful due to the quenching effect of the enzyme reaction by components of the metalworking fluids. The catalase microbial assay estimates microbial loads by indirectly measuring the concentration of catalase enzyme present in most bacteria and all fungi found in metalworking fluids. Good correlation between standard plate counts and catalase activity of microbial load were determined for all the three water-based metalworking fluid types.[133] Pressure detection used for measurement of microbial load was also confirmed to be proportional to oxygen evolution and microbial catalase concentration in the metalworking system.[134] This assay requires some operator training to yield reproducible results, purchase of expensive equipment and generation of individual standard calibration for each metalworking system, but the rapid results allow for real-time monitoring and control of microbial contamination in the system.

Another microbial assay included the development of the significance tests or sig-tests that address the Eh, pH, oxygen and nutrient gradients found in metalworking fluid sumps using a gel system to determine the overall contribution of the metalworking fluid microbial community to spoilage.[45,135]

Other potential microbial assays may be developed, such as using molecular DNA or rRNA probes, to quantify specific microbial populations like *Pseudomonas aeruginosa*, commonly associated with the biodeterioration of metalworking fluids.

There is no single best method for measuring and monitoring microbial load in a metalworking fluid system. Each metalworking system will establish its own characteristic operating profile. The best approach is to select a microbial assay that will provide the most economical and efficient routine microbial monitoring process for that particular metalworking fluid system.

Monitoring microbial loads using the different microbial assays is just one parameter for monitoring microbial contamination in metalworking systems. Other parameters that also impact the overall control of microbial contamination in metalworking systems include gross observations, system operating parameters, physical analysis and chemical analysis of the in-use metalworking fluids.[128,132,136,137]

11.8 Summary

If constructive efforts are made to monitor and control microbial contamination in metalworking fluid systems, with proper know-how and use of the selected preservative, marked improvement in the overall plant management of metalworking fluid systems will occur, with significant reduction in metalworking plant operation costs.

References

1. Springborn, R.K. (1967) *Cutting and Grinding Fluids: Selection and Application*, American Society of Tool Manufacturing Engineering, Dearborn, Michigan.
2. Nachtman, E.S. and Kalpakjian, S. (1985) *Lubricants and Lubrication in Metalworking Operations*, Marcel Dekker, New York.
3. Hill, E.C. (1967) The significance and control of microorganisms in rolling mill oils and emulsions. *Metals Mater.*, **1**, 294–297.
4. Bennett, E.O. (1974) The deterioration of metal cutting fluids. *Progr. Indust. Microbiol.*, **13**, 121–149.
5. Rossmoore, H.W. (1974) Microbiological causes of cutting fluid deterioration. *Soc. Manufact. Eng. Tech. Paper*, #MR74–169.
6. Hill, E.C. (1977) Microbial infection of cutting fluids. *Tribology Internat.*, **10**, 49–54.
7. Almen, R., Mantelli, G., McTeer, P. and Nakamaya, S. (1982) Application of high-performance liquid chromatography to the study of the effect of microorganisms in emulsifiable oils. *Lubrication Eng.*, **38**, 99–103.
8. Holtzman, G.H.M., Rossmoore, H.W., Holodnik, E. and Weintraub, M. (1982) Interrelationships between biodeterioration, chemical breakdown, and function in soluble oil emulsions. *Develop. Indust. Microbiol.*, **23**, 207–216.
9. Geiger, R.M., Bennett, E.O. and Gannon, J.E. (1984) Microbial degradation of selected water-soluble lubricants. *Develop. Indust. Microbiol.*, **25**, 567–573.
10. Tant, C.O. and Bennett, E.O. (1956) The isolation of pathogenic bacteria from used emulsion oils. *Appl. Microbiol.*, **4**, 332–338.
11. Rossmoore, H.W. and Williams, B.W. (1967) Survival of coagulase-positive staphylococci in soluble cutting oils, *Health Lab. Sci.*, **4**, 160–165.
12. Holdum, R.S. (1976) Microbial spoilage of engineering materials: Part 3 – Are infected oil emulsions a health hazard to workers and to the public? *Tribology Internat.*, **9**, 271–281.
13. Bennett, E.O. (1983) Water based cutting fluids and human health. *Tribology Internat.*, **16**, 133–136.
14. Hill, E.C. and Al-Zubaidy, T.S. (1979) Some health aspects of infections in oil and emulsions. *Tribology Internat.*, **12**, 161–164.
15. Hill, E.C. (1983) Microbial aspects of health hazards from water based metalworking working fluids. *Tribology Internat.*, **16**, 136–140.
16. Elsmore, R. (1989) The survival of *Legionella pneumophila* in dilute metalworking fluids. *Tribology Internat.*, **22**, 213–214.
17. Rossmoore, H.W. (1993) Biostatic fluids, friendly bacteria, and other myths in metalworking fluids©, *Lubrication Eng.*, **49**, 253–260.
18. Cookson, J.O. (1971) Machine tool design and use in relation to cutting fluids. *Ann. Occupational Hygiene*, **14**, 181–190.
19. Cookson, J.O. (1977) An introduction to cutting fluids. *Tribology Internat.*, **10**, 5–11.
20. Nicholson, R. (1977) Care and handling of cutting fluids. *Tribology Internat.*, **10**, 17–20.
21. Kotvis, P. (1983) A guide to metalworking coolant formulations and performance. *Fluid & Lubrication IDEAS*, **6**, 10–12.

22. Saunders, A.R. (1990) The metalworking lubricants market: Recent and future trends. *Indust. Lubrication Tribology*, 3–10.
23. Sutcliffe, T. and Barber, S.J. (1977) How to select a water-base coolant. *Am. Machinist*, 138–141.
24. Hunz, R.P. (1984) Water-based metalworking lubricants©. *Lubrication Eng.*, **40**, 549–553.
25. Eachus, A.C., Selleck, J.R. and Hunsucker, J.H. (1978) Antimicrobial aspects of metalworking fluids. *Chem. Times Trends*, **1**, 51–55.
26. Yang, C.C. (1979) The effects of water hardness on the lubricity of a semi-synthetic cutting fluid. *Lubrication Eng.*, **35**, 133–36.
27. Marinello, R.L. (1976) Industrial synthetic lubricants. *Plant Eng.*, 94–98.
28. Edwards, J. and Jones, E. (1977) Synthetic cutting fluids – A brief history and state of the art. *Tribology Internat.*, **10**, 977.
29. Bennett, E.O. (1973) The disposal of metal cutting fluids. *Lubrication Eng.*, **29**, 300–307.
30. Kipers, K.C., Shook, F.C. and DeBoer, R. (1983) Evaluation of emulsifier systems for metalworking fluids – Part I. *Lubrication Eng.*, **39**, 358–365.
31. Kipers, K.C., Shook, F.C. and DeBoer, R. (1983) Evaluation of emulsifier systems for metalworking fluids, Part II. *Lubrication Eng.*, **39**, 368–375.
32. Mahdi, S.M. and Sköld, R.O. (1990) Surface chemistry aspects on the use of ultrafiltration for the recycling of water-based synthetic metalworking fluids: – Component studies. *J. Dispersion Sci. Technol.*, **11**, 1–30.
33. Flick, E.W. (1988) *International Surfactants*, Noyes Publication, Park Ridge, New Jersey, USA.
34. Childers, J.C., Huang, S. and Romba, M. (1990) Metalworking fluid additives for waste minimization©. *Lubrication Eng.*, **46**, 349–358.
35. McCutcheon's (1991) *Emulsifiers & Detergents*, Volume I, McCutcheon Division, MC Publishing, New Jersey, USA.
36. Keefer, L. K., Goff, U., Stevens, J. and Bennett, E.O. (1990) Persistence of *n*-nitrosodiethanolamine contamination in American metal-working fluids. *Food Chem. Toxicol.*, **28**, 530–534.
37. Canter, N.M., Chaloupka, J.J. and Fischesser, G.J. (1988) The use of ethyleneoxide/ propylene oxide (EO/PO) esters as additives in semi-synthetic metalworking fluids. *Lubrication Eng.*, **44**, 257–261.
38. Sabina, L.R. and Pivnick, H. (1956) Oxidation of soluble oil emulsion and emuslifiers by *Pseudomonas oleovorans* and *Pseudomonas formicans*. *Appl. Microbiol.*, **4**, 171–175.
39. Hernandez, P.M., Larsen, I.C., and Peterson, C.R. (1985) The various factors involved in developing a microbial-resistant aqueous cutting fluids©. *Lubrication Eng.*, **41**, 425–429.
40. Ashjian, H., Giacobbe, T.J., Loveless, F.C., Mackerer, C.R., Novick, N.J. and O'Brian, T.P. (1992) Bioresistant surfactants and cutting oil formulations. PCT Patent Application WO 92/07925, 1–31.
41. Kane, E.L. and Pfhul, W. (1976) Preservation and preservatives in the aluminum hotrolling and beverage can processing industry. *Lubrication Eng.*, **32**, 249–253.
42. Bennett, E.O. (1972) The biology of metalworking fluids. *Lubrication Eng.*, **28**, 237–247.
43. Bennett, E.O. (1974) Water quality and coolant life. *Lubrication Eng.*, **30**, 549–555.
44. Kitze, E.D. and McGray, R.J. (1963) The occurrence of molds in modern industrial cutting fluids. *Lubrication Eng.*, **19**, 110–113.
45. Hill, E.C. (1984) Microorganisms: number, types, significance, detection, in *Monitoring and Maintenance of Aqueous Metal-Working Fluids*, Chater, K.W.A. and Hill, E.C. (eds), John Wiley, New York, pp. 97–113.
46. Passman, F.J. (1988) Microbial problems in metalworking fluids. *Lubrication Eng.*, **44**, 431–433.
47. Foxall-van Aken, S., Brown, J.A., Jr., Young, W., Salmeen, I., McLure, T., Napier, S., Jr. and Oslen, R.H. (1986) Common components of industrial metal-working fluids as sources of carbon for bacterial growth. *Appl. Environ. Microbiol.*, **51**, 1165–1169.
48. Fleming, C.D. and Baker, R.J. (1960) Controlling the spoilage of water soluble cutting fluids. *Lubrication Eng.*, **16**, 414–419.

49. Carlson, V. and Bennett, E.O. (1960) The relationship between the oil-water-ratio and the effectiveness of inhibitors in oil-soluble emulsions. *Lubrication Eng.*, **16**, 572–574.
50. Feisal, E.V. and Bennett, E.O. (1961) The effect of water hardness on the growth of *Pseudomonas aeruginosa* in metal cutting fluids. *J. Appl. Microbiol.*, **24**, 125–130.
51. Cook, P.E. and Gaylarde, C.C. (1988) Biofilm formation in aqueous metal working fluids. *Internat. Biodeterioration*, **24**, 265–270.
52. Sondossi, M., Rossmoore, H.W. and Wireman, J.W. (1985) Factors affecting the regrowth of *Pseudomonas aeruginosa* following biocide treatment. *Lubrication Eng.*, **41**, 366–369.
53. Fabian, F.W. and Pivnick, H. (1953) Growth of bacteria in soluble oil emulsion. *Appl. Microbial.*, **1**, 199–203.
54. Bennett, E.O. and Wheeler, H.O. (1954) Survival of bacteria in cutting oil. *Appl. Microbiol.*, **2**, 368–371.
55. Tant, C.O. and Bennett, E.O. (1958) The growth of aerobic bacteria in metal-cutting fluids. *Appl. Microbiol.*, **6**, 388–391.
56. Wort, M.D., Lloyd, G.I. and Schofield, J. (1976) Microbial examination of six industrial soluble oil emulsion samples. *Tribology Internat.*, **9**, 35–37.
57. Mattsby-Baltzer, I., Sandin, M., Ahlström, B., Allenmark, S., Edebo, M., Falsen, E., Pederson, K., Rodin, N., Thompson, R.A. and Edebo, L. (1989) Microbial growth and accumulation in industrial metal-working fluids. *Appl. Environ. Microbiol.*, **55**, 2681–2689.
58. Rossmoore, H.W., Rossmoore, L.A. and Young, C.E. (1987) Microbial ecology of an automative engine plant, in *Biodeterioration Research Vol. 1*, Llewellyn, G.C. and O'Rear, C.E. (eds), Plenum, New York, pp. 255–267.
59. Bennett, E.O. (1957) The role of sulfate-reducing bacteria in the deterioration of cutting emulsions. *Lubrication Eng.*, **13**, 215–219.
60. Guynes, G.J. and Bennett, E.O. (1959) Bacterial deterioration of emulsion oils I. Relationship between aerobes and sulfate-reducing bacteria in deterioration. *Appl. Microbiol.*, **7**, 117–121.
61. Isenberg, D.L. and Bennett, E.O. (1959) Bacterial deterioration of emulsion oils II. Relationship between aerobes and sulfate-reducing bacteria. *Appl. Microbiol.*, **7**, 121–125.
62. Abu Shaqra, Q.M. and Hill, E.C. (1984) Distribution of sulphate reducing bacteria in oil emulsion and its relationship to aeration and swarf. *Tribology Internat.*, **17**, 31–34.
63. DeMare, J., Rossmoore, H.W. and Smith, T.H.F. (1972) Comparative study of triazine biocides. *Develop. Indust. Microbiol.*, **13**, 341–347.
64. Rossmoore, H.W. and Holtzman, G.H. (1974) Growth of fungi in cutting fluids. *Develop. Indust. Microbiol.*, **15**, 273–280.
65. Rossmoore, H.W. and Treusch, P.J. (1975) Comparison of several media for enumeration of cutting fluid fungi. *Develop. Indust. Microbiol.*, **16**, 475–482.
66. Bennett, E.O. (1973) Formaldehyde preservatives for cutting fluids. *Internat. Biodeterioration Bull.*, **9**, 95–100.
67. Rossmoore, H.W., DeMare, J. and Smith, T.H.F. (1972) Anti- and pro-microbial activity of hexahydro-1,3,5-tris-(2-hydroxyethyl)-*s*-triazine in cutting fluid emulsion. In *Biodeterioration of Materials Vol. 2*, Walters, A.H. and Hueck-Van Der Plas, E.H. (eds), John Wiley, New York, pp. 286–293.
68. Rossmoore, H.W., Holtzman, G.H. and Kondek, L. (1976) Microbiology with a cutting edge. In *Proceedings of the Third International Biodeterioration Symposium*, Sharpley, J.M. and Kaplan, A.M. (eds), Applied Sciences, London, pp. 221–231.
69. Hill, E.C., Gibbon, O. and Davies, P. (1976) Biocides for use in oil emulsions. *Tribology Internat.*, **9**, 121–130.
70. Rossmoore, H.W. (1981) Antimicrobial agents for water-based metalworking fluids. *J. Occupational Med.*, **23**, 247–254.
71. Smith, T.H.F. (1969) Toxicological and microbiological aspects of cutting fluid preservatives. *Lubrication Eng.*, **25**, 313–320.
72. Keevil, C.W., MacKerness, C.W. and Colbourne, J.S. (1990) Biocide treatment of biofilms. *Internat. Biodeterioration*, **26**, 169–179.
73. Eagon, R.G. and Barnes, C.P. (1986) The mechanism of microbial resistance to hexahydro-1,3,5,-triethyl-*s*-triazine. *J. Indust. Microbiol.*, **1**, 113–118.

74. Sondossi, M., Rossmoore, H.W. and Wireman, J.W. (1986) The effect of fifteen biocides on formaldehyde-resistant strain of *Pseudomonas aeruginosa*. *J. Indust. Microbiol.*, **1**, 87–96.
75. Sondossi, M., Rossmoore, H.W. and Williams, R. (1989) Relative formaldehyde resistance among bacterial survivors of biocide-treated metalworking fluid. *Internat. Biodeterioration*, **25**, 423–437.
76. Onyekwelu, I.U., Bennett, E.O. and Gannon, J.E. (1981) The effective life of preservatives in cutting fluid concentrates. *Tribology Internat.*, **14**, 7–9.
77. Onyekwelu, I.U. and Bennett, E.O. (1979) The effects of filtering agents upon the activity of preservatives in cutting fluids. *Internat. Biodeterioration Bull.*, **15**, 88–95.
78. Bennett, E.O. (1982) The effects of metals upon the inhibitory activities of cutting fluid preservatives. *Internat. Biodeterioration Bull.*, 7–12.
79. Bennett, E.O., Oppong, D. and Lee, P. (1988) The effects of hydraulic fluids on cutting fluid rancidity control. *Soc. Manufact. Eng. Tech. Paper*, #MR88–182.
80. Barman, B.N. and Preston, H.G. (1992) The effects of pH on degradation of isothiazolone biocides. *Tribology Internat.*, **25**, 281–287.
81. Barman, B.N. (1994) Influence of temperature on the degradation of isothiazolone biocides in aqueous media and in metalworking fluid concentrate. *Lubrication Eng.*, **50**, 351–355.
82. Heenan, D.F., Burrell, R.E., Hajscar, E.E., Lauzon, D.C. and Dowdle, W.R. (1991) Isothiazolone microbiocide-mediated steel corrosion and its control in aluminum hot rolling emulsions. *Lubrication Eng.*, **47**, 545–548.
83. Engler, R. (1980) Regulations pertaining to the use of industrial biocides. *Develop. Indust. Microbiol.*, **22**, 117–122.
84. Halleux, P. (1990) Regulatory demands for biocides today. *Internat. Biodeterioration*, **26**, 251–258.
85. Barnes, C.P. and Eagon, R.G. (1986) The mechanism of action of hexahydro-1,3,5-triethyl-*s*-triazine. *J. Indust. Microbiol.*, **1**, 105–112.
86. Rossmoore, H.W. and Sondossi, M. (1988) Applications and mode of action of formaldehyde condensate biocides. *Adv. Appl. Microbiol.*, **33**, 223–277.
87. Wheeler, H.O. and Bennett, E.O. (1956) Bacterial inhibitors for cutting oil. *Appl. Microbiol.*, **4**, 122–126.
88. Singer, M. (1976) Laboratory procedures for assessing the potential of antimicrobial agents as industrial biocides. *Process Biochem.*, **11**, 30–35.
89. Rossmoore, H.W. (1979) Heterocyclic compounds as industrial biocides. *Develop. Microbiol.*, **20**, 41–71.
90. Collier, P.J., Ramsey, A.J., Austin, P. and Gilbert, P. (1990) Growth inhibitory and biocidal activity of some isothiazolone biocides. *J. Appl. Bacteriol.*, **69**, 569–577.
91. Bennett, E.O. and Futch, H.N. (1960) Nitroparafin inhibitors for cutting fluids. *Lubrication Eng.*, **16**, 228–230.
92. Bennett, E.O., Gannon, J.E. and Bennett, D.L. (1983) Inhibitory properties of 1,3-propanediols in cutting fluids. *Tribology Internat.*, **16**, 199–202.
93. Bennett, E.O., Onyekwelu, I.U., and Gannon, J.E. (1979) Corrosion inhibitors as preservatives for metalworking fluids-morpholine compounds. *Mater. Perform.*, **18**, 40–45.
94. Leder, J. (1987) Glutaraldehyde: A new microbiocide for metalworking©. *Lubrication Eng.*, **43**, 666–671.
95. Leder, J. and Russo, M.R. (1989) Biocide usage in metalworking fluids: The effect of treatment patterns on efficacy. *Lubrication Eng.*, **45**, 217–220.
96. Rossmoore, H.W. (1983) Nitrogen compounds. In *Disinfection, Sterilization and Preservation*, 3rd edn, Block, S.S. (ed.), Lea & Febiger, Philadelphia, pp. 271–307.
97. Bennett, E.O. (1978) The antimicrobial properties of diethylene triamines in metal-working fluids. *Internat. Biodeterioration Bull.*, **14**, 21–29.
98. Bennett, E.O., Onyekwelu, I.U., Bennett, D.L. and Gannon, J.E. (1980) Inhibitory activities of triazole compounds in metalworking fluids. *Lubrication Eng.*, **36**, 215–218.
99. Oppong, D. and Bennett, E.O. (1989) Use of a mixture of alcoholamines to potentiate the antimicrobial activities of certain cutting fluid preservatives. *Tribology Internat.*, **22**, 343–346.

100. Sandin, M., Mattsby-Baltzer, I. and Edebo, L. (1991) Control of microbial growth in water-based metal-working fluids. *Internat. Biodeterioration*, **27**, 61–74.
101. Izzat, I.N. and Bennett, E.O. (1978) The potentiation of the antimicrobial activities of cutting fluid preservatives by EDTA. *Lubrication Eng.*, **35**, 153–159.
102. Izzat, I.N., Bennett, E.O., Gannon, J.E. and Onyekwelu, I.U. (1981) Effects of EDTA on the antimicrobial properties of mixtures of cutting fluids preservatives. *Tribology Internat.*, **14**, 171–175.
103. Bennett, E.O., Gannon, J.E. and Bennett, D.L. (1982) Effects of EDTA on the antimicrobial properties of mixtures of cutting fluids preservatives. Part II. *Tribology Internat.*, **15**, 187–189.
104. Piet, L. and Rossmoore, H.W. (1985) A further study of the use of monocopper(II) citrate as an antimicrobial agent in metalworking fluid©. *Lubrication Eng.*, **41**, 103–105.
105. Riha, V.F., Sondossi, M. and Rossmoore, H.W. (1990) The potentiation of industrial biocide activity with Cu^{2+}. II. Synergistic effects with 5-chloro-2-methyl-4-isothiazolin-3-one. *Internat. Biodeterioration*, **26**, 303–313.
106. Sondossi, M., Riha, V.F. and Rossmoore, H.W. (1990) The potentiation of industrial biocide activity with Cu^{2+}. I. Synergistic effects with formaldehyde. *Internat. Biodeterioration*, **26**, 51–61.
107. Bennett, E.O. (1974) The biologial testing of cutting fluids. *Lubrication Eng.*, **30**, 128–135.
108. Hill, E.C. (1976) Evaluation of biocides for use with petroleum products. *Process Biochem.*, **11**, 36–38, 41.
109. Shennan, J.L. (1983) Selection and evaluation of biocides for aqueous metal-working fluids. *Tribology Internat.*, **16**, 317–330.
110. Rossmoore, H.W. (1977) Evaluation techniques for biodeterioration of water-miscible metalworking fluids. In *Biodeterioration Investigation Techniques*, Walters, A.H. (ed.), Applied Science, London, pp. 227–242.
111. Morton, L.H.G. (1987) Detection of significant spoilage microorganisms in soluble oil-in-water metal-working fluids. In *Industrial Microbiological Testing*, Hopton, J.W. and Hill, E.C. (eds.), Blackwell Scientific, Palo Alto, California, pp. 221–225.
112. Rossmoore, H.W., Seszny, P. and Rossmoore, L.A. (1977) Evaluation of source bacterial inoculum in development of a cutting fluid test procedure. *Lubrication Eng.*, **33**, 372–377.
113. Rossmoore, H.W. and Rossmoore, L.A. (1980) The identification of a defined microbial inoculum for the evaluation of biocides in water-based metalworking fluids. *Lubrication Eng.*, **36**, 6–20.
114. Rossmoore, H.W. (1982) The role of the inoculum in the microbiological evaluation of water-based metalworking fluids. In *International Yearbook of Tribology*, Bartz, W.J. (ed.), Expert Verlag, West Germany, pp. 791–797.
115. Pivnick, H. and Fabian, F.W. (1953) Methods for testing the germicidal value of chemical compounds for disinfecting soluble oil emulsions. *Appl. Microbiol.*, **1**, 204–207.
116. Himmelfarb, P. and Scott, A. (1968) Simple circulating tank test for evaluation of germicides in cutting fluid emulsions. *Appl. Microbiol.*, **16**, 1437–1438.
117. Shennan, J.L. and Chater, K.W.A. (1984) The philosophy and compatibility of biocide additions. In *Monitoring and Maintenance of Aqueous Metal-Working Fluids*, Chater, K.W.A. and Hill, E.C. (eds), John Wiley, New York, pp 113–129.
118. Sharpell, F.H (1979) Development of test protocols for antimicrobial agents by the ASTM. *Develop. Indust. Microbiol.*, **20**, 73–80.
119. ASTM (1993) Standard test method for evaluating the bioresistance of water-dilutable metalworking fluids. In *Standards on Materials and Environmental Microbiology*, 2nd edn, Philadelphia, pp. 642–644.
120. ASTM (1993) Standard test method for evaluation of antimicrobial agents in aqueous metalworking fluids. In *Standards on Materials and Environmental Microbiology*, 2nd edn, Philadelphia, pp. 277–279.
121. Rossmoore, H.W. and Rossmoore, L.A. (1991) Effect of microbial growth products on biocide activity in metalworking fluids. *Internat. Biodeterioration*, **27**, 145–156.

122. Brandeberry, L.J. and Myers, H.V., Jr. (1960) Test procedures for compounds used as preservatives in industrial coolants. *Lubrication Eng.*, **16**, 161–164.
123. Rossmoore, H.W. and Williams, B.W. (1971) An evaluation of a laboratory and plant procedure for preservation of cutting fluids. *Internat. Biodeterioration Bull.*, **7**, 55–60.
124. Rogers, M.R., Kaplan, A.M. and Beaumont, E. (1975) A laboratory in-plant analysis of a test procedure for biocides in metalworking fluids. *Lubrication Eng.*, **31**, 301–310.
125. Hill, E.C., Davies, I., Pritchard, J.A.V. and Byron, D. (1967) The estimation of microorganisms in petroleum products. *J. Inst. Petroleum*, **53**, 275–279.
126. Rossmoore, H.W. (1971) Methylene blue reduction for rapid inplant detection of coolant breakdown. *Internat., Biodeterioration Bull.*, **7**, 147–154.
127. Yanis, R.J. and Wolfe, G.F. (1960) Test procedures for the evaluation of cutting fluids. *Lubrication Eng.*, **16**, 164–170.
128. Passman, F.J. (1989) Monitoring microbial contamination in metalworking fluds. *Soc. Manufact. Eng. Tech. Paper*, #MR89–454.
129. Genner, C. (1976) Evaluation of the dip-slide technique for the mcirobiological testing of industrial fluids. *Process Biochem.*, **11**, 39–48.
130. Genner, C. and Hill, E.C. (1981) Evaluation of the 'dip-slide' technique for cutting oils. *Tribology Internat.*, **14**, 11–13.
131. Rossmoore, L.A., Wireman, J.W. and Rossmoore, H.W. (1986) Rapid field method for the detection and enumeration of sulfate reducers. In *Biodeterioration 6 – Proceedings of the Sixth International Biodeterioration Symposium*, Sheila, B. and Houghton, D.R. (eds), C.A.B. Intnl. Mycological Inst., The Biodeterioration Society, United Kingdom, pp. 413–419.
132. McCoy, J.S. (1978) A practical approach to central system control. *Lubrication Eng.*, **34**, 180–186.
133. Gannon, J.E. and Bennett, E.O. (1981) A rapid technique for determining microbial loads in metalworking fluids. *Tribology Internat.*, **14**, 3–6.
134. Passman, F.J. (1984) A catalase test for quickly estimating microbial loads in metalworking fluids. *Soc. Manufact. Eng. Tech. Paper*, #MR84–916.
135. Hill, E.C. (1987) Spatial distribution of spoilage microbes in fluids using the gel system. In *Industrial Microbiological Testing*, Hopton, J.W. and Hill, E.C. (eds), Blackwell Scientific, Palo Alto, California, pp. 195–199.
136. Holdum, R.S. (1977) Microbial spoilage of engineering materials: Part 6 – Improving monitoring and control. *Tribology Internat.*, **10**, 273–280.
137. Yust, P.R. and Becket, G.J.P. (1980) Microbiology theory and practice in metalwoking fludis. *Indust. Lubrication Tribology*, November/December, 220–225.

12 Toxicology of preservatives

R.A. RODFORD

12.1 Introduction

Preservatives are, by their very nature, biologically active molecules. Basically, they need to be aggressive to do a good job. Therefore, most biocides, particularly those used for product preservation, are subject to some form of regulation specifying scope and concentration for use. In today's tightly regulated environment, the development of a new preservative is a very expensive and time-consuming operation. The test requirements for regulatory approval of a new chemical vary depending on where it is made and marketed. In addition, many countries have positive lists for preservatives, particularly when they are used in cosmetics or foods. To obtain a positive listing, a package of toxicological studies needs to be presented to and approved by the relevant regulatory authority.

Exposure to a preservative may be by ingestion, inhalation or by skin absorption. Skin irritation and/or skin sensitization may result from direct contact of the preservative with the skin. All new preservatives should therefore be regarded as potential health hazards until the required toxicological evaluation has been completed and a risk assessment performed for the proposed use. In general, the manufacturer will need to supply data from a package of acute studies including oral, dermal and possibly inhalation, skin and eye irritation, sensitization and genotoxicity. A recommendation for an acceptable air-borne concentration in the working environment, termed an occupational exposure limit (OEL) in the UK, a threshold limit value (TLV) in the US or a maximum Arbeitskonzentration (MAK) in Germany, should be proposed. The particular application and extent of human exposure will indicate which subacute or chronic studies should be performed to provide an adequate safety support package.

In view of the intrinsic aggressiveness of preservatives towards biological material, it should come as no surprise that they often give positive results in toxicological studies especially *in vitro* mutagenicity assays and stringent skin sensitization studies such as the Magnusson and Kligman guinea pig maximization test. This is now better appreciated and any assessment of their safety should be based on the fullest possible understanding of their toxicity together with a risk assessment for the particular use.

Detergent formulations generally provide ideal media for the propagation of microorganisms. The Gram-negative type such as *Pseudomonas* are the most common although Gram-positive organisms, moulds and yeasts may also be found. These organisms not only spoil the product but can pose a serious health hazard to the consumer. Therefore, an effective preservative should be used to control the microorganisms and prevent spoilage of the product. For products needing an extended shelf life, regulations require the use of a preservative which is both effective and safe.

The use of nonionic surfactants in both personal care and detergent formulations have increased the problems of product preservation as it has been shown that the efficiency of certain preservatives is reduced when the ratio of the surfactant to the preservative exceeds a certain critical value. This ratio varies depending on the surfactant, the preservative and the test organism.[1] The mechanism of inactivation of preservatives by nonionic surfactants has been studied by many investigators including Higuchi and coworkers[2,3] who showed the complexing of phenolic preservatives with polyethylene glycols and Dyer[4] and Evans[5] who attributed inactivation to a partitioning of the preservative between the micelles of the surfactant and the aqueous phase. Anionic and cationic surfactants can also interact with preservatives and this has been reviewed by Coates.[6]

Many other factors, including the pH and partitioning in oil-in-water systems can also affect the activity of any particular preservative. The overall assessment of preservative performance can, therefore, only be made on the finished product.

12.2 Preservatives in general use

In considering a new product formulation, at least four points need to be addressed. The first of these is the spoilage potential of the product. Within this context, it is necessay to take into account the presence of chemical substrates for microbial growth, physical conditions, i.e. pH, water activity, etc., microbial challenge during manufacture, storage conditions, pattern of use and storage time and temperature. Normal in-use conditions, expected shelf working life and the probability of consumer abuse should also be considered.

The second point to consider is whether product preservation is possible without the use of a biocide. In the case of preservative-free formulations, the physical or chemical properties should be inimical to growth and the exclusion of microorganisms should be easy to achieve during manufacture. It would also be beneficial if one or more of the ingredients has bacteriostatic or bacteriocidal action.

Thirdly, it has to be considered which biocides are appropriate. Chemical compatibility with the product has to be taken into account along with suitability to meet the expected microbial challenge, legislative restrictions, costs and safety approval implications. Last but not least, the requirements of occupational health and environmental measures must be taken into account.

The preservatives listed below have found application generally in surfactant-containing formulations for cosmetic and detergent products. Other product types, for example, pharmaceuticals, paints and adhesives can also be preserved with these chemicals. Some compounds are also active against algae and have been used as slimicides, for example, quaternary ammonium salts or the isothiazolinones.

12.2.1 Organic acids

Included within this group of compounds are acetic, lactic, benzoic, sorbic and propionic acids. These preservatives generally have slight to moderate activity against both Gram-positive and Gram-negative bacteria. Benzoic and sorbic acids also have strong activity against yeasts whilst sorbic and propionic acids have a strong activity against moulds. Evidence has been produced that many organic acids act as uncoupling agents by preventing the uptake of substrates which depend on a protonmotive force for their entry into the cell.[7] The major incompatibility of organic acids is a high pH formulation.

12.2.2 Phenolics

This class of biocides includes *ortho*-phenyl phenol, hexachlorophene, chloroxylenol, chlorhexidene, phenoxyethanol, triclosan and dichlorophen. These compounds generally have a moderate to strong activity against Gram-positive and/or Gram-negative bacteria. *Ortho*-phenyl phenol and dichlorophen also have good activity against both yeasts and moulds. Various possible modes of action have been suggested for the different phenolic biocides.[8] Phenoxyethanol is thought to inhibit active transport by uncoupling oxidation from phosphorylation and protein coagulation. Chlorhexidene and hexachlorophene are suggested to act by general coagulation and disruption of the permeability barrier of the cell. Phenol itself causes generalized membrane damage by possible lysis of the cytoplasmic membrane which results in a leakage of amino acids, purines, pyrimidines, potassium ions and pentoses.

Ortho-phenyl phenol is poorly soluble in surfactants, chloroxylenol is incompatible with both cationic and nonionic surfactants and chlorhexidene is incompatible with anionics and gums. Phenoxyethanol is also incompatible with nonionics. Some of these phenolics, especially the chlorinated ones, do not have an acceptable toxicological or environmental profile for use in high volume products.

12.2.3 Quaternary ammonium salts

Benzalkonium chloride, benzethonium chloride and cetrimide are members of this group of materials which have their strongest biocidal activity against Gram-positive bacteria. The mode of action of quaternary ammonium salts has not yet been completely elucidated but certain accepted and well-defined processes have been reported.[9,10] Firstly, the preservative is adsorbed onto the bacterial cell surface. Diffusion then takes place through the cell wall and binding to, and disruption of, the cytoplasmic membrane follows. Next there is leakage of potassium ions and cytoplasmic constituents and finally, precipitation of the cell contents and cell death.

Quaternary ammonium compounds are useful at pH ranges above 5 but are incompatible with anionics, proteins and salts.

12.2.4 Mercury salts

Two examples of this group are thiomersal and phenyl mercuric acetate which both have broad spectrum biocidal activity and are also very effective against both yeasts and moulds. Mercury salts can combine with the –SH groups of enzymes.[11] This can occur with enzymes associated with the membrane or with enzymes and proteins found in the cytoplasm. Low concentrations of some mercuric salts can lyse growing cultures of *Escherichia coli*.

The effective in-use concentration is very low, typically below 0.05% w/w and the useful pH range is less than 8. Major incompatibilities are nitrogen- and sulphur-containing organics, plastics and halides. Mercury salts are not considered appropriate for consumer products from both toxicological and environmental standpoints.

12.2.5 Ureas

Imidazolidinyl-urea and diazolidinyl-urea are both preservatives in this class of compounds. The former has strong activity against Gram-positive bacteria and yeasts whilst the latter has strong activity against both Gram-positive and Gram-negative bacteria and moderate activity against both yeasts and moulds. It is presumed that this class of preservatives owes its antimicrobial activity to the release of formaldehyde. The useful pH range for these urea compounds is 3–9.

12.2.6 Isothiazolinones

The biocidal materials in this group are benzyl, methyl or chloro derivatives. They are broad spectrum preservatives active against both Gram-positive and Gram-negative bacteria. They are also very effective against yeasts and moulds with a wide useful pH range from 1–9. The mechanism of antibacterial action of 1,2-benzisothiazolin-3-one, the active ingredient in the Proxel range of preservatives from Zeneca, has been shown to involve reaction with the thiol groups associated with the organisms. Interaction at the cytoplasmic membrane surface is indicated. Proxel inhibits the oxidation of carbohydrate substrates which are transported across the cytoplasmic membrane by thiol-dependent enzymes. It inhibits the oxidation of glycerol which enters the cell by diffusion and also inhibits the utilization of the electron transport chain by bacteria, possibly by action on the dehydrogenase enzymes.[12] The mode of action of other isothiazolinone preservatives is still under investigation but a reasonable assumption would be that initially, a process of generalized damage to the cell membrane is involved, followed by interaction with proteins and enzymes probably via disulphide linkages. Major incompatibilities of this group of biocides are amines, sulphites and mercaptans.

12.2.7 Hydantoins (formaldehyde donors)

Dimethyloldimethyl hydantoin and monomethyloldimethyl hydantoin are most effective against Gram-positive bacteria. Activity against yeasts and moulds is slight and the useful pH range is around 4–9. It is assumed that the biocidal action of these chemicals is due to the release of formaldehyde. Thus, the use of these preservatives may be restricted in certain countries in which adverse publicity about formaldehyde has led to a retail ban.

12.2.8 Miscellaneous organics

Other notable examples of compounds with biocidal activity are, the methyl, propyl and butyl esters of parabens, 2-bromo-2-nitropropane-1,3-diol (Bronopol), formaldehyde, quaternium 15 (Dowicil 200), oxazolooxazole (Nuosept 95 or Nuosept C), 4,4-dimethyl-1,3-oxazolidine (Oxaban A), 1-azo-3,7-diox-5-ethylbicyclo(3.3.0)octane (Oxaban E), glutaraldehyde, *N*-(trichloromethylthio)cyclohex-4-ene-1,2-dicarboximide (Captan), glyceryl monolaurate and zinc pyrithione.

The strongest activity of the parabens esters is against Gram-positive bacteria and moulds. The mechanism of action of parabens esters is concentration dependent.[13] Low concentrations inhibit transport whilst higher concentrations affect membrane integrity. This has the effect of leakage of intracellular constituents and selective inhibition of ΔpH across the cell membrane. A major drawback of the parabens esters is their incompatibility with nonionics and cationics.

Both Bronopol and formaldehyde are effective preservatives with strong activity against both Gram-positive and Gram-negative bacteria with slightly less activity against yeasts and moulds. Bronopol acts by interfering with cytoplasmic membrane enzymes oxidizing thiol groups to disulphides.[11] Very low concentrations of formaldehyde cause lysis of the cell wall. In addition, formaldehyde interacts with amino groups in the cytoplasm.[11] Bronopol and formaldehyde are useful across a wide range of pH around 3/4–10. The former is incompatible with –SH compounds and aluminium and the latter is incompatible with proteins.

Dowicil 200 and Nuosept C are both strongly active against both Gram-positive and Gram-negative bacteria, moderately effective against moulds but only slightly active against yeasts. The mode of action of Dowicil 200, a hexamine derivative quaternized with *cis*-1,3-dichloropropene, has been studied by Rossmoore and Sondossi.[14] Hexamine breaks down by acid hydrolysis to release formaldehyde so the activity of Dowicil 200 is almost certainly due to this mechanism. The activity of Nuosept C is also assumed to be due to the release of formaldehyde. The useful pH range for both these preservatives is 4–9/10 but Nuosept C is incompatible with bisulphite at high or low pH.

Oxabans A and E and glutaraldehyde are effective against all bacteria, yeasts and moulds, the latter particularly so at pH levels of 7.5–8.5. Oxabans A and E are only useful at a pH of greater than 6 as below this they break down. The mode of action of glutaraldehyde is complex due to its diverse chemical reactivity. Its activity is, therefore, probably due to more than one process. The following effects on bacteria have been observed: crosslinking of the peptidoglycan in Gram-positive bacteria; interaction with protein in the cell walls of Gram-negative bacteria; reaction with spore cortex constituents; and increased resistance to lysis of protoplasts and spheroplasts.[15] Glutaraldehyde is inactivated by ammonia and primary amines at neutral to basic pH values.

Captan and glyceryl monolaurate are both moderately effective against bacteria, yeasts and moulds although glyceryl monolaurate is less effective against Gram-negative bacteria. Captan is an important agricultural fungicide. Its mode of action is centred on the reaction with cellular thiol groups.[16–18]

Zinc pyrithione is strongly active against Gram-positive bacteria, yeasts and moulds and moderately effective against Gram-negative bacteria. It is useful across a pH range of 4.5–9.5 but is incompatible with EDTA and some nonionics.

The toxicology of the materials discussed above obviously varies widely as there is such a range of chemical types. It is therefore proposed to limit the discussion on the various toxicological aspects to a selection of the most effective all-round preservatives, namely, the isothiazolinones (Proxel and Kathon), Bronopol, formaldehyde, Oxabans A and E and glutaraldehyde.

12.3 Acute toxicity

12.3.1 Formaldehyde

Formaldehyde is a normal metabolite in mammals and, in small quantities, is rapidly metabolized.[19] The major route of biotransformation appears to be oxidation to formic acid, followed by further oxidation to carbon dioxide and water.[20,21] Most of the formaldehyde absorbed by ingestion or inhalation is rapidly oxidized in the blood and excreted via the kidneys. The half-life of formaldehyde in blood is considered to be 1.5 min or less in rats, guinea pigs, rabbits and cats.[22] At concentrations considerably greater than those likely to be experienced by the general public from its use as a preservative, formaldehyde is severely irritating to the oral mucosa and cases of fatal ingestion of formaldehyde are known. Industrial exposure represents a potential worst case situation, and therefore maximum exposure limits are set by government bodies to ensure any inconvenience to the work force is minimized. The absolute concentrations of formaldehyde permitted industrially vary slightly from country to country. Examples of permissible exposure limits for formaldehyde are:

- UK – 2 ppm (MEL – TWA and STEL)[23]
- USA – 0.3 ppm (TLVc)[24]
- Germany – 0.5 ppm (MAK)[25]

where MEL = maximum exposure limit, TWA = 8 h time weighted average, STEL = 10 min short term exposure limit TLV(c) = threshold limit, value (ceiling) and MAK = maximum Arbeitskonzentration.

Acute ingestion of formaldehyde by humans has resulted in loss of consciousness, vascular collapse, pneumonia, haemorrhagic nephritis and abortion. Formaldehyde has occasionally injured the larynx and trachea, but damage to the gastrointestinal tract has occurred primarily in the stomach and lower oesophagus.[26]

12.3.2 Isothiazolinones

The two best known preservatives in this class are benzisothiazolinone (the Proxel range of biocides from Zeneca) and a 3:1 mixture of 5-chloro-2-methyl-4-isothiazolin-3-one and 2-methyl-4-isothiazolin-3-one (the Kathon range of biocides manufactured by Rohm and Haas). Neither of these products is acutely toxic by ingestion, inhalation or skin absorption. In tests on animals conducted many years ago, oral LD50s in the rat (the dose required to kill 50% of the animals) were 1100 mg kg^{-1} for Proxel GXL[27] which is 20% active to >2000 mg kg^{-1} for Proxel XL2[27] which is 9.5% active. For Kathon CG (1.5% active),[28] the rat oral LD50 was established at 3350 mg kg^{-1} (males) and 2630 mg kg^{-1} (females). The dermal LD50 in the rabbit was >5000 mg kg^{-1} for Kathon CG.[28] The LC50, four-hour exposure, to the rat (the air-borne concentration required to kill 50% of the animals) was about >13.7 mg l^{-1} of air for Kathon CG.[28]

12.3.3 2-Bromo-2-nitropropane-1,3-diol

The tradename of this compound is Bronopol manufactured by the Boots Company plc. Following acute administration, Bronopol was classified as moderately toxic to rats and mice by the oral route using the classification system of Hodge and Sterner[29] intraperitoneal, subcutaneous and percutaneous acute toxicity data were consistent with this assessment.[30] The oral LD50s to rats were 307 mg kg^{-1} (males) and 342 mg kg^{-1} (females) whilst the corresponding data for mice showed 374 mg kg^{-1} (males) and 327 mg kg^{-1} (females).[31]

12.3.4 4,4-Dimethyl-1,3-oxazolidine and 1-azo-3,7-diox-5-ethylbicyclo (3.3.0) octane

These preservatives are produced by Angus Chemicals and are known commercially as Oxaban A and Oxaban E, respectively. The acute oral LD50 of Oxban A to male rats was established at about 950 mg kg^{-1} [32] indicating that this chemical would be classified in the EC as harmful. The acute dermal LD50 of Oxaban A to rabbits was estimated to be around 1400 mg kg^{-4} with the skin response being moderate to severe.[32] In an acute inhalation study, the LC50 to rats was found to be about 11.7 mg l^{-1} for Oxaban A.[32] On the other hand, Oxaban E is less acutely toxic to rats than Oxaban A, particularly by the oral route where the LD50s were found to be 5250 mg kg^{-1} and 3680 mg kg^{-1} for males and females, respectively.[33] The acute dermal LD50 of Oxaban E in rabbits was found to be 1948 mg kg^{-1} and the LC50, following a four-hour mist inhalation study in rats was determined to be 3.1 mg l^{-1}.[33]

12.3.5 Glutaraldehyde

Data indicate that glutaraldehyde vapour may produce complaints of 'irritation' of the eyes and respiratory tract in man at around 0.3 ppm in air.[34] Therefore, in the UK[24] a STEL has been set at 0.2 ppm and, likewise in the USA[25] and Germany[26] a ceiling TLV-eight-hour TWA and a MAK, respectively, have been set at 0.2 ppm to control these effects. Animal studies[35] indicate that aqueous solutions of glutaraldehyde (Ucarcide) are of moderate acute toxicity. LD50s in the rat ranged from 410–733 mg kg^{-1} body weight for a concentration range of 25–50% (w/w). However, further studies with dilute solutions of Ucarcide show a very low order of acute oral toxicity with an LD50 of 17.5 ml kg^{-1} in the rat. Studies in rabbits have given dermal LD50s ranging from 897 mg kg^{-1} to 3045 mg kg^{-1} body weight for 50% to 25% aqueous solutions. The former would be classified as harmful by the EC. Concentrations of 5% or less were shown to have a very low order of acute dermal toxicity. Acute inhalation studies in rats show that atmospheres generated from saturated vapour from 25%, 45% or 50% could produce irritant effects but there were no effects to indicate they were harmful.

12.4 Skin and eye irritation

Most biocides in a concentrated form are corrosive or irritant to skin and/or eyes.

12.4.1 Formaldehyde

On contact with the skin formaldehyde in solution may cause primary irritation.[36,37] Although its irritancy potential is generally recognized, few quantitative data exist for formaldehyde. The only relative irritancy data are from an open epicutaneous test using guinea pigs where marked irritation occurred at 30% and 10% of formaldehyde. Some irritation was also produced at 3%, 0.1% and 0.03%, although there were no effects at 1% nor at 0.3% formaldehyde.[38] No comment was made on this apparent discrepancy; possibly it is an artefact of this type of test.

Human data indicate that formaldehyde is a severe irritant to the eyes and upper respiratory tract. Formaldehyde, at a concentration of 4 ppm, can produce lachrymation but the irritancy threshold may be much lower. At 10 ppm, formaldehyde causes severe eye irritation in man.[39]

12.4.2 Isothiazolinones

Single applications of 1,2-benzisothiazolin-3-one (BIT) in aqueous solution to rat skin were irritant or strongly irritant at concentrations above 8%, whereas a single application of an aqueous solution containing 1.65% BIT was not irritant.[40] Dilute aqueous solutions of BIT (<0.07%) were not irritant to the rabbit eye but more concentrated solutions (say >0.5%) were irritant or very irritant with the possibility of causing serious damage to eyes.[40]

Studies carried out according to OECD guidelines have indicated that Kathon CG (1.5% active ingredient (a.i.)) is irritant to rabbit eyes but not to rabbit skin.[41] Other animal studies, however, using the Draize eye irritation test showed Kathon 886 (14% a.i. as supplied), ranging in concentration from 1.1% to 14% a.i., and Kathon CG at 1.5% a.i. were corrosive when tested as supplied. Kathon CG was also found to be severely irritating to rabbit skin using occlusive patches on both intact and abraded test sites.[42]

12.4.3 Bronopol

In rabbit eye tests, Bronopol was not irritating at a concentration of 0.5% in normal saline applied once a day on four successive days. Following single applications, 5% Bronopol in polyethylene glycol 400 was irritant to rabbit eyes but a concentration of 2% was not.[31] Single applications of 0.5% Bronopol in 2.5% aqueous methyl cellulose were not irritant when applied to intact rabbit skin for six hours under occlusion but 2% Bronopol was irritant using the same test conditions. Since 5% Bronopol in polyetheylene glycol was not irritant under these test conditions, the irritancy potential of Bronopol would appear to be vehicle dependent. This was confirmed in human studies where Bronopol was not irritant to skin under occlusive conditions at 0.5% in petrolatum or 0.1% in aqueous buffer at pH 5.5 but concentrations of 1% in petrolatum or 0.25% in aqueous buffer were slightly irritant.[31] These data indicate that irritation reactions due to Bronopol are unlikely to occur in man even if an undiluted product containing 0.02% Bronopol were in contact with the skin or entered the eye.

12.4.4 Oxabans A and E

Due to its alkaline nature, Oxaban A can cause severe burns to eyes, although at a concentration of 5000 ppm in water, eye irritation potential is extremely low.[32] Using the US Department of Transport test method, Oxaban A is not corrosive to skin. However, the undiluted chemical is still severely irritant to skin. Animal studies also indicate that Oxaban E is severely irritating to skin and eyes but not corrosive.[33]

12.4.5 Glutaraldehyde

Undiluted glutaraldehyde caused severe irritation and corrosion in rabbit eye tests.[43,44] Whilst a 5% solution of glutaraldehyde also caused corneal damage, 0.1% and 0.5% did not cause inflammation or corneal damage to rabbit eyes.[45] In an accidental exposure, a 2% aqueous solution of glutaraldehyde has caused eye irritation in man.[46] In a four-hour skin covered patch test in rabbits, a 25% aqueous solution of glutaraldehyde caused severe inflammation and corrosion.[45] Whilst these investigators identified 1% as a 'threshold' for inflammation, other workers saw no local effects in rabbits at concentrations up to 7% and only mild inflammation and some crusting at 25% following a 24-hour covered patch test.[47] In man, nine out of 13 patients with contact dermatitis had irritant reactions to 1% glutaraldehyde in petrolatum following 48-hour covered patch tests, whilst at 0.1% no irritation was experienced by 884 patients.[48] Exposure to glutaraldehyde vapour at 2–11 ppm for 6–8 h was irritating to rats eyes,[45] as was exposure to 3 ppm for unknown daily periods for 90 days.[49]

12.5 Skin and respiratory sensitization

Many biocides can cause allergic contact dermatitis or respiratory sensitization.

12.5.1 Formaldehyde

Formaldehyde is a frequent cause of sensitization, as assessed in patients attending dermatology clinics. In a series of patch tests carried out on preservatives and fragrance materials used in consumer products, the incidence of positive patch test reactions ranged from 2.1% to 4.7%.[50–54] There have also been numerous reports of occupational contact dermatitis to formaldehyde.[55,56] In addition, some cases of formaldehyde-associated occupational asthma have been reported.[57–59] However, the number of cases is small and the evidence for an immune response is not clear.[60] This response of the respiratory tract to formaldehyde would appear to be considerably less prevalent than the cutaneous response to the substance. Formaldehyde is a sensitizer in predictive guinea pig tests, although the observed responses may differ between laboratories.[61] In human predictive tests, repeated applications of a 5% or 10% aqueous solution of formaldehyde followed by a challenge with a 1% aqueous formaldehyde solution led to 8% of the 154 subjects tested becoming sensitized.[62] Jordan *et al.*[63] determined a minimum eliciting level of formaldehyde by patch testing and observed that four out of nine formaldehyde-sensitive subjects reacted to 30 ppm formaldehyde after 120 continuous hours of testing. These tests indicate that levels of formaldehyde below 30 ppm should be tolerated by sensitive subjects when used repeatedly on normal skin. This conclusion is supported by the fact that contact dermatitis from formaldehyde in cosmetics is rare.[55]

12.5.2 Isothiazolinones

Cases of allergic contact dermatitis were reported in man following exposure to concentrations of Proxel greater than the recommended use level.[64–66]

Both human and animal studies have shown Kathon CG to have strong skin sensitization potential.[67,68] Much work was carried out to establish a use level for Kathon CG which would not pose any significant risk of allergic contact dermatitis to the consumer. The level determined was 15 ppm in 'rinse-off' products such as shampoos and shower gels and 7.5 ppm in 'leave-on' products such as skin creams and lotions.[42]

12.5.3 Bronopol

Overall, Bronopol has a low, but variable, sensitization in the Magnusson and Kligman guinea pig maximization test and is, at most, a weak sensitizer by this method. There are some inconsistencies in published literature concerning the skin sensitization potential of Bronopol in man. This has probably arisen from the use of irritant concentrations of Bronopol for patch tests. Initial studies indicated that 2.5% Bronopol in petrolatum was not irritant and that Bronopol was a potential sensitizer.[69] However, later studies[70] showed that Bronopol was irritant at concentrations above 1% and a test in 93 subjects using the Draize procedure, with an induction concentration of 5% and a non-irritant challenge concentration of 0.25%, showed no evidence of skin sensitization. A test using a single application of 0.25% Bronopol in petrolatum in 149 patients attending a dermatitis clinic also found no evidence of skin sensitization.[31] A study on cosmetic intolerance in over 5000 Belgian patients being tested for contact dermatitis found only 0.1% of the sample gave a positive reaction to Bronopol in a patch test.[71] The divergence in the reported incidence of reactions to Bronopol may result from the inclusion in some studies of subjects who have been occupationally exposed to high levels of Bronopol. These data indicate a very low sensitization potential for Bronopol; it is unlikely to induce allergic contact dermatitis in man at normal preservative use concentrations.

12.5.4 Oxabans A and E

In the Magnusson and Kligman guinea pig maximization test, Oxaban A was found to be a strong sensitizer at a challenge of 1% following intradermal and topical inductions of 0.1% and 2.5% a.i., respectively. In human skin irritation and sensitization studies, a 3% solution of Oxaban A in water was found to be irritant; 60% of the 101 test panel reacting by the end of the second week. The concentration of the test material was reduced to 0.3% for the third week and applied to a fresh test site. Following two weeks rest, a further fresh test site was challenged with the 0.3% solution of Oxaban A. Two of the panellists reacted, so that two weeks later these two were rechallenged with a concentration of 0.1% Oxaban A. There was no reaction to this concentration.[32] The conclusion may be drawn that whilst Oxaban A has been shown to possess a fairly strong sensitization potential in animal studies, the risk of consumers exhibiting allergic contact dermatitis from contact with products containing Oxaban A at the normal preservative use concentration of 0.1% is very low.[32]

In human studies, one out of 201 subjects showed sensitization to 3% Oxaban E although some 12% of the panellists showed varying degrees of skin irritation during the induction period.[33] However, Oxaban E was shown to be a strong sensitizer in the Magnusson and Kligman guinea pig maximization test at a challenge concentration of 2.5% following intradermal and topical application inductions of 2.5% and 5%, respectively. At a second challenge following a week's rest, a weak response was observed at a concentration of 0.1%, but there was no response at 0.01%. Strong cross reactions to formaldehyde were also demonstrated.

12.5.5 Glutaraldehyde

Glutaraldehyde is a well known skin contact allergen in animals and man. The lowest induction or challenge concentration producing skin sensitization in man was shown to be 0.5%. Concentrations of 0.2% and below have shown no potential for allergic contact dermatitis in man.[69,72,73] Glutaraldehyde is also on the prescribed list of agents known to cause occupational asthma.[74]

12.6 Subacute and chronic toxicity including carcinogenicity

12.6.1 Formaldehyde

In a chronic inhalation study, rats and mice were exposed to 0, 2.0, 5.6 or 14.3 ppm formaldehyde in air for six hours per day five days per week for two years.[75] Squamous cell carcinoma of the nasal passages was induced in some rats (44% at 14.3 ppm and 0.85% at 5.6 ppm). Polyploid adenomas were found in all exposed rats. Nasal lesions also occurred in mice, but at a much lower incidence of only 0.93% at the top dose. This difference in response may be related to a species difference in physiological responses to respiratory irritants. This study leads to the conclusion that formaldehyde does have carcinogenic potential in rats and mice.

The issue is to determine the relevance of these animal data to man. Many epidemiological studies have examined the effects of industrial exposure to formaldehyde. One such study has examined the mortality of men employed in the British chemical and plastics industry from 1920 to 1989.[76] There were slight increases in lung cancer, respiratory disease and stomach cancer, but no correlation was found with the estimated cumulative dose or the time since the first exposure. It was concluded that these findings do not provide a clear conclusion on the potential human carcinogenicity of formaldehyde. A previous study examined the mortality in the UK of pathologists and medical laboratory technicians occupationally exposed to formalin[77] and found no health problems related to formaldehyde exposure.

A case-control study of cancer of the nose and paranasal sinuses was conducted in France.[78] Out of 207 cases and 409 controls, no significant association was found between exposure to formaldehyde and squamous cell carcinomas of the sinonasal cavities. However, the study did suggest that exposure to both formaldehyde and wood dust may increase the risk of nasal adenocarcinoma, when compared with the risk due to wood dust alone. Another earlier case control study of cancer mortality among Du Pont workers exposed to formaldehyde[79] showed no differences between the company's formaldehyde-exposed workers and non-exposed co-workers.

A recent meta-analysis of the epidemiologic evidence of an association between formaldehyde and respiratory cancer[80] did not indicate an excess risk or an exposure–response gradient for lung cancer. An exposure–response gradient was seen for both sinonasal and nasopharyngeal cancers which suggested that at least substantial levels (which induce chronic irritation) of occupational exposure to formaldehyde are associated with a risk of these cancers. This information supports the contention that a threshold exists for formaldehyde carcinogenicity.

A review suggests that man is less susceptible to formaldehyde than rats and that the risk of carcinogenic effects from even occupational exposures is very low.[81] It is concluded that there is no risk presented to man from use of formaldehyde as a preservative.

12.6.2 Kathon CG

In a three-month drinking water study in rats, no treatment-related effects were observed with Kathon CG except for mild gastric irritation.[82] A dermal carcinogenicity study, in which Kathon CG was applied to the skin of mice three times a week for 30 months, showed no treatment-related tumours or other adverse effects apart from local tissue irritation.[82]

12.6.3 Bronopol

Repeated oral administration of Bronopol to rats[31] established the following no-effect levels:

1. 20 mg kg^{-1} body weight per day by gavage for 90 days.
2. 100 mg kg^{-1} body weight per day in the diet for 12 weeks.
3. 10 mg kg^{-1} body weight per day in the drinking water for two years.

Oral administration of Bronopol to dogs at 20 mg kg^{-1} body weight per day for 90 days produced no significant toxic reaction, although some vomiting was noted.[31] These data indicate that Bronopol is moderately toxic following oral administration to rats and dogs. Bronopol was not carcinogenic when applied dermally to mice at doses of up to 1.5 mg per day for 80 weeks.[31] When Bronopol was administered to rats in the drinking water at doses up to 160 mg kg^{-1} body weight per day for two years, there was no effect on the incidence of tumours over the controls.[31]

12.6.4 Oxaban A

In a 90-day dermal study in rats with Oxaban A, no deaths occurred and no treatment-related effects were observed in serum chemistry, haematology, urine or organ pathology. The top dose of 195 mg kg^{-1} body weight, representing a level of Oxaban A equal to 100 times normal preservative use concentration, produced moderate to severe skin irritation after the third week. At actual Oxaban A use concentrations, equivalent to 1.95 mg kg^{-1} body weight per day, no adverse effects were observed.[32]

12.6.5 Glutaraldehyde

In an 11-week rat drinking water study with glutaraldehyde at concentrations up to 0.5% there were no clinical signs of toxicity during the live phase of the study. Subsequent histopathology revealed no morphological abnormality in central or peripheral nervous tissue.[83] Rats exposed to up to 10 ppm glutaraldehyde vapour for an unknown period daily for 90 days, showed no overt neuromuscular abnormalities.[49] A six-week dermal study in rabbits using 5 mg kg^{-1} body weight per day of 2% aqueous glutaraldehyde in unoccluded applications, produced no observable adverse effects other than skin irritation.[84] Carcinogenicity studies in rats and mice by the inhalation and dermal routes were in progress in 1988 as part of the National Toxicology Program in the USA.

12.7 Genetic toxicology

Genotoxicity studies are conducted to provide information on the capability of a chemical to mutate, or change, genes or cause disruption of chromosomes. If it can be demonstrated that a chemical can cause gene mutations, then it may have the capability to cause cancer or changes to the heritable gene line of man.

12.7.1 Formaldehyde

Studies by various investigators have concluded that formaldehyde is not mutagenic in the Ames test.[85–88] However, other studies have indicated that formaldehyde is mutagenic in this test but that the mutagenicity is masked by its high toxicity at low doses.[89–93] In addition, formaldehyde was shown to be mutagenic in male larvae of *Drosophila melanogaster*, when added to the feed medium whereas female larvae and adults were unaffected. Injection of formaldehyde into male flies was also mutagenic but to a lesser degree. Formaldehyde vapour was not mutagenic to *Drosophila melanogaster*.[94,95] Other studies have shown that formaldehyde is mutagenic in prokaryotic and in lower eukaryotic systems, in *Drosophila* and in mammalian cells *in vitro*.[94–98] The results of mutagenicity tests in mammals *in vivo* are, however, negative or equivocal.[96,99] It is considered that the mutagenic activity of formaldehyde is comparatively weak, except under very specific conditions in *Drosophila*. In addition, it is possible that it may be detoxified by metabolism before it causes damage to cellular DNA.

12.7.2 Isothiazolinones

The genotoxicity of Kathon CG has been assessed in many different assays, both *in vitro* and *in vivo*; viz. Ames, in a variety of bacterial strains, mouse micronucleus, *in vitro* and *in vivo* cytogenetics, *Drosophila melanogaster*, unscheduled DNA synthesis, cell transformation, mouse lymphoma and DNA binding.[42] Positive results were obtained in the Ames test[100–103] and the mouse lymphoma assay.[103] The mutagenicity data have been reviewed by a variety of experts and, taking into account negative results from a valid dermal carcinogenicity study (see previous section) the conclusion has been drawn that Kathon CG does not represent a genotoxic hazard for man.[42]

12.7.3 Bronopol

Limited reported mutagenicity testing has shown that Bronopol was not mutagenic in the Ames test (*in vitro*) nor in the mouse dominant lethal assay (*in vivo*) when administered at the maximum tolerated dose.[31]

12.7.4 Oxabans A and E

The mutagenic potentials of Oxaban A and E have been assessed in a variety of *in vitro* and *in vivo* assays. There are some positive mutagenicity data *in vitro* for Oxaban A, but *in vivo* assays have been negative. Oxaban E does not appear to have mutagenic potential either *in vitro* or *in vivo*.[33] Oxaban A and E do not, therefore, pose a genotoxic hazard to man.

12.7.5 *Glutaraldehyde*

The genotoxicity of glutaraldehyde has been investigated in *in vitro* tests which have given both positive and negative results and *in vivo* where only negative results have been shown. Glutaraldehyde has demonstrated a positive response in Ames tests,[91,93,104–109] and also in other tests using the same bacterium *Salmonella typhimurium*.[110,111] Both positive[108] and negative[112,113] results have been obtained in mutagenicity assays using the bacterium *Escherichia coli*. Glutaraldehyde also gave a positive result in a *Bacillus subtilis* rec assay which measures DNA damage indirectly.[106] Negative results have been achieved in *Drosophila melanogaster* assays.[114–117] In *in vitro* tests using mammalian cells, glutaraldehyde induced mutations in mouse cells[117,118] and caused DNA damage in hamster cells.[119] Equivocal results were obtained in other assays for mutagenicity, chromosome damage and sister chromatid exchange.[107,120] Glutaraldehyde was also negative in an unscheduled DNA test using rat liver cells[107] and a cell transformation assay using hamster cells.[121] Glutaraldehyde showed no evidence of mutagenicity in two *in vivo* tests, viz. a dominant lethal assay using female mice[122] and an unscheduled DNA synthesis assay using male rats.[123] Evaluation of these latter data indicates that glutaraldehyde is not an *in vivo* genotoxin, and therefore is probably not a risk to man from its use as a preservative.

12.8 Reproductive and developmental effects

12.8.1 *Formaldehyde*

The teratogenic potential of formaldehyde has been examined in many studies, but none has demonstrated a teratogenic effect.[124]

12.8.2 *Isothiazolinones*

In a rabbit study, Kathon CG was not teratogenic but was embryotoxic and foetotoxic at high doses.[42] Kathon CG was not teratogenic to rats at doses up to 15 mg kg^{-1} body weight per day.[42] In addition, in a 15-week rat drinking water study, no adverse effects were observed on fertility, reproduction, foetal survival or health or maternal health up to a dose of 20 mg kg^{-1} body weight per day.[42]

12.8.3 *Bronopol*

Bronopol produced no significant embryotoxic or teratogenic effects when administered by oral intubation to rats throughout pregnancy at doses up to 100 mg kg^{-1} body weight per day.[31] Bronopol had no effect on parturition, litter size, post-natal survival or development of the young when given orally to rat dams from day 15 of gestation and throughout lactation at doses up to 40 mg kg^{-1} body weight per day. There was also no effect on reproductive performance or fertility in both male and female rats up to this dose level.[31] Bronopol had no embryotoxic or teratogenic effects when administered orally to rabbits from days 8–16 of pregnancy at doses up to 10 mg kg^{-1} body weight per day.[31]

12.8.4 Glutaraldehyde

Glutraldehyde produced no evidence of foetotoxicity or teratogenic effects in a rat study up to a dose of 50 mg kg^{-1} body weight per day.[19] In mice, glutaraldehyde had no effects on maternal survival and weight gain, the number of live foetuses, foetal weight or foetal malformation up to 24 mg kg^{-1} body weight per day. Maternal survival and weight gain were decreased at 40 mg kg^{-1} body weight per day and above, and at 50 mg kg^{-1} body weight per day a slight increase in the incidence of foetal malformations was observed. At 100 mg kg^{-1} body weight per day, glutaraldehyde produced reduced foetal weight and an increase in the number of stunted foetuses and in the incidence of foetal malformations.[125] At 15 mg kg^{-1} body weight per day, glutaraldehyde had no effects on rabbits when administered by stomach tube for 13 days during pregnancy. Even at 40 mg kg^{-1} body weight per day there were no foetal malformations but there were indications of maternal toxicity as evidenced by early foetal deaths and reduced foetal weights.[126]

12.9 Environmental considerations

Formaldehyde is a naturally occurring substance which is a normal metabolite in mammals. It is reported to be present in all fruit and vegetables in the diet at a concentration of 4 mg kg^{-1}; 3–7.5 mg of formaldehyde are formed when the 50–150 mg of caffeine in a cup of coffee are metabolized in the human body.[127] Formaldehyde occurs in the atmosphere by the photochemical decomposition of organic products (e.g. methane). The natural background concentration of formaldehyde in clean maritime air is 0.0001 ppm and in clean continental air 0.001 ppm. Formaldehyde in solution does not pose a hazard to the environment as it has a low octanol:water partition coefficient, which means that it will remain in the aqueous phase not significantly adsorbing or bioaccumulating and, at the levels likely to be released to the sewage treatment works, it can biodegrade both aerobically and anaerobically.

Proxel is also biodegradable and has a low potential for bioaccumulation.

Also, it has a relatively low toxicity to birds and mammals. Proxel is, however, acutely toxic to aquatic organisms and to sewage treatment operations at concentrations similar to those at which surfactants are toxic.[40] The concentrations of Proxel in the environment likely to arise from its use as a preservative in surfactants are extremely low (a few ppb and below) and are several (>3) orders of magnitude lower than concentrations at which adverse effects might arise.

The environmental fate of Kathon CG has been extensively investigated.[128,129] Studies have shown that although the active ingredients of Kathon CG are toxic to aquatic species and wildlife, they are biodegradable into simple, non-toxic products which are rapidly dissipated from soil, natural waters and sewage treatment systems. It is known that the biodegradation of Kathon CG does not lead to a chlorinated organic species; the 5-chloro constituent breaks down and releases the chloride ion. Therefore, Kathon CG does not persist in the environment. However, from a factory handling standpoint, Kathon CG-containing wastes must not be discharged into public waters. Such wastes should either be deactivated or adequately diluted before discharge into any public water or sewage treatment system.[41] Deactivation can be effected by the addition of a slightly acidic sodium metabisulphite or sodium bisulphite solution.

Bronopol does not biodegrade at concentrations as low as 3 mg l^{-1}. Hydrolysis at environmental pH values is unlikely to reduce the Bronopol level during sewage treatment, but photolysis could completely degrade it in seven days. The degradation products formed, whilst potentially more toxic than the parent material, will not persist in the environment. It is expected that Bronopol will tend to remain in the aqueous phase during sewage treatment and subsequent discharge to rivers. The bacterial minimum inhibitory concentrations of Bronopol (12.5–50 ppm) are several orders of magnitude higher than the levels likely to be present in sewage so that adverse effects on the microbiological processes of sewage treatment are unlikely. The concentrations of Bronopol likely to occur in rivers are orders of magnitude below the levels which produce toxic effects in daphnia (48-hour EC50 = 1.4 mg l^{-1}), fish (96-hour LC50 = 20 mg l^{-1}) or algae (EC50 = 0.2 mg l^{-1}). The potential for bioconcentration of Bronopol in aquatic organisms should be low.

12.10 Conclusion

It is important in the preparation of cosmetic, toiletry or speciality chemical products that there is adequate protection from microbiogical contamination. This contamination can arise from the consumer, the environment during the shelf life of the product, the manufacturing plant or the raw materials, including water. Good preservatives, such as those which have been discussed in this chapter, can provide peace of mind to the manufacturer that the product will be effectively protected from microbiological attack. This in turn provides the consumer with a good quality product free from any microbiological contamination which can lead to serious adverse health effects. However, it must be remembered that preservatives are biologically active molecules, they have to be to kill or prevent the growth of the organisms responsible for microbiological attack. The assessment of the risk of potential adverse health effects on man from the use of these preservatives is a complex process in which the known animal and human toxicity data available have to be assessed against the likely human exposure. Because preservatives are so biologically active, their suppliers should ensure that there is an adequate data base from which a risk assessment can be made. In addition, suppliers should adopt the good practice of providing recommended use levels which provide effective preservation but which do not represent unacceptable risks to either man or the environment.

Generally, preservatives are used at such low levels that any irritation or skin sensitization potential is irrelevant at the recommended use concentration. This being said, there are still certain applications, such as in leave-on cosmetics in which some preservatives close to their recommended use levels may cause either irritation or, in particular, sensitization. There are many published papers which testify to this effect. Some manufacturers have worked extremely hard to make sure that their preservatives are adequately tested to provide the information which they need to set suitable limits for use and which will ensure that the risk of adverse health effects during normal use of the product are minimized. Indeed, some manufacturers even monitor their customers' products in the marketplace to ensure that their preservatives are being used at the recommended levels.

One important and often forgotten aspect of preservative management is to communicate effectively with the consumer. Formaldehyde is a good example of a technically excellent, cost-effective preservative which has now been banned in several countries because the risk assessment process was not effectively managed and communicated to the general public. Bad public relations has led to these marketing restrictions for formaldehyde despite regulatory approval for the use of this preservative in many other countries. Extensive marketing experience, together with critical evaluation of many epidemiological studies on workers exposed industrially to concentrations of formaldehyde which are considerably higher than those which would occur following the use of products preserved with formaldehyde, show that formaldehyde in its use as a preservative is safe. Formaldehyde is a very effective preservative for personal products. It is effective at very low concentrations and is functional with most ingredients across a wide pH range. It is generally only incorporated in products which are diluted during use and subsequently rinsed from the skin. Thus consumer exposure, and the risk in terms of skin irritation and sensitization, from these sources is only minimal.

References

1. Wedderburn, D.L. (1970) Interactions in cosmetic preservation. *Am. Perfum Cosmet.*, **85**, 49–53.
2. Higuchi, T. and Lach, J.L. (1954) Study of the possible complex formation between macromolecules and certain pharmaceuticals. *J. Am. Pharm. Assoc., Sci. Ed.*, **44**, 465–470.
3. Guttman, D. and Higuchi, T. (1956) Possible complex formation between macromolecules and certain pharmaceuticals. X. *J. Am. Pharm. Assoc., Sci. Ed.*, **45**, 659–664.
4. Dyer, D.L. (1958) A review of detergent–germicide interactions. *Soap Chem. Specialities*, **34**, 53–55, 139–141.
5. Evans, W.P. (1964) The solubilization and inactivation of preservatives by nonionic detergents. *J. Pharm. Pharmacol.*, **16**, 323–331.
6. Coates, D. (1973) Interaction between preservatives and surfactants. *Mfg. Chem. Aerosol News*, **44**, 41–42.
7. Freese, E. and Levin, B.C. (1978) Action mechanisms of preservatives and antiseptics. *Develop. Indust. Microbiol.*, **19**, 207–227.
8. Merianos, J.J. (1991) Quaternary ammonium antimicrobial compounds. In *Disinfection, Sterilization and Preservation*, 4th edn, Block, S.S. (ed), Lea & Febiger, Philadelphia and London, p. 250.
9. Ikeda, T. and Tazuke, S. (1985) Biocidal polycations. *Polym. Prep.*, **26**, 226–227.
10. Franklin, T.J. and Snow, G.A. (1989) *Biochemistry of Antimicrobial Action*, 4th edn, Chapman and Hall, London.
11. Hugo, W.B. (1992) Disinfectant mechanisms. In *Principles and Practice of Disinfection, Preservation and Sterilization*, Russell, A.D., Hugo, W.B. and Aycliffe, G.A.J. (eds), Blackwell Scientific, Oxford, pp. 187–210.
12. Fuller, S.J. *et al.* (1985) The mode of action of 1,2-benzisothiazolin-3-one on *Staphylococcus aureus*. *Lett. Appl. Microbiol.*, **1**, 13–15.
13. Russell, A.D. (1991) Principles of antimicrobial activity. In *Disinfection, Sterilization and Preservation*, Block, S.S. (ed.), 4th edn, Lea & Febiger, Philadelphia and London, pp. 29.
14. Rossmoore, H.W. and Sondossi, M. (1988) Applications and mode of action of formaldehyde condensate biocides. *Adv. Appl. Microbiol.*, **33**, 223–277.
15. Gorman, S.P. *et al.* (1980) Antimicrobial activity, uses and mechanisms of action of glutaraldehyde. *J. Appl. Bacteriol.*, **48**, 161–190.
16. Richmond, D.V. and Somers, E. (1962) Studies on the fungitoxicity of captan. I. The structure specificity of captan and six *n*-trichloromethylthio analogues. *Ann. Appl. Biol.*, **50**, 33–43.
17. Richmond, D.V. and Somers, E. (1962) Studies on the fungitoxicity of captan. II. The uptake of captan by conidia of *Neurospora crassa*. *Ann. Appl. Biol.*, **50**, 45–56.
18. Richmond, D.V. and Somers, E. (1963) Studies on the fungitoxicity of captan. III. Relation between sulfhydryl content of fungal spores and their uptake of captan. *Ann. Appl. Biol.*, **52**, 327–336.
19. Akabane, J. (1970) Aldehydes and related compounds. *Internat. Encyclopaedia Pharmacol. Therapy*, **2**, 523–560.
20. Buss, J. *et al.* (1964) Enterale Resorption von Formaldehyd. *Naunyn-Schmiedebergs Archiv., Exp. Pathol. Pharmakol.*, **247**, 380–381.
21. Williams, R.T. (1959) Detoxification mechanisms. *The Metabolism and Detoxification of Drugs, Toxic Substances and other Organic Compounds*, 2nd edn, J. Wiley and Sons, New York.
22. Dutch Ministry of Health and Environmental Protection (1982). *Formaldehyde and*

Cancer in Humans, Advice of the Special Advisory Committee on Formaldehyde, Presented at OECD *ad hoc* meeting on formaldehyde, June 7th, 1985, Paris, France.
23. Health & Safety Executive (1993) *Occupational Exposure Limits 1993*. EH-40/93.
24. American Conference of Governmental Industrial Hygienists (1993/1994) *Threshold Limit Values for Chemical Substances and Physical Agents and Biological Exposure Indices*, ISBN: 1–882417–03–8.
25. Deutsche Forschungsgemeinschaft (1992) MAK- and BAT-Values 1992. *Maximum concentrations at the workplace and biological tolerance values for working materials*, Commission for the Investigation of Health Hazards of Chemical Compounds in the Work Area, Report No. 28.
26. Newall, G.W. (1983) Overview of formaldehyde. In *Formaldehyde Toxicity*, Gibson, J.E. (ed.), Hemisphere, Washington, New York and London, p. 7.
27. ICI (1982) *Biocides Technical Information Sheet B18.82*. Proxel biocides for the preservation of synthetic polymer emulsions. Products: Proxel BD, Proxel GXL, Proxel CRL, Proxel XL2 and Proxel TN.
28. Rohm & Haas (1988) *European Product Safety Sheet for Kathon CG*. Rohm & Haas Company, European Region, Product Integrity Department, Sophia Antipolis 06565, Valbonne Cedex, France.
29. Hodge, H.E. and Sterner, J.H. (1949) Tabulation of toxicity classes. *Am. Ind. Hyg. Assoc. Quart.*, **10**, 93–96.
30. Cosmetic Ingredient Review (1980) Final report of the safety assessment for 2-bromo-2-nitropropane-1,3–diol. *J. Environ. Pathol. Toxicology.*, **4**, 48–61.
31. Bryce, D.M., *et al.* (1978) The activity and safety of the antimicrobial agent Bronopol (2-bromo-2-nitropropan-1,3-diol). *J. Soc. Cosmet. Chem.*, **29**, 3–24.
32. Angus Chemical Company (1986) *Oxaban A Technical Bulletin TB 54*. Norbrook, Illinois, USA.
33. Angus Chemical Company (1985) *Oxaban E Preliminary Technical Data Sheet and Material Safety Data Sheet*. Angus Chemie GmbH, Ibbenbüren, Germany.
34. Health & Safety Executive (1993) *Occupational Exposure Limits: Criteria Document Summaries*. HMSO, London, UK.
35. Union Carbide (1982) *Ucarcide Antimicrobials Technical Data Sheet*. Union Carbide Corporation, Speciality Chemicals and Plastics Division, Danbury, Connecticut, USA.
36. Glass, W.I. (1961) An outbreak of formaldehyde dermatitis. *N.Z. Med. J.*, **60**, 423–427.
37. Pirila, V. and Kilpio, O. (1949) On dermatitis caused by formaldehyde and its compounds. *Ann. Med. Intern. Fenn.*, **38**, 38–51.
38. Maibach, H. (1983) Formaldehyde: effects on animal and human skin. In *Formaldehyde toxicity*, Gibson, J.E. (ed.), Hemisphere, USA.
39. HSE (1993) *Occupational Exposure Limits: Criteria Document Summaries*. Synopses of the data used in setting occupational exposure limits, Health and Safety Executive.
40. ICI Technical Information Sheet B16.81. (1981) *ICI Biocides for the Preservation of Miscellaneous Aqueous Solutions and Dispersions*. ICI plc, Organics Division, Application Research & Technical Service Dept., Manchester, UK.
41. Rohm & Haas (1989) *Kathon CG Cosmetic and Toiletry Preservative Safety Guidelines*. Rohm & Haas Company, European Region, Product Integrity Dept., Sophia Antipolis 06565, Valbonne Cedex, France.
42. CIR. (1992). Final report on the safety assessment of methylisothiazolinone/methylchloroisothiazolinone. *J. Am. Coll. Toxicol.*, **11**, (1), 75–128.
43. Smyth, H. F. Jr. *et al.* (1962). Range-finding toxicity data: list VI. *Am. Ind. Hyg. Assoc. J.*, **23**, 95–107.
44. Union Carbide (1970) *Data Sheet*. 30th January (cited in NIOSH, 1987).
45. Ballantyne, B. *et al.* (1985) Acute toxicity and irritancy of glutaraldehyde. *Toxicologist*, **5**, 204.
46. Murray, W.J. and Ruddy, M.P. (1985) Toxic eye injury during induction of anesthesia. *Southern Med. J.*, **78**, (8), 1012–1013.
47. Uemitsa, N. *et al.* (1976) *Oyo Yakuri*, **12**, 1 (from translation 81–12518–06T carried out by the National Translation Centre and cited in the BIBRA Toxicity Profile on Glutaraldehyde, 3rd edn, 1991).
48. Hansen, E.M. and Menne, T. (1990) Glutaraldehyde: patch test, vehicle and concentration. *Contact Dermatitis*, **23**, (5), 369–370.

49. Gad, S.C. (1982) A neuromuscular screen for use in industrial toxicology. *J. Toxicol. Environ. Health*, **9**, 691–704.
50. Rudner, E.J. (1977) North American group results. *Contact Dermatitis*, **3**, 208.
51. BCDG (1982) *British Contact Dermatitis Group*, unpublished data.
52. NACDRG (1982) North American Contact Dermatitis Research Group. Eiermann, M.J., Larson, W., Maibach, H.I. and Taylor, J.S. Prospective study of cosmetic reactions: 1977–1980. *J. Am. Acad. Dermatol.*, **6**, 909–917.
53. Meynadier, J.-M., *et al.* (1982) Allergie aux conservaturs. *Ann. Dermatol. Venereol. (Paris),*, **109**, 1017–1023.
54. De Groot, A.C. *et al.* (1985.) Patch tests with fragrance materials and preservatives. *Contact Dermatitis*, **12**, 87–92.
55. Cronin, E. (1980) *Contact Dermatitis*, Churchill Livingstone, Edinburgh, London and New York.
56. Foussereau, J. *et al.* (1982). *Occupational Contact Dermatitis. Clinical and Chemical Aspects*, Munksgaard, Copenhagen.
57. Popa, V. *et al.* (1969) Bronchial asthma and asthmatic bronchitis determined by simple chemicals. *Diseases. Chest*, **56**, 395–404.
58. Hendrick, D.J. and Lane, D.J. (1975) Formalin asthma in hospital staff. *Br. Med. J.*, **1**, (5958), 607–608.
59. Hendrick, D.J. and Lane, D.J. (1977) Occupational formalin asthma. *Br. J. Indust. Med.*, **34**, 11–18.
60. ECETOC. Technical Report No. 1. (1981) *Assessment of Data on the Effects of Formaldehyde on Humans*, European Chemical Ecology and Toxicology Centre, Brussels, Belgium.
61. Andersen, K.E., *et al.* (1984). Guinea pig maximization tests with formaldehyde releasers. Results from two laboratories. *Contact Dermatitis*, **10**, 257–266.
62. Marzulli, F.N. and Maibach, H.I. (1974) Antimicrobials: experimental contact sensitization in man. *J. Soc. Cosmet. Chem.*, **24**, 399–421.
63. Jordan, W.P., Jr., *et al.* (1979) Threshold response in formaldehyde-sensitive subjects. *J. Am. Acad. Dermatol.*, **1**, 44–48.
64. Pederson, N.B. (1976) Occupational allergy from 1,2-benzisothiazolin-3-one and other preservatives in plastic emulsions. *Contact Dermatitis*, **2**, (6), 340–342.
65. Slovak, A.J.M. (1980) Contact dermatitis due to benzisothiazolone in a works analytical team. *Contact Dermatitis*, **6**, (3), 187–190.
66. Roberts, D.L., *et al.* (1981) Occupational dermatitis due to 1,2-benzisothiazolin-3-one in the pottery industry. *Contact Dermatitis*, **7**, (3), 145–147.
67. De Groot, A.C. (1990) Labelling cosmetics with their ingredients. *Br. Med. J.*, **300**, 1636–1638.
68. Chan, P.K., *et al.* (1983). Kathon biocide: manifestation of delayed contact dermatitis in guinea pigs is dependent on the concentration for induction and challenge. *J. Invest. Dermatol.*, **81**, (5), 409–411.
69. Marzulli, F.N. and Maibach, H.I. (1974) The use of graded concentrations in studies of skin sensitizers: experimental contact dermatitis. *Food Cosmet. Toxicol.*, **12**, 219–227.
70. Maibach, H.I. (1977) Dermal sensitization potential of 2-bromo-2-nitropropan-1,3-diol (Bronopol). *Contact Dermatitis*, **3**, 99.
71. Broeckx, W., *et al.* (1987). Cosmetic intolerance. *Contact Dermatitis*, **16**, 189–194.
72. Union Carbide Corporation (1980) *Repeated Insult Patch Test*. Conducted for Union Carbide by Testkit Laboratories, Inc. (Cited in Union Carbide Technical Brochure - Ucarcide Antimicrobials as Preservatives).
73. Skog, E. (1968) Sensitivity to glutaraldehyde. *Contact Dermatitis Newsletter*, **4**, 79.
74. UK Industrial Injuries Advisory Committee (1990) *Occupational Asthma*. Report by the Industrial Injuries Advisory Committee in accordance with Section 141 of the Social Security Act 1975 on the question of whether sensitizing agents should be prescribed under the Act. Cm 1244. Department of Social Security, HMSO, London.
75. Kerns, W.D. *et al.* (1983) Carcinogenicity of formaldehyde in rats and mice after long-term inhalation exposure. *Cancer Res.*, **43**, 4382–4392.
76. Gardner, M.J. *et al.* (1993) A cohort study of workers exposed to formaldehyde in the British chemical industry: an update. *Brit. J. Indust. Med.*, **50**, (9), 827–834.

77. Harrington, J.M. and Shannon, H.S. (1975) Mortality study of pathologists and medical laboratory technicians. *Brit. Med. J.*, **4**, (5992), 329–332.
78. Luce, D. *et al.* (1993) Sinonasal cancer and occupational exposure to formaldehyde and other substances. *Internat. J. Cancer*, **53**, (2), 224–231.
79. Fayerweather, W.E. *et al.* (1982). Case-control study of cancer deaths in Du Pont workers with potential exposure to formaldehyde. *ECETOC Technical Report No. 6.*
80. Partanen, T. (1993) Formaldehyde exposure and respiratory cancer – a meta-analysis of the epidemiologic evidence. *Scand. J. Work Environ. Health*, **19**, (1), 8–15.
81. Purchase, I.F.H. and Paddle, G.M. (1989) Does formaldehyde cause nasopharyngeal cancer in man? *Cancer Lett.*, **46**, (2), 79–85.
82. Rohm & Haas (1989) *The Kathon CG Report.* An expert briefing on use, safety and toxicity. Rohm & Haas Company, European Region, Product Integrity Department, Sophia Antipolis 06565, Valbonne Cedex, France.
83. Spencer, P.S., *et al.* (1978) On the specific molecular configuration of neurotoxic aliphatic hexacarbon compounds causing central-peripheral distal axonopathy. *Toxicol. Appl. Pharmacol.*, **44**, 17–28.
84. Stonehill, A.A., *et al.* (1963) Buffered glutaraldehyde. A new chemical sterilizing solution. *Am. J. Hosp. Pharm.*, **20**, 458–465.
85. Florin, I. *et al.* (1980) Screening of tobacco smoke constituents for mutagenicity using the Ames test. *Toxicology*, **18**, 219–232.
86. Koops, A. and Butterworth, B.E. (1976) *In Vitro Mutagenicity Studies of Formaldehyde*, E.I. du Pont de Nemours, Haskell Laboratory for Toxicology and Industrial Medicine, Newark, DE.
87. Gocke, E. *et al.* (1981) Mutagenicity of cosmetic ingredients licensed by the European Communities. *Mutat. Res.*, **90**, 91–109.
88. Couch, D.B. *et al.* (1982) The mutagenicity of formaldehyde to *Salmonella typhimurium. Environ. Mutagen.*, **4**, 336.
89. Connor, T.H. *et al.* (1983) Mutagenicity of formalin in the Ames test. *Mutat. Res.*, **119**, 145–149.
90. Donovan, S.M. *et al.* (1983). Mutagenic activities of formaldehyde and hexamethylphosphoramide in reverse and forward *Salmonella typhimurium* mutation assays. *Environ. Mutagen.*, **5**, 476.
91. Haworth, S. *et al.* (1983) Salmonella mutagenicity test results for 250 chemicals. *Environ. Mutagen.*, **5**, Suppl. 1, 3–142.
92. Saski, Y. and Endo, R. (1978) Mutagenicity of aldehydes in *Salmonella. Mutat. Res.*, **54**, 251–252.
93. Levin, D.E. *et al.* (1982) A new Salmonella tester strain (TA 102) with A-T base pairs at the site of mutation detects oxidative mutagens. *Proc. Natl. Acad. Sci. USA*, **79**, 7445–7449.
94. Rapoport, I. A. (1946) Carbonyl compounds and the chemical mechanisms of mutations. *Compt. Rend. Acad. Sci. USSR*, **54**, 65–67.
95. Auerbach, C. *et al.* (1977) Genetic and cytogenetical effect of formaldehyde and related compounds. *Mutat. Res.*, **39**, 317–362.
96. Brusick, D. J. (1980) In vivo/in vitro mutagenesis assays with formaldehyde. In *Third CIIT Conference on Toxicology: Formaldehyde Toxicology*, Chemical Industry Institute on Toxicology, 20–21 November, 1980, Raleigh, NC., USA.
97. Obe, G. and Beek, B. (1979) Mutagenic activity of aldehydes. *Drug and Alcohol Dependence*, **4**, 91–94.
98. Grafstrom, R.C. *et al.* (1985). Genotoxicity of formaldehyde in cultured human bronchial fibroblasts. *Science*, **228**, 89–91.
99. Fontignie-Houbrechts, N. (1981) Genetic effects of formaldehyde in the mouse. *Mutat. Res.*, **88**, 109–114.
100. Wright, C. *et al.* (1983). Mutagenic activity of Kathon, an industrial biocide and cosmetics preservative containing 5-chloro-2-methyl-4-isothiazolin-3-one and 2-methyl-4-isothiazolin-3-one. *Mutat. Res.*, **119**, 35–43.
101. Ashby, J. *et al.* (1985). Chloracetamide-*N*-metholol: an example of an *in vitro* and *in vivo* clastogen which is non-mutagenic to *Salmonella. Mutat. Res.*, **156**, (1–2), 19–32.

102. Monte, W.C. *et al.* (1983) Mutagenicity of two nonformaldehyde-forming antimicrobial agents. *Food Chem. Toxicol.*, **21**, (5), 695–696.
103. Scribner, H.E. *et al.* (1983) The genetic toxicology of Kathon biocide, a mixture of 5-chloro-2-methyl-4-isothiazolin-3-one and 2-methyl-4-isothiazolin-3-one. *Mutat. Res.*, **118**, (3), 129–152.
104. Marnett, L.J. *et al.* (1985) Naturally occurring carbonyl compounds are mutagens in Salmonella tester strain TA104. *Mutat. Res.*, **148**, (1–2), 25–34.
105. NTP (1989) *Review of Current DHHS, DOE and EPA Research Related to Toxicology*, Fiscal year 1989, National Toxicology Program, NTP-89-168. US Department of Health and Human Services, Research Triangle Park, North Carolina.
106. Sakagami, Y. *et al.* (1988) DNA repair test of disinfectants by liquid rec-assay. *Mutat. Res.*, **193**, (1), 21–30.
107. Slesinski, R.S. *et al.* (1983) Mutagenicity evaluation of glutaraldehyde in a battery of *in vitro* bacterial and mammalian test systems. *Food Chem. Toxicol.*, **21**, (5), 621–629.
108. Wilcox, P. *et al.* (1990) Comparison of *Salmonella typhimurium* TA102 with *Escherichia coli* WP2 tester strains. *Mutagenesis*, **5**, (3), 285–291.
109. Zeiger, E. (1980) Unpublished data (cited in Wade *et al.* (1982) *FDA By-lines* No. 1, p. 7).
110. Ruiz-Rubio, M. *et al.* (1985) Oxidative mutagens specific for A.T. base pairs induce forward mutations to L-arabinose resistance in *Salmonella typhimurium. Mutat. Res.* **147**, 153–163.
111. Sakagami, Y. *et al.* (1988) The evaluation of genotoxic activities of disinfectants and their metabolites by umu test. *Mutat. Res.*, **209**, (3/4), 155–160.
112. Hemminki, K. *et al.* (1980) Comparison of alkylation rates and mutagenicity of directly acting industrial and laboratory chemicals. *Arch. Toxicol.*, **46**, 277–285.
113. von der Hude, W. *et al.* (1988) Evaluation of the SOS chromotest. *Mutat. Res.*, **203**, (2), 81–94.
114. NTP (1982) *National Toxicology Program Technical Bulletin*, No. 8, p. 11.
115. NTP (1985) *Fiscal Year 1985 Annual Plan*, National Toxicology Program, NTP-85–055, US Department of Health and Human Services, Research Triangle Park, North Carolina.
116. Yoon, J.S. *et al.* (1985) Chemical mutagenesis testing in Drosophila. IV. Results of 45 coded compounds tested for the National Toxicology Program. *Environ. Mutagen.*, **7**, (3), 349–367.
117. Zimmering, S. *et al.* (1989) Chemical mutagenesis testing in Drosophila. VII. Results of 22 coded compounds tested in larval feeding experiments. *Environ. Molec. Mutagen.*, **14**, (4), 245–251.
118. McGregor, D.B. *et al.* (1988). Responses of the L5178 tK+/tK-mouse lymphoma cell forward mutation assay. II: 18 coded chemicals. *Environ. Molec. Mutagen.* **11**, (1), 91–118.
119. Anon (1970) *Symp. Fund Cancer Res.*, **23**, 346 (cited in NIOSH, 1987. Registry of Toxic Effects of Chemical Substances, 1987–1988 edition, DHHS (NIOSH) Pub. No. 87–114. National Institute for Occupational Safety and Health, Cincinnati, Ohio, 1987).
120. Galloway, S.M. *et al.* (1985) Development of a standard protocol for *in vitro* cytogenetic testing with Chinese hamster ovary cells: comparison of results for 22 compounds in two laboratories. *Environ. Mutagen.*, **7**, 1–51.
121. Tsutsui, T. *et al.* (1987) Assessment of the carcinogenic hazard of 27 substances used in dental practices. *Jap. J. Pharmacol.*, **43**, 132P.
122. Tamada, M. *et al.* (1978) Mutagenicity of glutaraldehyde in mice. *Bokin Bobai*, **6**, (2), 62–68 (Japan). (*Chem. Abstr.* **91**, 135063X).
123. Mirsalis, J.C. *et al.* (1989) Measurement of unscheduled DNA synthesis and S-phase synthesis in rodent hepatocytes following *in vivo* treatment: testing of 24 compounds. *Environ. Molec. Mutagen.*, **14**, (3), 155–164.
124. Staples, R.E. (1983) Teratogenicity of formaldehyde. In *Formaldehyde Toxicity*, Gibson, J.E. (ed.), Hemisphere New York, Chap. 6, pp. 51–59.
125. Marks, T.A. *et al.* (1980) Influence of formaldehyde and Sonacide (potentiated acid glutaraldehyde) on embryo and fetal development in mice. *Teratology*, **22**, (1), 51–58.

126. Anon. (1990) (Preliminary results of a gavage study of prenatal toxicity of glutaraldehyde in rabbits). *Pestic. Toxic. Chem. News.* **18**, (33), 28.
127. Schmahl, D. (1985) Was the fuss about formaldehyde necessary? No risk of cancer to the ordinary citizen. *Rhein-Neckar-Zeitung*, 22 February, 1985.
128. Krzeminski, S.F., *et al.* (1975) Fate of microbicidal 3-isothiazolone compounds in the environment: modes and rates of dissipation. *Agric. Food Chem.*, **23**, 1060–1068.
129. Krzeminski, S.F., *et al.* (1975) Fate of microbicidal 3-isothiazolone compounds in the environment: products of degradation. *Agric. Food Chem.*, **23**, 1068–1075.

13 The safe use of preservatives

D.N. MUNRO

The safe use of preservatives represents a specific example of a more general need. That is, to ensure that in harnessing chemical and other technologies for the benefit of man, the potential risks to health, to the environment, and to plant and equipment, are all adequately assessed and managed.

This chapter will focus on the avoidance of adverse occupational health effects. This encompasses both the prevention of any such effects and avoidance of significant interference with the wellbeing of those involved in the manufacture or use of preservatives. It should not however be forgotten that there will be a need, and often specific legal duty, to assess and control adequately risks of physical and environmental damage.

Predominant amongst the ill health effects associated with the use of preservatives are skin effects such as irritant and allergic dermatitis[1,2] which may occasionally also include photoallergenic effects.[3] There may also be the potential for serious respiratory disease, 'biocide lung', which comprises pneumonia and chronic pulmonary fibrosis[4] arising from exposure to certain biocides.

Cases of ill health have been seen in manufacture of preservatives, but they occur more commonly at the user stage. Many applications have been implicated, including the textile finishing industry,[5] paint manufacturing and use,[6,7] metal working,[8] agriculture,[9] the pottery industry,[10] and health care and food industry workers.[11] It is not always necessary to have contact with the preservative itself for such problems to arise. There are reports for example of symptoms arising from accidental exposure to formulations containing only small quantities of preservative,[12] and from exposure to glues and pastes containing biocides.[13] Exposure may not in fact be immediately obvious, as in the unfortunate case of the deep sea diver who developed generalised allergic dermatitis following the disinfection of his wet suit with a 'preservative-containing' cleaning solution.[14]

How then are we to ensure that ill health effects do not arise? A first principle to establish is that there is nothing which has no inherent toxicological hazard. This is hardly a novel observation, but one which is frequently overlooked. Clearly even substances vital to life such as salt and sugar will have a significant toxic effect if the dose is sufficient, although they are usually regarded as being 'safe'. Paracelsus, the 'founding father' of toxicology summarised this concept over four hundred years ago!

Alle Dinge sind Gift
und nicht ohn Gift;
allein die Dosis macht,
dass ein Ding kein Gift ist.

Paracelsus (1493–1541).

Loosely translated this states that 'All things are poisonous, it is the dose which separates the medicine from the poison'. This insight established for the first time the concept of a 'dose response curve' and more importantly that of a 'safe dose'.

This leads us to the question of what it is that we regard as 'safe'. The title of this chapter is 'Safe use of preservatives', yet there can be no such thing as absolute safety. Beware all who demand or promise absolute safety; they will either be naive or corrupt!

The concept of safety is better regarded in terms of risk. When we talk about managing safety it is actually risk that we are concerned to manage; to reduce risk to an acceptable or accepted level. That these two levels are not necessarily the same becomes immediately obvious when one considers everyday risk. Consider the risk of premature death associated with cigarette smoking or motorbike riding in comparison with the requirements imposed on the nuclear and aviation industries!

The first stage in managing risk is to identify the hazard with which we are dealing. This, in the case of a preservative, or any other chemical, will require us to identify the inherent toxicological hazard. For substances in commercial use this will be facilitated by regulatory requirements for toxicological testing and product registration. For example, in the USA the Environmental Protection Agency (EPA) initiates and administers requirements, derived primarily from the Toxic Substances Control Act, for the registration and use of such substances. In Europe, various European directives and member state regulations, largely based on the Dangerous Substances Directive (7th Amendment), cover the issues such as the notification of new substances and the marketing, labelling and transport of hazardous materials. Similar specific regulatory provisions apply in countries such as Canada, Australia, Japan and elsewhere.

The consequence for the user is that the preservative will be supplied with information relating to its hazardous properties. This information will be in the form of a manufacturer's safety data sheet, labels and possibly supporting technical information. This documentation will include data derived from standardised toxicological testing packages specified by the regulatory authorities. It will often also be supplied in a standard format and be interpreted as standard phrases and symbols. These data are the key that unlocks the door leading to risk assessment and ultimately to 'safe working'.

Although the hazard data is key, it will not in fact be the hazard that necessarily predominates in the level of risk that arises in use. It must be *risk* = *f* × *hazard*

function *f* is largely a measure of *exposure*

f is dependent upon	*task* *intensity* *duration* *frequency* etc.

risk = *exposure* × *hazard*

Figure 13.1 Relationship between hazard and risk.

recognised that whilst the toxicity of a material is an 'inherent property' of the substance, the risk will also be influenced by many other factors. Amongst these we must include all those factors which can contribute to the actual or potential exposure of workers to the material in question. This concept is illustrated in Figure 13.1.

If an effective strategy to minimise 'risk' is to be defined then clearly this can be usefully divided into two key elements. First we can seek to minimise the hazard. Thereafter we must concentrate on managing those risks which might arise.

Considering then the first element, the hazard of the material, how should this be approached? Clearly it would be advantageous to select the least hazardous preservative practicable. In practice however the choice of a particular agent is going to be based on a range of criteria. Regulatory status, availability, technical suitability, cost and other factors will all obviously be crucial and usually predominant influences in the selection of a particular agent. It is important however to evaluate the hazard profile of short-listed candidate preservatives before a final selection is confirmed. The nature of the hazard is likely to have consequences throughout the product lifecycle, from manufacture through to use and ultimate disposal, and thus it merits serious consideration.

The starting point for such consideration will usually be the manufacturer's safety data sheet and possibly supporting technical literature. Detailed specific guidance on 'interpreting biocide health and safety data sheets' has been published.[15] A guide to the key elements of a typical safety data sheet format and its interpretation and implications is given in Table 13.1.

The second stage in the strategy is to manage those risks which might arise as a consequence of use of the chosen agent. This process will essentially seek to eliminate or at least minimise actual and potential exposures to a level below those at which we can expect to see evidence of any ill health effects or significant nuisance.

The risk management process will be dependent upon the context. For a new workplace or process there are opportunities to influence effectively

Table 13.1 'Typical' hazard data sheet format

1. Identification	What is the name? Are there alternatives?
2. Product description	Solid or liquid? What does it look like? Appearance/odour?
3. Summary	Is the material hazardous? Exposure standard?
4. Physicochemical data	Liquid/solid? Boiling point? Flash point? Other properties?
5. Stability and reactivity	Avoid contact with other materials?
6. Storage	How to store? Special requirements?
7. Handling	Special handling precautions?
8. Personal protection	Is PPE generally needed? What about my specific use?
9. Spillage/accidental release	What do I need to do?
10. Waste disposal	Any special advice? Any specific requirements?
11. First aid measures	What procedures? Antidotes etc?
12. Fire and explosion	Will it burn/explode? How do I deal with it?
13. Health hazard	What if I breathe it? It gets on skin/eyes? Is swallowed? Long term effects?
14. Environmental information	Effects?
15. Regulatory information	Labelling? Rules for transport

PPE = personal protective equipment.

the level of risk right from the initial conceptual level. The later the points of influence in implementing change to process, to equipment or to procedures are effected the more difficult, more expensive and less effective they will be. The major stages where input on the control of risk can be effected are illustrated in Table 13.2.

Table 13.2 Strategic opportunities to control risk

Conceptual Strategic Process design	Best use of expertise. Expense largely time only. Opportunity for step changes and continuous improvement (CI).
Project specification Project implementation Workplace modification	Expensive in time, materials and cash. Can achieve sustained improvements.
Audit of control Workplace monitoring	Involves local expertise. Ongoing requirement. Mainly preventive but can be used as a tool for CI.
Worker exposure monitoring Health surveillance	Expensive in time, expertise and cash. Ongoing requirement. Essentially quantification of level of failures.

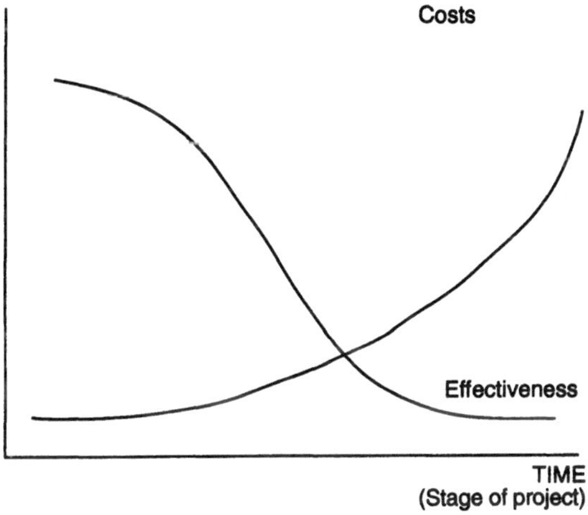

Figure 13.2 Cost and effectiveness of risk control measures.

The relative costs and effectiveness of intervention at various stages are highlighted in Figure 13.2.

The more common scenario unfortunately is that an existing application or facility exists, and it is within these constraints that it is required to establish and maintain safe working. We will consider in some detail how the strategy outlined above can be applied in these circumstances.

The initial stage should be to carry out an appropriate risk assessment. This assessment will not necessarily invoke sophisticated analysis or quantitative techniques. The objective of the risk assessment is to determine whether or not it is probable that there will be exposure sufficient to trigger an adverse effect.

The initial stage of such a risk assessment will be the identification of the hazard. What is the potential to induce adverse effects? The answer can be based on an understanding of the toxic and physicochemical properties of the substance gleaned from the safety data sheet or other sources.

The next stage involves an assessment of exposure potential. The routes and patterns of all potential exposures should be established. The nature, duration, frequency and magnitude of exposures should be estimated based on knowledge of the specific operation. It is important that all work patterns are considered, including abnormal situations like plant start up and shut down, routine and 'breakdown' maintenance, and factors such as shift and seasonal work variations.

A judgement then needs to be made as to whether or not exposure is adequately controlled. Very often the answer will be obvious. Either the exposure will be insignificant, or, alternatively so gross that it is clear that measures should be taken to mitigate the exposure. In other cases there may be established exposure standards such as threshold limit values (TLV) published by the American Conference of Governmental Industrial Hygienists (ACGIH),[16] permitted exposure limits (PEL) published by the National Institute for Occupational Safety and Health (NIOSH) or other mandatory national standards such as the occupational exposure standards (OES) and maximum exposure limits (MEL) which apply in Great Britain.[17] In the absence of such limits for the substance in question it may be necessary to refer to industry practice, advice from trade associations or from the supplier.

Alternatively it may be possible to classify the substance into a particular toxic hazard category, and then assume some form of generic exposure limit towards which control of exposure should be targeted. An example of one such approach is illustrated in Figure 13.3. Such approaches have been widely debated,[18,19] have been adopted by a number of leading chemical manufacturers[20] and have been published in various industry guidance documents.[21] The prime objective is to enable the relative hazards of substances to be readily compared and to suggest generic exposure limits against which the requirements for, and the effectiveness of, control measures can be determined.

In order to sensibly estimate exposure potential, and subsequently possibly facilitate its elimination or minimisation, it is first important to understand the nature of this exposure. An individual's total occupational exposure to a substance in use represents the sum of the individual's task exposures, incident-related exposures and background exposures. These contributory exposures may be defined as follows.

Task-related exposures arise from planned process tasks, where the nature of the task may lead to a release to atmosphere or to direct contact

COSHH TOXICITY CLASSIFICATION GRID

Class	Properties	O.E.S. Range	Risk Phrases
H1	Very Toxic Respiratory Sensitiser Potent Carcinogen (Human)	< 0.1 mg/M^3 < 0.1 ppm	R26,27,28 R42 R45
H2	Toxic Corrosive Animal Carcinogen Skin Sensitiser "Unknown" Toxicity	> 0.1 ppm < 10 ppm < 1 mg/M^3	R23,24,25 R34,35,41 R45 R43
M	Harmful Skin/Eye Irritants Low Potency Carcinogene (E.C. Cat 3.)	> 10 ppm < 500 ppm < 10 mg/M^3	R20,21,22 R36,37,38 R40
L	Non Hazardous Non Irritant Non Genotoxic Non Sensitising	> 500 ppm	

R20 Harmful by inhalation
R21 Harmful by skin contact
R22 Harmful if swallowed
R23 Toxic by inhalation
R24 Toxic by skin contact
R25 Toxic if swallowed
R26 Very toxic by inhalation
R27 Very toxic byskin
R28 Very toxic if swallowed
R34 Causes burns
R35 Causes severe burns
R36 Irritating to eyes
R37 Irritating to respiratory system
R38 Irritating to skin
R39 Danger of very serious irreversible effects
R40 Risk of irreversible effects
R41 Risk of serious eye damage
R42 May cause sensitisation by inhalation
R43 May cause skin sensitisation
R45 May cause cancer

Figure 13.3 Illustration of a scheme to classify substances by toxicological hazard. Redrawn from Zeneca Specialties 'Control of Hazardous Substances'.

with the substance. Such tasks typically include filling or emptying vessels or packages, maintenance work etc.

Incident-related exposures result from unplanned, although possibly foreseen, events such as spillages or a planned task going wrong. Minor frequent events may contribute to personal exposure either directly, e.g. someone exposed locally to a leak, or by increasing the background concentration. Infrequent events may cause acute problems but are unlikely to contribute significantly to routine exposures.

Background exposure is that exposure which would be experienced by someone being present in the workplace, but not directly carrying out any operations involving the specific substance. Background exposures will arise from fugitive emissions, such as continual leaks from equipment, vents and drains and so forth and additionally from emissions from tasks and incidents.

Moving on to the practical application of these understandings it is usual to subdivide a workplace into a succession of smaller elements in order to facilitate the risk assessment process. This involves splitting operational activities down into a series of 'unit operations' such as charging materials, sampling, packing off and so forth. The exposure potential arising from each of these elements of the work can then easily be assessed taking into account factors such as the methods employed, the scale and intensity of

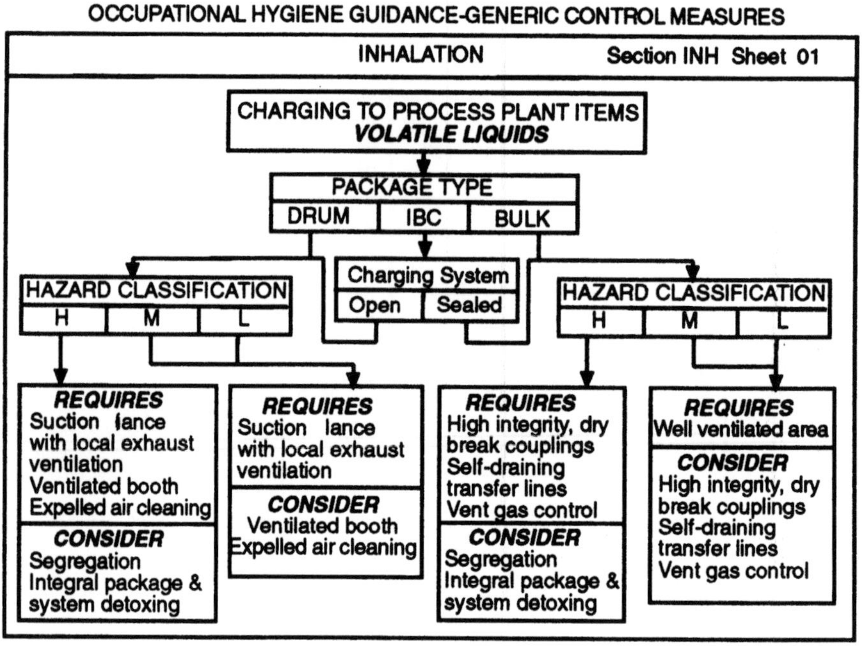

Figure 13.4 Example of generic guidance on exposure controls. Redrawn from Zeneca Specialties 'Control of Hazardous Substances'.

operations, and the physical properties of the material being handled. The level of risk arising from the exposure potential can then be matched to an appropriate control regime. It is possible to develop 'generic guidance' in this area (an illustrative example is given in Figure 13.4).

In practice it will be necessary to consider each task element in turn and then apply any necessary control, taking into account the hierarchy of available occupational hygiene control measures. This hierarchy represents both 'best practice' and, in certain cases, specific legal requirements.[22] It is based on the simple premise that control at source, resulting in a safe workplace, will be more effective and easier to sustain, than a strategy which relies upon an individual routinely to use personal protective equipment (PPE), or to follow rigorous procedures in order to prevent excessive exposure.

The hierarchy of occupational hygiene controls is as follows:

- elimination
- substitution
- total enclosure
- segregation

- partial enclosure (with LEV)
- local exhaust ventilation (LEV)
- general ventilation (dilution)
- personal protective equipment
- procedural controls
- provision of amenities
- training.

Considering the hierarchy of available controls the first option is to substitute a toxic hazard with a material which is either non-toxic or, at least, is of lesser toxicity. As referred to earlier, other technical considerations may determine the agent of choice. They may not however preclude changes in formulation, e.g. aqueous rather than solvent-based formulations, changes in physical form, e.g. pellets, granules or flakes rather than powders or changes in particle size such that adequate control may be more readily achieved.

Where use of a 'non-hazardous' material is impracticable, consideration should be given to enclosing the process totally, preferably under slight negative pressure, so as totally to separate the operator from the process. Consideration must however also be given to ease of maintenance and the potential exposures that might arise from such activities including any cleaning.

Where complete enclosure is not practicable then incomplete enclosure, supplemented by effective local exhaust ventilation, can often achieve similar standards of containment. It will of course be necessary to ensure that the local exhaust ventilation is properly designed, installed and regularly tested in order to achieve, demonstrate and sustain acceptable standards of control. Fortunately authoritative guidance on these topics is readily available.[23–25] Examples of this approach range from fume cupboards, through various types of flow booths (e.g. horizontal laminar flow booths and downdraught booths), to the use of 'through pan draughting' on chemical reactors.

If, due to the nature of the operation, even partial enclosure is impracticable, local exhaust ventilation can still be used to good effect. It should be noted however that effectiveness will be critically dependent on the position of the exhaust inlet in comparison with the point(s) of emission of the substance being controlled. Effectiveness will also be dependent upon the volume and velocity of the exhaust air supplied. The velocity, at the furthest point of capture, will need to be sufficient to capture the pollutant, and to convey it in the desired direction, despite any interference from thermal effects or other induced draughts. It should be noted as a rule of thumb that the velocity of air will have fallen to one tenth of the duct velocity at a distance 'one duct diameter' away from the orifice. If that seems hard to believe then blow out the flame from a match at a distance of approximately six inches. Now try to suck out the flame at the same distance!

Good general ventilation, even where supplemented by positive forced ventilation, should not normally be relied upon for control of exposure. Although there will be dilution of the contaminant this may be inadequate in the case of an operator close to the source of exposure. It will also lead to general contamination of the workplace, permit contamination of others who would not otherwise be affected, and may well be expensive in terms of volumes of air extracted and input.

Where an effective standard of control cannot otherwise be achieved it may be necessary to rely upon personal protective equipment (PPE). This should generally be seen as a temporary measure whilst the standard of control is improved. It must be recognised however that there will be some situations, e.g. certain maintenance activities, where PPE may be the only viable alternative. In these situations, as in all others where there is reliance on PPE as a prime means of protection, consideration needs to be given to a range of consequent issues. These will include selection of appropriate equipment, issues such as nominal protection factors, permeation and degradation rates, useability for the tasks in question, regulatory compliance etc. There will be a need to provide for suitable maintenance, cleaning, storage, repair and replacements. Employees must be suitably trained, validated and supervised in the use of PPE. These considerations are only indicative of the issues to be addressed and are not exhaustive. They do however begin to illustrate that, whilst PPE is never the preferred solution, neither is it likely to be a cheap option in the long term.

Whatever the control mechanisms implemented they will need to be supported by effective training and implementation of appropriate procedures. The effectiveness of the whole should never be assumed. The effectiveness must be demonstrated in some way and potential enhancements identified and incorporated into improvement plans. An illustrative example is given in Table 13.3.

Monitoring of the effectiveness of control may involve checks on performance of the equipment, monitoring of exposures to contamination, health surveillance of workers, or a mixture of all three techniques. In the absence of any specific overriding legal requirement the emphasis should be on the former technique. Ideally there will be some easily measured parameter, face velocity, duct air velocity, pressure drop etc., that equates to previously established effective control. Such data are usually required for design purposes, but if not already documented can readily be obtained at commissioning or later. This affords the opportunity to monitor the success of control on an ongoing basis, and to detect and remedy any trend to reduced effectiveness before it becomes otherwise obvious.

A second alternative is to implement a comprehensive monitoring

Table 13.3 Task assessment record

Risk assessment	
Task	Transfer of biocide (50 kg) from 200 l drum to a mixer tank.
Method	Pumped via flexible hose. Weight measured 'by difference' from drum on weighscale.
Material hazard	Causes burns. Harmful if swallowed. A potential skin sensitiser. Non-volatile viscous liquid.
Risk	Primarily arises from skin contact. Protection needed to avoid burns will be adequate to prevent skin sensitisation arising.
Control	Must avoid skin and eye contact and the inhalation of any mists or aerosols which might be generated. (Vapour will not be a problem – unless the material is heated)
Conclusion	Current controls are unacceptable.
Improvement/action plan	
Immediate	(1) Transfer material by vacuum (applied to the mixer tank) rather than by pump – eliminates risk of mists, minimises the exposure from maintenance/cleaning. (2) Provide drip can + detox solution in order to collect/neutralise drainings from the transfer line and prevent any contamination of the area. (3) Introduce a PPE regime (gloves, face shield and apron or disposable overall) and check adequacy of washing and showering facilities. (4) Formalise written procedure in consultation with the operatives. Include all foreseeable circumstances. (5) Train and validate, supervise and audit.
Medium term	As above + (1) Seek alternative less viscous/hazardous biocide. (2) Modify charging facility – e.g. 'drt-break' couplings, bunded/sumped area, easy clean finishes etc.
Longer term	As above + (1) Automated enclosed dosing system. (2) Integral detox of equipment and packages.

programme to establish workplace and personal exposures. Thus we quantify the dose delivered to the workers, but probably will not detect any loss of control until after the event! The need for a comprehensive monitoring programme is referred to here as, unfortunately, 'spot samples' are unlikely to have any real significance. Atmospheric concentrations will vary rapidly with time, distance and circumstance. Plots of data tend to follow a distribution not unlike that of a log normal population.[26] Thus a single or a few isolated samples are unlikely to have statistical significance and hence will defy scientifically valid and defensible interpretation. Monitoring does of course have its role to play, but it is enough to state here that it may as often be abused as used! References are given to introductory and other texts for those who would wish to investigate further.[27,28]

A third alternative is to rely upon some form of health surveillance.

Again, although health surveillance has many useful roles to play,[22] it is not ideally placed as a measure of effective workplace control. A whole chain of events will need to be in place before health surveillance identifies a problem. First, we will need to have a failure of control, secondly this will need to have led to exposure, at a concentration, time, duration etc. leading to symptoms. Thirdly this must be coincident with health surveillance, and finally the link between condition and exposure must be made before any queries about control will be raised for investigation. Clearly this is a somewhat insensitive and unreliable feedback loop!

In summary then the key to safe working with preservatives is as follows. To ensure that all hazards are identified. From the hazard data, taking into account all the factors which might contribute to exposure, assess the risks which may arise. Where this risk is not acceptable it should be controlled by direct and effective means to a level representing an insignificant risk.

Thus matching and sustaining control measures appropriate to the level of risk is the answer. This may not require sophisticated techniques but must encompass all actual and foreseeable work activities and circumstances.

I referred earlier to Paracelsus who expounded the truism that all substances are hazardous to a greater or lesser extent. I would also suggest a corollary – that there are none so hazardous that they cannot be handled safely! Certainly preservatives have the demonstrated capacity to cause harm if they are not adequately controlled. They are not however, by any means the most aggressively toxic materials handled by industry. Ill health and discomfort must never be accepted as an inevitable consequence of their use. The application of the approach outlined above can go a long way towards preventing ill health effects being realised in those manufacturing and using preservatives.

References

1. Goh, C.L. (1985) Irritant dermatitis from tri-*N*-butyl tin oxide in paint. *Contact Dermatitis*, **12** (3), 161–163.
2. O'Driscoll, J.B. and Beck, M.H. (1988) Occupational allergic contact dermatitis from Kathon WT. *Contact Dermatitis*, **19** (1), 63.
3. Norris, P.G., Hawk, L.J.M. and White, I.R. (1987) Photoallergic contact dermatitis from Fentichlor. *Contact Dermatitis*, **18** (5), 318–320.
4. Lings, S. (1982) Pesticide lung: a pilot investigation of fruit growers and farmers during the spraying season. *Br. J. Indust. Med.*, **39** (4), 370–376.
5. Kawai, K., Nakagawa, M., Sasaki, Y. and Kawai, K. (1993) Contact dermatitis from Kathon 930. *Contact Dermatitis*, **28** (2), 117–118.
6. Sanz-Gallen, P., Planas, J., *et al.* (1992) Allergic contact dermatitis due to 1,2-benzisothiazolin-3-one in paint manufacture. *Contact Dermatitis*, **27** (4) 271–271.
7. Oleage, J.M., Aguirre, A. *et al.* (1992) Allergic contact dermatitis from Kathon 893. *Contact Dermatitis*, **27** (5), 345–346.
8. de-Boer, E.M., van-Ketel, W.G. and Bruynzeel, D.P. (1989) Dermatoses in metal workers. *Contact Dermatitis*, **20** (4), 280–286.

9. Anon. (1983) Precautions to be taken when cooking narcissus bulbs in formalin. *Veiligheid*, **59** (10), 503.
10. Roberts, D.L., Messenger, A.G. and Summerly, R. (1981) Occupational dermatitis due to 1,2-benzisothiazalin-3-one in the pottery industry. *Contact Dermatitis*, **7** (3), 145–147.
11. Wilkinson, S.M., Schouten, P. *et al.* (1991) Allergic dermatitis from Tego 51. *Contact Dermatitis*, **24** (1), 74, 75.
12. Clark, E.G. (1987) Risk of isothiazolines. *J. Soc. Occupational Med.*, **37** (1), 30–31.
13. Mathias, C.G.T. (1983) Contact dermatitis to a new biocide (Tektamer 38) used in a paste glue formulation. *Contact Dermatitis*, **9** (5), 418.
14. Munro, C.S., Shields, T.G. and Lawrence, C.M. (1989) Contact allergy to Tego 103G in a deep sea diver. *Contact Dermatitis*, **21** (4), 278–279.
15. Health and Safety Commission Paper and Board Advisory Committee (1990) *Interpreting Biocide Health and Safety Data Sheets*.
16. ACGIH (1993) *Threshold Limit Values for Chemical Substances and Physical Agents and Biological Exposure Limits*, 1993–1994.
17. Health and Safety Executive London (1994) *EH40/94 Occupational Exposure Limits* 1994.
18. Gardner, R.J. and Oldershaw, P.J. (1991) Development of pragmatic exposure control concentrations based on regulatory risk phrases. *Ann. Occupational Hygiene*, **35**, 57–59.
19. Money, C.D. and Munro, D.N. (1992) An integrated system to assist in ensuring satisfactory occupational hygiene performance within a diverse business. *Proceedings of the First International Occupational Hygiene Association Meeting*, Brussels 1992.
20. Zeneca Specialties (1990) *Control of Substances Hazardous to Health – Assessment Training Manual*.
21. Chemical Industries Association London (1993) *Safe Handling of Colourants (2) – Hazard classification and the selection of occupational hygiene strategies*.
22. HMSO London (1989) *Control of Substances Hazardous to Health Regulations* 1988.
23. Health and Safety Executive London (1993) HS(G)37 *An Introduction to Local Exhaust Ventilation*, HSE Books.
24. ACGIH (1992) *Industrial Ventilation – A Manual of Recommended Practice*, 21st edn, ACGIH Ann Arbor, Michigan, USA: Edward Brothers.
25. Health and Safety Executive London (1990) HS(G)54 *The Maintenance, Examination and Testing of Local Exhaust Ventilation*, HSE Books.
26. Rappaport, S.M. *et al.* (1981) Sampling in the assessment of continuous exposure to acutely toxic chemicals. Part 1 Strategy. *J. Am. Indust. Hygiene Assoc.*, **42**, 831–888.
27. Health and Safety Executive London (1989) *EH42 Monitoring Strategies for Toxic Substances*.
28. Corn, M. (1987) Strategies of air sampling. In *Recent Advances in Occupational Health*, McDonald (ed.), Churchill Livingstone, London, pp. 199–202.

14 Regulation of preservatives in the USA

M.E. BURT

14.1 Introduction

Depending on their use, antimicrobials or preservatives are regulated in the United States by the US Environmental Protection Agency (EPA) and the US Food and Drug Administration (FDA). The use of preservatives is regulated by the EPA under the Federal Insecticide, Fungicide and Rodenticide Act, as amended (FIFRA). This law regulates the use of all pesticides, including antimicrobials, in the USA. FIFRA was first enacted in 1947 and has been amended several times, most recently in 1988. The use of preservatives in indirect food additive situations is regulated by the FDA, operating under the Federal Food, Drug and Cosmetic Act (FFDCA). The FFDCA was first enacted in 1938 and the Food Additive Amendment was passed in 1958. The purpose of this chapter is to provide an overview of the regulation of preservatives in the USA by the EPA, FDA, and other relative schemes.

14.2 Industrial preservatives

The use of antimicrobials or preservatives in industrial use patterns and as preservatives in agrochemicals is regulated by the EPA under FIFRA. The EPA is charged with the promulgation of regulations, policy and procedures so that the use of antimicrobials can be registered or accepted for use. Prior to registering the use of any antimicrobial, the EPA requires that certain data requirements be fulfilled. These data requirements address the areas of product chemistry, environmental fate, toxicology, and wildlife and aquatic organism testing. Data requirements are specified in EPA's pesticide regulations[1] and are reviewed herein.

The applicant for registration is required to provide complete information on the composition of the antimicrobial active ingredient, including certified upper and lower limits of concentration for both the technical product and end-use formulations, CAS registry number, and chemical identity. The applicant must also provide a description of the manufacturing or formulation process, a discussion on the formation of impurities, the results of chemical analysis and an analytical method, suitable for use by regulatory authorities to analyse samples collected in the marketplace. Additionally, certain physical/chemical properties data are required for both the technical and end-use formulations.

Environmental fate data requirements for most industrial antimicrobial uses are fulfilled with the submission of a hydrolysis study. Wildlife and aquatic organism data requirements are as follows: avian oral LD50 (mallard duck or bobwhite quail); avian dietary LC50 (mallard duck and bobwhite quail); rainbow trout and bluegill LC50; *Daphnia* LC50.

The registration of all pesticides must be supported by a complete acute toxicity data base for both the technical grade of the active ingredient and the end-use formulations. These requirements include: acute oral toxicity – rat; acute dermal toxicity; primary eye irritation; primary dermal irritation; and dermal sensitization. The results of these studies are used to determine appropriate label warning and precautionary statements as specified in the regulations.[1] Also, inhalation toxicity testing will be required if the end-use pattern specifies spray or aerosol application methods. The EPA toxicity categories and related precautionary statements are reviewed in Table 14.1.

Table 14.1 US EPA toxicity category definitions and precautionary statements by toxicity category, Title 40 CFR Part 156.[1]

Hazard indicators	Toxicity categories			
	I	II	III	IV
Oral LD_{50}	Up to and including 50 mg kg^{-1}	From 50 to 500 mg kg^{-1}	From 500 to 5000 mg kg^{-1}	>5000 mg kg^{-1}
Inhalation LD_{50}	Up to and including 0.2 mg l^{-1}	From 0.2 to 2 mg l^{-1}	From 2 to 20 mg l^{-1}	>20 mg l^{-1}
Dermal LD_{50}	Up to and including 200 mg kg^{-1}	From 200 to 2000	From 2000 to 20 000	>20 000
Eye effects	Corrosive: corneal opacity not reversible within 7 days	Corneal opacity reversible within 7 days; irritation persisting for 7 days	No corneal opacity; irritation reversible within 7 days	No irritation
Skin effects	Corrosive	Severe irritation at 72 h	Moderate irritation at 72 h	Mild or slight irritation at 72 h

Table 14.1 *Continued*

Toxicity category	Precautionary statements by toxicity category	
	Oral, inhalation or dermal toxicity	Skin and eye local effects
I	Fatal (poisonous) if swallowed (inhaled or absorbed through skin). Do not breathe vapor (dust or spray mist). Do not get in eyes, on skin, or on clothing. Front panel statement of practical treatment required.	Corrosive, causes eye and skin damage (or skin irritation). Do not get in eyes, on skin, or on clothing. Wear goggles or face shield and rubber gloves when handling. Harmful or fatal if swallowed. Appropriate first aid statement required.
II	May be fatal if swallowed (inhaled or absorbed through the skin). Do not breathe vapors (dust or spray mist). Do not get in eyes, on skin, or on clothing. Appropriate first aid statements required.	Causes eye (and skin) irritation. Do not get in eyes, on skin, or on clothing. Harmful if swallowed. Appropriate first aid statement required.
III	Harmful if swallowed (inhaled or absorbed through the skin). Avoid breathing vapors (dust or spray mist). Avoid contact with skin (eyes or clothing). Appropriate first aid statement required.	Avoid contact with skin, eyes or clothing. In case of contact immediately flush eyes or skin with plenty of water. Get medical attention if irritation persists.
IV	No precautionary statements required.	No precautionary statements required.

The requirement for subchronic and chronic toxicity testing varies, depending on whether or not the proposed use is of low, medium or high exposure (1987 EPA antimicrobial data call-in). For the most part, industrial uses of antimicrobials are considered to be low exposure uses and a minimum set of subchronic data are required. These studies include a 90-day study (oral or dermal, as appropriate) and a teratology study in one species. Also, a genotoxicity test battery to assess the active ingredient's potential for gene mutation, structural chromosomal aberration or other genotoxicity effects are required for all antimicrobials. EPA will require exposure assessment information. This information would confirm the low potential for human exposure to industrial use antimicrobials following the labeled use directions, thus obviating the requirement for chronic toxicity testing.

Data submitted to EPA in support of product registrations are subject to exclusive use by the original registrant for a period of 10 years after initial registration, as provided by FIFRA. Also, FIFRA entitles the original data submitter to data ownership protection for a period of 15 years after submission to EPA. During this time, the original data submitter can claim compensation for the use of data by any other party citing that data at EPA in order to gain product registration.

Although the use of antimicrobials for agrochemical preservatives is registered by the EPA, the EPA must also be petitioned for an exemption from the requirement to provide a tolerance for residues of the antimicrobial when used as an inert ingredient in pesticide formulations applied to growing crops (or raw agricultural commodities after harvest, or to animals). The amount of preservative in agrochemicals is at a low level and residues in or on crops would usually not be of concern to the Agency. Generally, data requirements to support a petition for the exemption from the requirement of a tolerance are similar to those for antimicrobial registration.

In addition to obtaining EPA registration, each state or US territory, with few exceptions, requires that antimicrobials be registered in the state prior to marketing. Registration requirements vary from state to state but most require that an application for registration be supported by a copy of the product label, proof of EPA registration (copy of EPA stamped accepted label), a material safety data sheet and the appropriate registration fee. Only the state of California requires that applications for registration be supported by a complete set of product safety data and efficacy data. Products are then registered by the California Environmental Protection Agency after a complete and thorough data review.

If an antimicrobial is to be registered by EPA as a preservative in adhesives used for food packaging or in components of paper and paperboard (i.e. paper coating compositions), appropriate indirect food contact clearances must first be obtained from the FDA. The clearances are granted upon the submission and review of a petition containing appropriate chemistry, product safety and efficacy data.

14.3 Indirect food additives

The FDA's chemistry requirements for indirect food additives are reviewed.[2] In addition to basic chemical identity information, the FDA recommendations address the requirement for food migration data. Food migration studies are designed to estimate the amount of test substance (i.e. preservative) that would migrate from food packaging material into food. The results of the food migration study are used in conjunction with the product safety data to estimate human exposure. An acceptable daily intake (ADI) of the test substance is derived from product safety data and an estimated daily intake (EDI) is derived from the food migration study. Simply put, if the EDI does not exceed the ADI, the FDA would approve the proposed use and publish the regulation.[3]

The FDA's requirements for product safety data for direct food additives or color additives are outlined in the Agency's 'Redbook'.[4] However, the Redbook states that these guidelines do not specifically apply to indirect food additives since they are not intentionally added to food but are articles or components of articles used for food packaging. Any presence of the indirect food additive in food items would then be caused by the migration or extraction from the food contact surface into food. FDA indicates that product safety data recommendations for indirect food additives differ from those of direct additives and recommends that the Agency be contacted for additional information.

Briefly, the FDA's requirement for product safety or toxicity testing for indirect food additives vary, depending on dietary exposure potential (determined by migration studies). For 'virtually nil exposure' (less than 0.05 ppm), only an acute oral rodent study would be required. Toxicity tests for indirect additives with insignificant exposure (greater than 0.05 ppm but less than 1.0 ppm) would include subchronic studies and genotoxicity studies. Petitions for indirect food additives with significant dietary exposure (greater than 1.0 ppm) would be supported by a complete chronic data base, including rodent lifetime feeding studies and possibly a metabolism study.

Preservatives permitted by FDA for use in adhesives and as paper and paperboard components in contact with food items are listed in Tables 14.2, 14.3 and 14.4. In some cases, the regulations[6] provide cross references between specific areas (e.g. preservatives permitted for use as components of paper and paperboard in contact with fatty or aqueous food may be safely used as components of paper and paperboard in contact with dry food).

14.4 Cosmetic preservatives

In the United States, cosmetics have been self regulated since 1938. A voluntary registration program was initiated by the Cosmetics, Toiletries and Fragrance Association (CTFA) with the cooperation and support of the US FDA in the early 1970s. Only finished cosmetics are registered. Cosmetic ingredients, *per se*, are no longer included in the voluntary registration program. However, it is important for ingredients to be listed in the CTFA Ingredient Dictionary[5] as part of the self-regulation process.

There are three levels in the FDA voluntary cosmetic registration program. They are: registration of the manufacturing facility; registration of the cosmetic product; and submission of cosmetic experience reports.[6] Cosmetic registration includes complete disclosure of all ingredients, including preservatives. The experience reports deal only with verified medical complaints and not with product performance complaints.

Table 14.2 Preservatives permitted for use as components of adhesives intended for use in packaging, transporting, or holding food. Title 21 CFR Part 175.105.[6]

n-Alkyl (C_{12}, C_{14}, or C_{18}) dimethyl (ethylbenzyl) ammonium cycloehxylsulfamate
Ammonium benzoate
1,2-Benzisothiazolin-3-one
p-Benzoxyphenol
Benzyl bromoacetate
p-Benzoxyphenol
Bis(tri-*n*-butylin) oxide
Bis(trichloromethyl)sulfone
2-Bromo-2-nitro-1,3-propanediol
Chlorinated pyridine mixture with active ingredients consisting of 2,3,5,6-tetrachloro-4-(methylsulfonyl) pyridine, 2,3,5,6-tetrachloro-4-(methylsulfinyl) pyridine and pentachloropyridine
1-(3-Chloroallyl)-3,5,7-triaza-1-azoniaadamantane chloride
4-Chloro-3,5-dimethylphenol (*p*-chloro-*m*-xylenol)
4-Chloro-3-methylphenol
5-Chloro-2-methyl-4-isothiazolin-3-one and 2-methyl-4-isothiazolin-3-one mixture at a ratio of 3 parts to 1 part, manufactured from methyl-3-mercaptopropionate. The mixture may contain magnesium nitrate at a concentration equivalent to the isothiazolone active ingredients (weight/weight).
Coconut fatty acid amine salt of tetrachlorophenol
Copper 8-quinolate
1,2-Dibromo-2,4-dicyanobutane
2,6-Di-*t*-butyl-4-methylphenol
4-(Diiodomethylsulfonyl) toluene
3,5-Dimethyl-1,3,5,2*H* tetrahydrothiadiazine-2-thione
Ethyl hydroxyethylcellulose
Glutaraldehyde
5-Hydroxymethoxymethyl-1-aza-3,7-dioxabicyclo[3.3.0]octane, 5-hydroxymethyl-1-aza-3,7-dioxabicyclo[3.3.0]octane, and 5-hydroxypoly-[methyleneoxy]methyl-1-aza-3,7-dioxybicyclo[3.3.0]octane mixture
4-4′-Isopropylidenediphenol, polybutylated mixture
Magnesium fluoride
2-Mercaptobenzothiazole and dimethyl dithiocarbamic acid mixture, sodium salt
2-Mercaptobenzothiazole, sodium or zinc salt
Pentaerythritol monostearate
Phenol
o-Phenylphenol
Potassium pentachlorophenate
Quaternary ammonium chloride (hexadecyl, octadecyl derivative)
Salicylic acid
Sodium dehydroacetate
Sodium fluoride
Sodium pentachlorophenate
Sodium μ-phenylphenate
Sodium salicylate
Sodium salt of 1-hydroxy 2(1*H*)-pyridine thione
Thymol
Tri-*t*-butyl-*p*-phenyl phenol
Tributylin chloride complex of ethylene oxide condensate of dehydroabietylamine
Tri-*n*-butyltin acetate
Tri-*n*-bytyltin neodecanoate

Table 14.3 Preservatives permitted for use as components of paper and paperboard in contact with aqueous and fatty foods. Title 21 CFR Part 176.170.[6]

Part 176.170 (a)(5)
1,2-Benzisothiazolin-3-one
n-Dodecylguanidine acetate
n-Dodecylguanidine hydrochloride
Glutaraldehyde
1,3,5-Triethylhexahydro-1,3,5-triazine
Part 176.170 (b)(2)
1-(3-Chloroallyl)-3,5,7-triaza-1-azoniaadamantane chloride
5-Chloro-2-methyl-4-isothiazolin-3-one and 2-methyl-4-isothiazolin-3-one mixture at a ratio of 3 parts to 1 part, manufactured from methyl-3-mercaptopropionate. The mixture may contain magnesium nitrate at a concentration equivalent to the isothiazolone active ingredients (weight/weight).
1,2-Dibromo-2,4-dicyanobutane
Dihydroxy dichlorophenyl methane
Formaldehyde
Sodium pentachlorophenate
Sodium *o*-phenylphenate

Table 14.4 Preservatives permitted for use as components of paper and paperboard in contact with dry food. Title 21 CFR Part 176.180.[6]

Barium metaborate
1,2-Benzisothiazolin-3-one
Bis(trichloromethyl) sulfone
Borax
Boric acid
5-Hydroxymethoxymethyl-1-aza-3,7-dioxabicyclo[3.3.0]octane, 5-hydroxymethyl-1-aza-3,7-dioxabicyclo[3.3.0]octane, and 5-hydroxypoly-(methyleneoxy)methyl-1-aza-3,7-dioxabicyclo[3.3.0]octane mixture

When a cosmetic manufacturer files a cosmetic registration form, FDA reviews the ingredient statement in order to determine if any new ingredients are listed. A new ingredient would be one not previously listed and not included in the CTFA Cosmetic Ingredient Dictionary. FDA will contact the manufacturer if there are illegal or deleted products listed on the cosmetic ingredient statement. Preservatives labeled only for use in cosmetics or personal care products are not subject to EPA registration as pesticides.

14.5 Summary

Preservatives are regulated in the USA by the US EPA, the US FDA, the various states and are self regulated in the case of cosmetic preservatives. Depending on the use pattern, preservatives can be regulated by more than one Federal agency.

References

1. Title 40 Code of Federal Regulations (updated annually) *Protection of Environment*, Parts 150 to 189. Office of the Federal Register, National Archives and Records Administration, Washington, DC.
2. *Recommendations for Chemistry Data for Indirect Food Additive Petitions* 1.2 (1993) Chemistry Review Branch, Office of Premarket Approval, Center for Food Safety and Applied Nutrition, US Food and Drug Administration, Washington, DC, March.
3. Title 21 Code of Federal Regulations (updated annually) *Food and Drugs*, Parts 170 to 199, Office of the Federal Register, National Archives and Records Administration, Washington, DC.
4. *Toxicological Principles for the Safety Assessment of Direct Additives and Color Additives Used in Food* (1993) Draft, Center for Food Safety and Applied Nutrition, US Food and Drug Administration, Washington, DC.
5. Nikitakis, J.M., McEwen, G.N. and Wenninger, J.A. (eds) (1991) *CTFA International Cosmetic Ingredient Dictionary*, The Cosmetic, Toiletry and Fragrance Association, Washington, DC.
6. Title 21 Code of Federal Regulations (updated annually), *Food and Drugs*, Parts 600 to 799, Office of the Federal Register, National Archives and Records Administration, Washington, DC.

15 European preservative legislation

G. LLOYD

15.1 Introduction

Legislation governing the use of preservatives within Europe is highly complex, involving a number of sub-legislative areas. For example, before consideration of the regulatory requirements (Table 15.1) for preservatives can be made, all chemicals used in the European Community must appear on either the European Inventory of Existing Commercial Substances (EINECS) or the European List of New Chemical Substances (ELINCS).

There are also specific applications where an alternative approval must first be sought before entry into the marketplace, for instance in the manufacture of food contact paper and the exploration and production of oil.

The approval of industrial preservatives in Europe can best be described as fragmented. There are countries where it is mandatory to gain full approval for all preservative applications, others require the approval for use in a limited number of applications, and there are countries where no approval is required at all (Appendix 1).

Approval is normally granted for the end use of formulated product, however before an approval can be granted the authority will require access to a data file covering all active ingredients used in the formulation.

The registration of a formulated, or end use product, requires a much smaller data package and is restricted to information on the product to be submitted; physical/chemical data and a series of acutes. The application must also include a letter of access from the owner of the data on each and every active used in the formulation, as these data are necessary for the calculation of the overall risk assessment.[1]

Again consideration to the fees charged, and language to be used must be taken. In addition to the overall registration position in Europe being fragmented, we also have other government departments involved in the granting of an approval for the use of preservatives.

15.2 Preservatives in the oil industry

The registration of preservatives for use in the oil industry (see Appendix 2), in the main, only applies to those member states with an oil production

Table 15.1 Regulatory compliance in Europe – typical data requirements (active ingredients)

Data	Typical registration[a]			Proposed Biocidal Products Directive[b]		
	Paint – incan	Metalworking fluids	Paper	Paint – incan	Metalworking fluids	Paper
General information						
Name of substance	×	×	×	×	×	×
EINECS/ELINCS/CAS No	×	×	×	×	×	×
Synonyms	×	×	×	×	×	×
Purity	×	×	×	×	×	×
Impurities	×	×	×	×	×	×
Molecular weight	×	×	×	×	×	×
Structural formula	×	×	×	×	×	×
Type of substance	×	×	×	×	×	×
Who is submitting the data set?	×	×	×	×	×	×
Quantity produced/ imported	×	×	×	×	×	×
Produced during last 12 months?	×	×	×	×	×	×
Classification and labelling	×	×	×	×	×	×
Use pattern	×	×	×	×	×	×
Amount of additives				×	×	×
Method of analysis	×	×	×	×	×	×
Spectral data	×	×	×	×	×	×
Means of personal protection	×	×	×	×	×	×
Side effects				×	×	×
Safety precautions	×	×	×	×	×	×
Methods to decontaminate			×	×	×	×
Incineration conditions				×	×	×
Other remarks	×	×	×		×	×
Efficacy	×	×	×	×	×	×
Physico-chemical data						
Physical state	×	×	×	×	×	×
Particle size	×	×		×	×	×
Melting point	×	×	×	×	×	×
Boiling point	×	×	×	×	×	×
Thermal stability	×	×	×	×	×	×
Vapour pressure	×	×	×	×	×	×
Octanol/water partition coefficient	×	×	×	×	×	×
Water solubility	×	×	×	×	×	×
Solvent solubility				×	×	×
Fat solubility	×	×	×	×	×	×
Flammability	×	×	×	×	×	×
Crystallisation	×	×	×	×	×	×
Flash point	×	×	×	×	×	×
Autoflammability	×	×	×	×	×	×
Explosivity	×	×	×	×	×	×
Surface tension	×	×	×	×	×	×
Oxidising properties				×	×	×
Other data and remarks	×	×	×	×	×	×

Table 15.1 *continued*

Data	Typical registration[a]			Proposed Biocidal Products Directive[b]		
	Paint – incan	Metalworking fluids	Paper	Paint – incan	Metalworking fluids	Paper
Toxicity						
Acute oral	×	×	×	×	×	×
Acute dermal	(×)	(×)	(×)	×	×	×
Acute inhalation	×	×	×	×	×	×
Skin irritation	×	×	×	×	×	×
Eye irritation	×	×	×	×	×	×
Sensitisation	×	×	×	×	×	×
Mutagenicity – Ames	×	×	×	×	×	×
Mutagenicity – (2)	×	×	×	×	×	×
Mutagenicity – (3)		×	×	×	×	×
Reproductive toxicity screen				×	×	×
28 day toxicity	×	×	×	×	×	×
90 day sub acute[a]	(×)	×	×	(×)	(×)	(×)
90 day sub acute[b]		×	×			(×)
Neurotoxicity		×	×			×
Cholinesterse inhibition		×	×			×
Teratogenicity		×	×		×	×
One generation reproduction		×	×			
Two generation reproduction			×			(×)
Toxicokinetics		×	×	×	×	×
Carcinogenicity screen	(×)	(×)	(×)			
Carcinogenicity	(×)	(×)	(×)		×	×
Other relevant information	×	×	×		×	×
Experience with human exposure	×	×	×	×	×	×
Toxicological mechanism				×	×	×
Worker health records				×	×	×
Epidemiology	×	×	×	×	×	×
Ecotoxicology						
Acute toxicity to birds			×			×
Subacute toxicity to birds			×			×
Acute toxicity to fish	×	×	×	×	×	×
Fish reproduction			×			×
Prolonged toxicity to fish		×	×			×
Bioaccumulation if P_{ow} >1000	×	×	×	×	×	×
Acute toxicity to Daphnia	×	×	×	×	×	×
Prolonged toxicity to Daphnia			×			×
Toxicity to algae	×	×	×	×	×	×
Toxicity to bacteria	×	×	×		×	
Toxicity to terrestrial organisms			×		×	
Toxicity to soil organisms (e.g. worms)					×	
Toxicity to higher plants			×			×
Toxicity to non target organisms			×	×	×	×
Other remarks	×	×	×		×	×

Table 15.1 *continued*

Data	Typical registration[a]			Proposed Biocidal Products Directive[b]		
	Paint – incan	Metalworking fluids	Paper	Paint – incan	Metalworking fluids	Paper
Environmental fate						
Biodegradation	×	×	×	×	×	×
BOD/COD	×	×	×			
Activated sludge respiration inhibition			×		×	×
Degradation/transformation, water/soil			×	×	×	×
Transport and distribution betweeen environmental compartments					×	
Adsorption/desorption soil screen			×	×	×	×
Distribution in the environment				×	×	×
Mobility			×		×	
Photodegradation			×		×	×
Full soil adsorption/ desorption study			×			×
Hydrolysis		×	×	×	×	×
Other remarks	×	×	×	×	×	×

Parentheses indicate conditional requirement. [a]Data requirements from more than one industry. [b]Data requirement is a proposal at this stage.

industry, although countries which border the North Sea do have legitimate concerns with regard to pollution control, and therefore take an active role in the activities of PARCOM (the Paris and Oslo Commission). This independent organisation, based in London, was set to advise Ministers on the best way to control, minimise or eliminate pollution in the North Sea.

At the present time, even with the increasing activities of PARCOM, there is little harmonisation between the various regulatory schemes operating in the offshore oil industry.[2] An alliance between the main contracting parties of PARCOM and industry has resulted in the development of a risk assessment programme known as CHARM (Chemical Hazard Assessment and Risk Management).[3] Once fully adopted it will form the basis of the approval schemes for all offshore chemicals of which preservatives are an important sector.

CHARM calls for a significant amount of marine toxicity and environmental fate data, plus information on the potential for the product to cause taint in fish. In the UK, approval is granted under the Offshore Chemical Notification Scheme (OCNS) administered by the Department of Trade and Industry, with technical assistance from the Ministry of Agriculture Fisheries and Foods. The scheme, introduced in 1979, has undergone a number of changes with the more recent ones being influenced by the activities of PARCOM.

In The Netherlands approval should be sought from the State Supervisor of Mines, whilst in Norway contact must be made with the State Pollution Control Authority. Both organisations will require conformation to the requirements of PARCOM including an assessment carried out using the CHARM model.

15.3 Preservatives in the paper industry

Regulatory conformance for preservatives used in the manufacture of paper is required by countries such as Sweden, the Netherlands,[4] Finland[5] and Denmark. There are national authorities set up to administer the approval schemes for chemicals, including preservatives, used in the manufacture of food contact paper, although approval from the FDA in the USA and German BGA (Bundesgesundheitamtes) are the most influential in the industry.

Approval by the BGA requires a considerable amount of information, similar but not quite as demanding as that for the registration of preservative in The Netherlands. There are however requirements specific to the paper industry such as proof of efficacy against slime-formers, migration data and field trial data.

Inevitably we have the influence of the EU in this area.

A number of directives have been published and adopted under a 'frame directive' established to control the use of material which might come into contact with foodstuffs. Within this frame directive is a proposal to produce a Community Directive that will regulate preservatives used in the manufacture of food contact paper. It is anticipated that this directive will replace all national legislation, e.g. the BGA.

15.4 The proposed Biocidal Products Directive (93/239/03)[6]

Following on from the introduction of the Crop Protection Directive 91/414/EEC,[7] this proposal, submitted by the commission in 1993, is designed to harmonise the regulatory requirements for biocides throughout the European market. The proposed scope is very wide covering preservatives in all applications, and as the directive is intended to harmonise the regulatory procedures throughout Europe it is essential that it takes into consideration the systems currently in place now.

Data required to register a preservative, fall under the heading of the common principles. These stipulate the data required for each application area, with a core data set of acutes, subacutes, wildlife and environmental data, supplemented by additional data of relevance to the particular application.

Like the established programmes the registration of end use products has a much smaller data requirement the cost of which is likely to be in the order of £60 000–£100 000 (sterling).

It may be possible to reduce the cost of a registration by data sharing. Data used in the registration process can be protected for up to 15 years; no member state will be allowed to use a company's data in support of a second company's registration without the permission of the original registrant.

The assessment of all active ingredients will be the responsibility of the European Commission with member states carrying out the assessment and approval of end use products. Once a member state has completed its assessment and granted an approval, all other member states will be required to grant approval for the preservative in their country under the principles of mutual acceptance, unless there are very good reasons not to. After adoption of the directive by the commission and the member states there will be a 10 year transitional period, during which all active ingredients and formulated products will be put through a health and environmental risk assessment. Any product classified as very toxic, a category 1 or 2 carcinogen, a mutagen or is toxic for reproduction (category 1 or 2) will not be authorised for sale to the general public.

At the present time the Biocidal Products Directive is still a proposal, it still has some way to go before it becomes law in the member states.

To conclude, legislation concerning preservatives currently lies with a variety of government departments within the boundaries of Europe. Although there are some countries with long standing registration procedures the introduction of the biocidal products directive will bring regulatory harmonisation to the industry.

References

1. *Product Development of Data Requirements and Common Principles for the Evaluation and Risk Assessment of Biocidal Products*. Ministry of Environment National Agency of Environment Protection. April 1994.
2. *Harmonised Offshore Chemical Notification Scheme*. 6 September 1994.
3. *CHARM Wizard* (V2.1), Chemical Hazard Assessment and Risk Management of Offshore Exploration & Production Chemicals. 21 December 1994.
4. *Guidelines for Submitting Applications for the Licensing of Pesticides*. Commissie Toelating, Bestrijdingsmiddelen, Wageningen, The Netherlands.
5. *Applications for the Advance Approval of Wood Preservative and Slimicides*. 13 April 1994. National Board of Waters and the Environment, Chemical Control Unit, PO Box 250, Helsinki, Finland.

6. Proposal for a Council Directive concerning the placing of biocidal products on the market. 93/C/239/03. *Official Journal of the European Community*. 3.9.93.
7. Council Directive 91/414/EEC (concerning the placing of plant protection products on the market). *Official Journal of the European Commission*. 15.7.91.

Appendix 1 – Countries with registration procedures for preservatives

All applications
The Netherlands
Belgium

Limited applications
UK
Denmark
Spain
Sweden
Norway
Finland

Food contact paper (e.g. BGA)
All member states of the EU

Oilfield applications (Majors)
Denmark
The Netherlands
Norway
UK

Appendix 2 – Typical data requirements for preservatives used in the offshore oil drilling and production industry

General information
Tradename
Supplier
Use/volumes
Composition
Is it on annex A?
Label
Physical/chemical

Ecotoxicology
Bioacummulation
Biodegradability
Taint potential
Marine LC_{50} values:
  Fish (optional)
  Algae
  Crustacea
Sediment Reworker
Bioconcentration factor

Animal toxicology
Acute LD_{50} mammal

Index

MIX
Papier aus verantwortungsvollen Quellen
Paper from responsible sources
FSC® C105338

If you have any concerns about our products,
you can contact us on
ProductSafety@springernature.com

In case Publisher is established outside the EU,
the EU authorized representative is:
Springer Nature Customer Service Center GmbH
Europaplatz 3, 69115 Heidelberg, Germany

Printed by Libri Plureos GmbH
in Hamburg, Germany